Ordering Life

ORDERING LIFE

KARL JORDAN
and the Naturalist Tradition

KRISTIN JOHNSON

The Johns Hopkins University Press
Baltimore

© 2012 The Johns Hopkins University Press
All rights reserved. Published 2012
Printed in the United States of America on acid-free paper
2 4 6 8 9 7 5 3 1

The Johns Hopkins University Press
2715 North Charles Street
Baltimore, Maryland 21218-4363
www.press.jhu.edu

Library of Congress Cataloging-in-Publication Data

Johnson, Kristin, 1973–
Ordering life : Karl Jordan and the naturalist tradition / Kristin Johnson.
    p.   cm.
Includes bibliographical references and index.
ISBN 978-1-4214-0600-8 (hdbk. : alk. paper) — ISBN 978-1-4214-0650-3 (electronic) —
ISBN 1-4214-0600-4 (hdbk. : alk. paper) — ISBN 1-4214-0650-0 (electronic)
1. Jordan, Karl, 1861–1959. 2. Entomologists—Biography. 3. Taxonomists—Biography.
4. Biology—Classification. I. Title.
QL31.J65J33 2012
595.7092—dc23
    [B]   2011048239

A catalog record for this book is available from the British Library.

*Special discounts are available for bulk purchases of this book.
For more information, please contact Special Sales at 410-516-6936 or
specialsales@press.jhu.edu.*

The Johns Hopkins University Press uses environmentally friendly book
materials, including recycled text paper that is composed of at least
30 percent post-consumer waste, whenever possible.

By using a single butterfly as a special text, one may discourse at pleasure of many.

—*Samuel H. Scudder,* The Life of a Butterfly, *1893*

Contents

Introduction 1

1
Joining the Naturalist Tradition 9
"Beetles. Beautiful beetles" 10 | Becoming a Zoologist 14 |
The Cosmopolitan Naturalists 21 | The "nice berth":
Curating a Zoological Museum 26 |
Mobilizing the Naturalist Tradition 33

2
Reforming Entomology 40
The "strange mixture" of Entomologists 41 | How to Do Entomology 48 |
The "making" of Species 55 | A New Type of Collection 62 |
Retraining the Natural History Network 67

3
Ordering Beetles, Butterflies, and Moths 74
"The great desideratum" 75 | Revising the Swallowtails 79 |
Making Systematics Scientific 86 | Crossing over to Biology 94 |
Amassing the Concreta 103

4
Ordering Naturalists 112
Men of Two Classes 113 | Organizing Entomologists 118 |
The End of Tring's Heyday 126 | "Science knows no country" 134 |
A "nation of Entomologists" 140

5
A Descent into Disorder 155
Telling "which way the wind blows" 156 | The Balance of Europe Is Upset 165 |
The Standstill 172 | Recovering Friends, Committees, and Congresses I 179 |
"The requirements for a thorough investigation" 185

6

Taxonomy in a Changed World   191

*The Rise of Applied Entomology 192 | "Something amiss" 199 |
Various Utopias I: The Ithaca Congress 206 | Various Utopias II:
The International Entomological Institute 212 | A Lad's Last Marble 219*

7

The Ruin of War and the Synthesis of Biology   229

*The Edges of Empire 230 | Where Subspecies Meet 235 |
"The end of Tring as we have known and cherished it" 242 |
"Provided Europe does not get quite mad" 249 |
"Without the collection I am hopeless" 257*

8

Naturalists in a New Landscape   265

*Recovering Friends, Committees, and Congresses II 266 |
The Quest to "clear up the chaos" in Weevils and Fleas 272 |
Avoiding the Snake in the Grass 280 | Glorified Office Boys 286 |
Late for a Knighthood 291*

Conclusion   300

Acknowledgments   313

Notes   315

Essay on Sources   357

Index   367

Illustrations appear following page 146

Ordering Life

Introduction

In the spring of 2000, a group of technology leaders gathered at a dinner party in San Francisco. Flush with money from the technology boom, former Microsoft chief technology officer Nathan Myhrvold wished to discuss what sorts of projects needed funding. Kevin Kelly, the founder of *Wired* magazine joined in, as did Stewart Brand, the creator of *The Whole Earth Catalog*. During the course of the evening, Kelly suggested supporting a global inventory of all living animals and plants, but the group assumed it was already being done. After all, the task had been mentioned in the first pages of Genesis, no less.

Yet the group soon found that not only did the global biological inventory remain unfinished but those who had given themselves the task—taxonomists—had completed only 1.5 to 30 percent of the job (depending on one's estimate of the number of creatures living on the planet). Having discovered these shocking facts, Kelly and others founded the All Species Foundation (ASF). The foundation's goal can be simply stated: catalog and describe every species on Earth within twenty-five years. They formulated the project, not surprisingly, with modern technology in mind, imagining a Web page devoted to each species. Proponents estimated that completing the list would require around $20 billion. Some of the biggest names in biology signed on to the project, including Edward O. Wilson of Harvard and Peter Raven of the Missouri Botanical Garden.[1]

Ultimately, the dot-com bust ended the ASF's extraordinarily ambitious project. But its founders' confrontation with the state of the "catalog of life" on this planet raises important questions about the status of the primary scientific tradition to which anyone with such goals must turn: the naturalist tradition, and, more specifically, the endeavor of taxonomy (*taxonomy* is here used interchangeably with the term *systematics*). In their astonishment that a world inventory had not yet been done, Kelly and his colleagues must have been wondering what taxonomists—those who name, describe, and classify living beings—had been doing with all their time and resources. Hadn't they been working on the "global biological inventory" (albeit under different names) since at least the eighteenth century? And didn't they have gigantic institutions—natural history museums—in every national capital in which to complete the project? After all, how difficult could naming organisms be? In other words, the ASF's vision highlights an interesting problem concerning the study of biodiversity; namely, how do we know what we know about biodiversity and, conversely, why do we seem to know so little?

Traditions by definition have histories. To help answer these questions in a historical way, this book takes the life and work of the entomologist Karl Jordan as a guide through the history of the naturalist tradition in the twentieth century. Jordan spent nearly seven decades endeavoring to name, describe, and order a small subset of the world's biodiversity. As a curator of insects employed first by Walter Rothschild's zoological museum in Tring, England, and then by the Natural History Museum in London, he described 2,575 species of Coleoptera, Lepidoptera, and Siphonaptera as well as another 851 species while working with Walter or Charles Rothschild. He thus would seem to have brought us a few steps closer to the fulfillment of the ASF's goal. The total number (3,426 species), while impressive, amounts to about 0.3 percent of the known animal world. Examining the backstory of that number provides a fascinating entree into the complex reasons for why completing the inventory of our planet's biodiversity remains elusive.

The fact that Jordan's works were (and still are) praised as models of taxonomic investigation (one lepidopterist remembers that when Jordan visited the halls of the Natural History Museum in London as an old man in the 1950s, one did not say hello to him: "You sort of clicked your heels and stood at attention!")[2] is an important element of why in these chapters he is chosen as a guide. For this book ultimately asks, how and in what kind of environment could Jordan do work that others found so impressive and sound? What prevented him from doing more? Using Jordan as a guide through the taxonomic wing of the naturalist tradition, one can examine the various opportunities and challenges confronting those working to describe, order, and explain biodiversity in the twentieth century.

As a specialist on various groups of insects, Jordan, or KJ, as he was known to his friends, was himself known by many names during the almost seven decades in which he carried out the meticulous work of ordering the chaos of diversity within nature. To medical entomologists he was the "Dean of Siphonapterists," owing to his pioneering work on flea classification and his role in identifying the flea responsible for transmitting the bubonic plague. To those who knew him best for the works on butterflies and moths he completed at Rothschild's command, he was an excellent lepidopterist; and to his closest friends, who knew his true love was beetles, he was a meticulous coleopterist. He was "one of the world's most international and catholic entomologists" to fellow specialists and allies of the International Congresses of Entomology that he founded in 1910.[3] And he was "a great zoologist and naturalist" and "first and foremost a biologist" to those who valued his rare comments on evolution.[4] The range of epithets illustrates the range of criteria by which naturalists have themselves been ordered as they have carried out the old project of ordering life.

Jordan called himself betimes a taxonomist, entomologist, systematist, or, in his more ironic moods, a "species-maker."[5] This term *species-maker* was an inside joke for his colleagues and a tongue-in-cheek reference to the fact that by the time Jordan began his life in systematics, many could see little science and less skill in naturalists' endeavor to name, describe, and order the living world. The reasons for such perceptions are various, but one can start with the implications of evolution theory for describing and ordering species. In a nonevolutionary paradigm governed by the theory of special creation, naturalists may have differed regarding their ability to discern the boundaries of those God-made species. But their difficulties certainly did not arise from the fact that species changed over time. Indeed, the naturalist tradition was often justified on the ground that naming, describing, and ordering animal and plant life revealed the plan of God's creation, thereby illustrating both his wisdom and beneficence.

The intellectual revolution inspired by Charles Darwin's proposals in the *Origin of Species*, at least as it relates to natural history, can be briefly outlined as follows: As the naturalist tradition flourished in the eighteenth and nineteenth centuries, both classification and description proved challenging under the deluge of specimens, especially for those studying insects. One entomologist toward the end of the nineteenth century imagined that when Adam had been given the task of naming all the animals, "his flagging energy grew lame" when he came to the insects. "So many species, and so small! It was no joke to name them all; And Adam said, 'I'll do no more; I think I'll leave the rest to Noah.'"[6] Indeed, the enormous variation in nature plagued naturalists' efforts to order the natural world. In particular, they had an increasingly difficult time sorting out whether new specimens with slight variations in form represented new species or were (less important) slight variations on the typical form, to be designated as varieties. Within a paradigm of static species, explaining the existence of such variation within species had taken various forms. Some, for example, held that species were each the product of separate creation by God, while varieties arose from the influence of the environment. Others decided that the varieties, too, must have been specially created.

Prior to 1859, these naturalists were united in their nonevolutionary view of species. But some naturalists, including most famously Darwin, grew dissatisfied with the supernatural (and therefore unscientific) element of such explanations. Ultimately, Darwin concluded that varieties and species differ only in their degree of divergence, rather than their distinct origins. Both, he argued, were products of descent with modification, with natural selection the predominant mechanism of change. When commenting on the implications of this tremendous shift in explanatory framework for practicing taxonomists, Darwin famously wrote: "We shall have to treat species in the same manner as those naturalists treat genera, who admit that genera are merely artificial combinations made for convenience. This may not be a cheering prospect; but we shall at least be freed from the vain search for the undiscovered and undiscoverable essence of the term species."[7]

Darwin insisted that a theory of descent with modification explained the tremendous difficulties taxonomists faced when trying to distinguish between species and varieties (and when classifying animals and plants more generally). He did not conclude from this that the entire endeavor should be abandoned. Yet the acceptance of an evolutionary history for species seemed, at least to some, to call the whole practice of naming and describing species into question. The "species-maker" label captured the belief of some that the thousands of species described by taxonomists were figments of their imagination because they doubted taxonomists' ability to discern the real boundaries between species. Others assumed Darwin's destruction of the boundary between varieties and species had dispensed with the need to consider species as real in the first place. In either case, the status of taxonomists' decisions as reflecting true entities in nature was called into question.

Jordan most certainly did not believe the phrase *species-makers* captured what taxonomists ideally did, and he developed meticulous methods, sometimes called Jordanian systematics, to ensure that the species taxonomists described and ordered reflected real units in nature. The taxonomist could then proceed to the "higher goal" of ordering those units into robust classifications upon which generalizations about the origin of biological diversity depended. The meticulous methods he developed in order to carry out that "higher goal" explain in part why the total number of species he named is not higher. But the methods Jordan developed to ensure the "species-makers" jibe was undeserved are, it turns out, only one part of the story.

Taxonomists' attempts to navigate the challenges and opportunities opened by Darwin's theory for their work were compounded by shifts in science regarding the criteria by which work would be called scientific. In a century in which physics, chemistry, and the experimental method set the standard for developing true knowledge about the natural world, many viewed taxonomy's strong emphasis on observations and dead specimens as at best archaic and at worst unscientific. Furthermore, proponents of a new, problem-based, theoretically driven discipline of "biology" (or, as it was sometimes called, "philosophical natural history") often posed biologists' approach as in contrast to the descriptive empiricism that had provided the methodological framework for taxonomy and natural history museums. By the time Jordan began his life in taxonomy, some even questioned whether taxonomists could say anything about the central questions facing life scientists; namely, how the tremendous diversity of living things had come about.

The taxonomists offered rejoinders, of course. After all, Darwin had famously devoted eight years of his life to a taxonomic project on barnacles in the firm belief that any theory regarding the origin of species must be based on the hard-won knowledge gained by "species work." And there were always "philosophical naturalists," such as the codiscoverer of natural selection, Alfred Russel Wallace, who paid tribute to the important work of those describing, cataloging, and classifying in the museum. Though Wallace conceded that he left the description of new species to others, he knew "the great generalizations" of biology had been rendered possible by "the laborious work of species-describers."[8] Working taxonomists composed similar manifestos: "When the dust-heaps of science come to be riddled and the dross separated from the ore," wrote one conchologist, "I have no fear that as large a proportion of the work done by the species-maker and the variety-monger will prove to be as good metal as that brought forward by the pedigree-maker and the so-called philosophical naturalist."[9]

Still, taxonomists like Jordan constantly lamented how extraordinarily difficult producing "good metal" could be. Jordan once joked that if only Noah had imprinted an inheritable name on the bodies of the millions of species he saved in the Ark, taxonomists would have been saved an enormous amount of drudgery.[10] The lament is also evidence that he did not see taxonomists' main job as giving names. Rather, Jordan described himself as having spent more than seven decades working "to bring the immense multitude of diverse forms into natural order."[11] The naming was of course crucial to this project. Jordan's ornithological colleague at Rothschild's museum, Ernst Hartert, once protested against characterizations of naming species as purely descriptive by pointing out that sorting organisms into Linnaean groups inevitably entailed an expression of knowledge about those organisms' natural affinities.[12] Indeed, Darwin himself had pointed out that he had simply formed a new and improved explanatory framework in which to make sense of the fact naturalists could place organisms in hierarchical groups, such as species, genera, families, and orders. Driven by this new, evolutionary interpretation of the Linnaean endeavor, Jordan firmly believed that in naming one was ordering. If the order devised was indeed to be natural, then the naming had to be meticulously guided by certain principles explicitly aimed at adjusting taxonomists' work to the implications of evolution.

During the course of his work ordering insect life, Jordan realized that for his particular community of naturalists—the entomologists—to truly fulfill their self-appointed task of finding the order in nature, they needed to be ordered as well. Thus, at the beginning of the century, he embarked on a series of efforts to organize those naturalists who were studying insects. This part of Jordan's life highlights how the enormous diversity of the naturalist tradition created enormous challenges for scientists intent on describing, ordering, and explaining the living world. All of this diversity influenced the ability of naturalists to develop research programs, create robust knowledge, and convince others that natural history, systematics, and museums deserved both respect and resources. Jordan blamed naturalists themselves for much of the apparent decline in the status of systematics, even as he composed eloquent replies to experimental biologists who ridiculed taxonomists as pigeonholing or stamp-collecting clerks. His constant endeavor to improve entomologists' organization, most evident in his foundation in 1910 of the international congresses of entomology, was based on a recognition that, while one could develop ideal methods aimed at deflecting the species-maker jibe, the taunt often captured what taxonomists were forced to do in the face of both unorganized priorities and methods and limited time and material.

Jordan's efforts to order both insects and entomologists were profoundly influenced by broader changes and events taking place in science and society during the twentieth century. It has been noted in a biography of the Nobel Prize–winning American geneticist Thomas Hunt Morgan that the rise of genetics cannot be understood outside of the context of changing economic patterns in the world at the end of the nineteenth and beginning of the twentieth centuries, including the development of European empires and the rapid growth of American science, industry, and agriculture.[13] The same can be said regarding the history (sometimes described, in contrast to genetics, as the decline) of systematics. Jordan became fascinated by the diversity of insects as a boy in the 1870s, when the study of such diversity was still the province of wealthy owners of private collections and hundreds of enthusiasts. It was also a time during which national museums of the major European powers amassed natural history objects in the name of demonstrating imperial might. Naturalists took advantage of capital, ease of communication, cheap postage, and access to far off regions to pursue their interests.

By the time Jordan died, in 1959, immense social and economic transformations had taken place. The governments of nations both old and new took over the role of financing science, ironically (in retrospect) on the grounds that they only could provide the long-term support required for research. Economic depression, an increased demand for the sciences to be accountable to the public, competition between nation-states and the rise in nationalism, and, ultimately, the devastation of world wars and the demise of Empire, wrought changes that continuously and profoundly altered the ability of naturalists to amass facts, complete research, and garner resources. Meanwhile, within science, Jordan witnessed the rise of Mendelian genetics and an associated emphasis on the central role of experimental methods in biology. The growth of applied entomology transformed the institutional framework of entomology in Europe, and with it, the task of the taxonomist. How naturalists worked, their status in the life sciences, and what kinds of knowledge they could produce were all influenced by these transitions.

As we shall see, Jordan launched vigorous defenses of both entomology and taxonomy as he joined others in attempting to successfully navigate the naturalist tradition into a new age. Thus, Jordan serves as a useful guide not only to understanding how knowledge about biodiversity is obtained but how the answer to that question has changed over time and why. For while composing those more than three thousand descriptions and choosing the names, he responded to the challenges of both diversity and a changing world with conscientious examinations of the tradition's strengths and weaknesses. He developed both methodological and organizational reforms that included careful assessments of not only how knowledge about species was amassed but why anyone should be naming and describing species in the first place.

Ultimately, the fate of both Jordan's vision for systematics and his organizational efforts illustrates how the difficulties involved in establishing synthetic, productive research programs within the life sciences depend greatly on what is happening outside the walls of laboratories and museums. Most importantly, society-changing events and shifts in values have influenced the justifications given for trying "to bring the immense multitude of diverse forms into natural order." Jordan's story shows how the diversity of the naturalist tradition itself has helped maintain that tradition as a vibrant, flexible part of modern science, even as the varied views, methods, and priorities of naturalists have created some of the tradition's most difficult challenges. Given the changing contexts in which the naturalist tradition has been pursued, this diversity has placed at least some naturalists in positions to take advantage of rapidly changing trends in the biological sciences and society.

Even as they adjusted to changing criteria as to what counts as useful science, Jordan occasionally admitted that the interest found by taxonomic entomologists in describing and ordering insects was somewhat intangible. Late in life he recalled how "when once a female relation asked me, in a somewhat disdainful spirit, why I had taken to the study of such creatures, I could only retaliate by asking why her husband had chosen her instead of somebody else." This had been in the days, Jordan explained years later, "when the economic and hygienic importance of Entomology was not yet realized and when one could not, as nowadays, astonish people by referring to the tens of millions of pounds of damage done every year by insects, or frighten the questioner by pointing to the multitude of diseases carried by arthropods."[14]

It was also a time when one could not, as nowadays, astonish people by referring to the "looming extinction crisis" and our "collective peril" should we fail to name and preserve biodiversity.[15] In recent years, many taxonomists have placed themselves in the frontlines of efforts to slow the extinction of the diversity of life on our planet. As that effort's most famous spokesman, E. O. Wilson, put it when explaining the ASF's call for a complete inventory of life during an interview on National Public Radio, "Obviously in the realm of conservation we can't save what we don't know."[16] But, of course, the tradition through which this extraordinary task is to be done arose long before anyone knew any "saving" needed to be done. This is not a trivial point. The endeavor of taxonomy arose from different goals, values, and concepts of science than those that drive modern taxonomists. Yet that past has conferred distinct trajectories on not only how taxonomy is done but why and where. Inevitably, these trajectories have posed challenges and opportunities that continue to both constrain and sustain taxonomists' various projects in important ways, including all modern endeavors to save what remains of the inhabitants of our Ark.

ONE

Joining the Naturalist Tradition

Karl Jordan became the curator of insects of the Walter Rothschild Zoological Museum—eventually one of the largest private natural history collections in the world—in 1893. He entered this museum's gilded doors as one of a lucky few who made a successful career out of a fascination with natural diversity. In 1898, a fellow entomologist wrote achingly that "the nice berths in the Natural History line, curatorships, secretaryships, etc., are so comparatively few and far between that I expect I shall have to forsake these paths for others less congenial."[1] Jordan's position was a "nice berth" indeed. Sitting before Rothschild's many rows of butterflies and beetles, he possessed both a salary and a directive to "sort, name, and describe" the captivating organisms before him.

But why would an individual have been trained to name, describe, and order animals—not to mention, be paid to do so? Desire for such an occupation on the part of the individual is not enough, no matter how much interest someone might have in learning about insects. A profession, a museum, a specialty, a discipline, and a tradition such as natural history cannot be pursued separate from particular social and economic contexts. And a museum cannot be billed as a center of science separate from a consensus on the part of at least some scientists that the description fits. When Jordan became an entomologist, distinct institutional and intellectual contexts determined what activities counted as science. Specific cultural movements deemed the endeavor an acceptable use of time and resources. And various economic developments provided the capital necessary for a relatively small group of experts to pay attention to nature in certain ways. Meanwhile, the fact that each of these factors—institutional, intellectual, cultural, and economic—were in the midst of often contentious change at the turn of the century would have profound influences on the particular vision Jordan developed of the naturalist tradition.

"BEETLES. BEAUTIFUL BEETLES"

Jordan once reflected on how, at some time in the 1860s, his elder brother, a schoolteacher, took him beetle collecting in the woods around his family's farm in the village of Almstedt, Hanover. Like all good schoolteachers in the nineteenth century, the brother knew some natural history. Jordan eagerly attended to the Latin names and focused observation that his brother applied to the flora and fauna they encountered, which included several specimens of a longicorn beetle, *Strangalia armata*. Jordan later described his brother's identification of this tiny insect using a pocket-size catalog of Coleoptera as commencing a lifelong fascination with beetles. Asked as an old man what he had tried to catch as a schoolboy, Jordan replied instantly, "Beetles. Beautiful beetles with tough elytra, crawling near the base of the trees in the woods at Almstedt."[2]

On rainy days, Jordan spent hours studying the lists contained in one of his favorite treasures, the *Catalogus Coleopterorum* (Catalog of Beetles), by Stein and Weise.[3] A system already existed for grouping the insects' names. More than a century earlier, the famous Swedish naturalist Carl Linnaeus had provided the two-word (or binomial) system of nomenclature that made remembering the beetles' names more simple. Building on the work of naturalists before him, Linnaeus had also developed the system of hierarchical ordering that allowed Jordan and his brother to group all beetles under the order Coleoptera, the class Insecta, and the kingdom Animalia. Linnaeus had cataloged 574 kinds of beetles in 1758. By 1788, that number stood at 4,000, and by 1868 a catalog by Gemminger and Harold listed more than 77,026 species.[4]

Linnaeus had justified naming, describing, and ordering organisms on various grounds, from the importance of cataloging Sweden's natural resources to the need to fulfill God's command to Adam to name all the animals. Both these justifications could be meshed with the imperatives of "natural theology," a field of knowledge devoted to demonstrating the existence and character of God through the study of his Creation. Indeed, that justification explains why Jordan's brother had a catalog of beetles in his pocket. Teaching natural history formed a central part of a schoolteacher's job, since many believed that demonstrating the wisdom and power of Providence from the intricate, purposeful structures and interactions of flora and fauna formed a primary means of protecting the moral and social order. Educators of the time argued that the study of nature would "awaken religious feeling" and serve as an antidote to various materialist philosophies that threatened both individual souls and political stability.[5]

Collecting specimens and observing living animals and plants represented accepted methods of studying natural history in this context. Such lofty justifications made it quite respectable, if not imperative, for Karl Jordan's own schoolmaster to have a terrarium in his classroom. Protected by this social sanction for studying animals and plants, Jordan recalled that, like many schoolboys, he had often collected various animals, keeping them in glass jars filled with schnapps until they began to smell. Once he even carried a live viper concealed in an umbrella to school—"not out of mischief, but because a master had casually mentioned that he had no viper in his terrarium."[6]

In that famously secularizing age, though, Jordan and other naturalists' ability to pursue natural history beyond a boyhood hobby depended on much more than demonstrations from nature of the attributes of God. Natural history also benefited from the ambitions of industrializing nation-states, intent on consolidating boundaries and unifying peoples. For example, the Prussian government used support for naturalists' institutions and organizations as a lever by which to gain loyalty in annexed territories. Intent on winning over the hearts of Jordan's fellow Hanoverians after annexation in 1866, Otto von Bismarck sent funds to both the Natural History Society in Hanover and the museum in Hildesheim.[7] So long as the pride in local flora and fauna could be subsumed within the goal of unifying Germany, the naturalist tradition was extraordinarily valuable since the work of naturalists could be seen as part of a broader endeavor to catalog the nation's natural resources and heritage.

Not surprisingly, the issuance of new maps provided one of the first tools of Germany's unification program. There, on the printed page, the abstract concept of the German nation took its first concrete form as new borders appeared, encompassing a diverse range of principalities. As a boy, Jordan devoted hours to poring over the family's small collection of atlases, and he learned early on how the political boundaries could shift from edition to edition. Still, the boundaries that most influenced the life of a young boy on the family farm were those of the hedgerow, field, and forest. As an old man, Jordan described how, from the highest point on the hill separating the valley of the Alme from the valley of the River Innerste, he could see distant villages with green orchards, meadows, and roofs of brick red—"forest, fields, and villages contrasting sharply."[8] As he looked down upon the varied landscape and played in the fields and woods, Jordan learned of the diverse flora and fauna of his homeland. He could even map certain plants to distinct areas, as, for example, when he found that bilberries occurred only on the sandstone of the Griesberg.

The two kinds of maps, political and natural, may seem quite different, but finding out what plants and animals occurred in one's homeland or a foreign land had often been part and parcel of the highly political endeavor of geography. Indeed, geography often seemed inseparable from natural history, much to the chagrin of proponents of mathematical and theoretical geography who bitingly characterized descriptive geography as based on the belief "To name is to know: this was the dictum of the unenlightened pedagogue."[9] To name could also be to possess, of course. Martin Lister wrote in the seventeenth century of the effort to describe England's flora as setting "forth exactly what she has of her own."[10] Such long-standing ties between natural history and geography explain why it made sense for naturalists to accompany surveying ships such as the HMS *Beagle*, as their captains and crews carved out spheres of political and economic influence by drawing up accurate maps. In other words, a love of atlases could fit very well with an interest in observing and describing flora and fauna.

Supported by the respectability conferred by either demonstrating the existence of God or cataloging a town or a nation's natural products, the naturalist tradition provided a happy pastime for many by the time Jordan collected his first beetle. Towns and even villages had natural history societies, and at meetings men (and a few women) contributed thousands of notes containing observations of the natural world. Advertisements, placed by the dozens of natural history agents who capitalized on this interest by selling specimens and books, peppered the journals. But dreaming of a career in natural history hardly provided a secure ambition, particularly for a boy of Jordan's social and economic status. Jordan's parents, Wilhelm and Johanne Jordan, were small-scale farmers: they owned thirty-five acres of ploughland and supported a household of fourteen men, women, and children. It was not a childhood certain to deliver on dreams of becoming a naturalist of any kind. But when Jordan's parents died with Karl still only a young boy, his older brothers—amid the careful reorganizing that accompanies such tragedies—took note of his love of learning and, with the help of a generous uncle, arranged for him to attend the Hildesheim High School. This pursuit of more education set Jordan apart from many of his fellow schoolmates. Most families of Jordan's class did not expect their children to attend more than eight years of school. Indeed, teachers rarely encouraged the sons of farmers to pursue secondary school, much less attend university.[11]

In such circumstances, the trajectory that would turn Jordan's boyhood interest in natural history into a livelihood depended on profound social and economic changes taking place in nineteenth-century Europe. Jordan's ability to attend a *Realschule* (or *Realgymnasium*, after the authorities added more Latin to the curriculum in 1882) in Hildesheim reflected such changes. Founded as an alternative to the *Gymnasien,* or traditional, classically oriented high schools that served the children of elites, the Realgymnasien provided the more "practical" education increasingly demanded by the rising middle classes. The advocates of these more modern schools argued that "We can no longer permit the schools to ignore the great works of our century by Bunsen, Helmholtz, Siemens, Faraday and Darwin!"[12] Science would be central to the new schools, which concentrated on modern languages and the natural sciences rather than on Latin and Greek. Not all found arguments for the value of such training convincing. For many years, universities refused to admit the new schools' graduates, and their certificates carried less prestige among those who valued the "pure," well-rounded intellectualism of the traditional schools. Indeed, some pupils from the traditional schools called the students from the modern schools "practical hacks" or "illiterates."[13] But the new emphasis on science provided an encouraging context for a boy fascinated by natural history.

During Jordan's years at the high school, an inspiring teacher named Wilken channeled his interest in beetles into the more methodical study of both botany and entomology. Jordan later recalled that this entailed learning "a fair amount of knowledge of detail in the life history of many beetles, concerning their environment, food and variation." He particularly remembered one incident when in the forests south of Hildesheim he found closely related species of *Bembidion* beetles living within a few centimeters of each other at the side of a brook. One lived in sand saturated with water; the other inhabited the sand above.[14] Similar observations had inspired some of the central questions of the naturalist tradition. Where did such diversity of form and place come from? and how? Why did two similar yet noticeably different species of beetles inhabit each their own space in nature?

Within the tradition of natural theology, the accepted answer to such questions entailed a recognition of God's design and providence. But as natural explanations became the hallmark of science, this kind of answer was increasingly unacceptable. By the time Jordan cast about for explanations of biological diversity, an all-encompassing naturalism had famously provided Darwin's driving methodological standard, ultimately inspiring his theory of descent with modification by natural selection. For Darwin, the purely natural process of differential survival acting upon chance variation molded species into seemingly perfectly adapted forms. Incidentally, Darwin had noted that to provide a truly convincing case for natural selection, naturalists required far more information on the occurrence of variation in nature. And so, whatever one believed about the origin of species, many naturalists proceeded in good conscience with their appointed task of describing nature's diversity in great detail. Someone like Jordan could quite respectably begin learning what others had described and dream of adding new observations to naturalists' "atlas" of the living world.

BECOMING A ZOOLOGIST

The strong tradition of natural history in the school curricula bolstered this dream. But the ability of a student of Jordan's background to imagine studying animals and plants at a university depended, once again, on transformations taking place in both Germany and European society in general. By attending a Realgymnasium where his interest in science could flourish, Jordan had profited from the enormous expansion of secondary education that had accompanied German industrialization and the associated rise of the middle class. But a highly entrenched academic status quo had barred such students from enrolling in the universities. Around the time Jordan graduated from high school, a shortage of schoolteachers created by the growth in secondary schools convinced administrators and politicians to open a few universities to graduates of the more "modern" schools. Such students would be allowed to take degrees in the philosophical faculty (which included political and rural economy, forestry, pharmacy, dentistry, and other more scientific subjects).[15] Jordan took advantage of this new policy and enrolled in the University of Göttingen.

His timing could not have been better, for eventually so many students from the middle and lower classes took advantage of the new policies that, as the positions for schoolteachers filled and jobs became scarce, some feared that an "academic proletariat" would arise from such liberal admission practices.[16] Chancellor Otto von Bismarck and Prince Wilhelm soon worried that the small contingent of students from the new, modern high schools would turn out useful only as recruits for socialism, while others feared that "the discontent of so many young men with ambitions beyond their station in life, would lead to a situation like the one that gave rise to nihilism in Russia."[17] Men acting on such anxieties closed the universities to students of the Realgymnasien in 1890 for a decade. Even at the time of Jordan's admission to the University of Göttingen, less than 10 percent of the twenty-one thousand students attending German universities came from the Mittelstand.[18] Jordan benefited, then, from both sporadic symptoms of social change and the unpredictable luck of good timing.

Similar factors determined his available options for taking a degree. For much of the nineteenth century, youths interested in pursuing natural history at university had little choice but to study medicine or enter the clergy. But by the 1880s, a young German naturalist could obtain a formal degree in zoology for its own sake, the result of a successful campaign by self-proclaimed "scientific zoologists" in the 1850s. Like other disciplines intent on carving out professional space, those insisting that zoology be an independent, academic discipline had defined themselves by comparison to others. They contrasted "scientific" zoologists, for example, with "old-fashioned" naturalists who spent their time describing new species and constructing classifications. In contrast to this "systematic zoology," the new zoology would earn the coveted epithet "scientific" by studying the *natural laws* governing animals' internal organization, development, and generation. Furthermore, this new zoology would not be based in natural history museums or on collections of thousands of dead specimens. Although the following generation of zoologists toned down this dichotomy between "scientific zoology" and "systematics," it inevitably reappeared when fights over status and institutional space occurred.[19]

Latent criticisms of systematics as unscientific, combined with the mentorship of Ernst Ehlers (director of the Zoological Institute at the University of Göttingen), would profoundly influence Jordan's life and work. The vision he would eventually outline for the role of systematics in biology, and how it should be done, can ultimately be traced to Ehlers's influence. Though Ehlers ridiculed the description of new species for its own sake, he insisted on the important role of careful descriptive and classification work as a foundation for biology. Careful efforts to trace the lines of common descent and divergence, for example, could establish the actual path of evolution. For many German zoologists only an explicit incorporation of evolutionary theory within the work of classification would establish zoology as a science. Yet amid this new theoretical framework for the centuries-old effort to order the natural world, Ehlers maintained an important role for natural history collections.

This approach to systematics guided Ehlers's tenure as editor of one of the most important zoological journals in Germany, the *Zeitschrift für wissenschaftliche Zoologie*. While authors wishing to publish isolated descriptions of new species could go elsewhere, Ehlers considered detailed monographs on particular groups legitimate, scientific contributions to the journal. Such works brought zoologists small steps closer to the grander goal of finding the overall order in nature. Ehlers had developed specific answers to zoologists' vociferous debates over how to find that order. In deciding, for example, which characters to use to build a "natural" classification, or phylogeny, Ehlers insisted that one must use as many characters as possible, including inconspicuous and microscopic characters.[20] Under Ehlers's tutelage, Jordan learned to have a microscope close at hand and to examine internal morphology as a rule rather than last resort. He also became committed to the idea that through meticulous attention to morphological detail, naturalists could indeed produce more accurate phylogenies.

The methodological discipline reflected in such careful work also inspired Ehlers, and eventually Jordan, to give cautious responses to theories purporting to explain how evolution actually happened. The completion of a good monograph called for such detailed attention to specimens that it forced the writer to set aside premature ambitions to make theoretical breakthroughs. The mind might inevitably wonder how differences within a group may have come about, but Ehlers frowned on what he considered premature discussions of mechanisms, and he criticized the speculative efforts of his compatriots August Weismann and Ernst Haeckel. Most of Ehlers's own research, and that of his students, focused on systematics, descriptive anatomy and embryology and steered away from speculation regarding mechanisms.[21] Ehlers also reproved those who claimed experimental research provided the only avenue to scientific knowledge, a skepticism that would stand Jordan in good stead when faced by critics of taxonomy and natural history museums.[22]

Guided by Ehlers's belief in the importance of careful monographs and detailed, comparative morphological work, Jordan carried out a careful study on the anatomy and biology (a term designating the life history of an organism) of a group then known as the Physapoda, or thrips, during his years at the zoological institute.[23] The resulting work exemplified both the challenges of finding the order in nature and the usefulness of Ehlers's meticulous methods. Jordan chose the Physapoda on the grounds that where zoologists placed them in classifications varied enormously, depending on the author; some placed them among the Orthoptera (grasshoppers and crickets), others among Pseudoneuroptera (a group then encompassing dragonflies and mayflies). Still others moved them to the Rhynchota (the true bugs, cicadas, etc.) or placed them in their own order. Ehlers thought the study of such confusing groups particularly important. Here the neat categories of classification broke down, transitional forms seemed to abound, and the clear-cut boundaries of groups faded.

After careful study, Jordan found that where an author placed the group depended on the characters chosen, a lesson he would remember when tempted to pronounce any character a foolproof source for determining evolutionary relationships. Given his animal's apparent combination of the characters of different insect orders, Jordan thus learned that zoologists must not rely on a single or even a few characters in order to classify animals. To broaden the range of characters examined, he took up the microscope, adding minute characters to the list of facts available. Jordan even considered the "biology" of the group at length, announcing that the conclusions reached from life history data confirmed his findings based on anatomy. The study proved an early lesson in the importance of using various kinds of data in assessing the evolutionary relationships between forms.

Jordan's meticulous microscopical work on the Physapoda no doubt seemed a far cry from tramping the woods in search of beetles. But Ehlers's ecumenical approach to the various (sometimes competing) types of zoology maintained a path back to those woods. For rather than neglecting the university's natural history museum in favor of more modern laboratories, Ehlers brought his students and their microscopes into the natural history collection. When Jordan showed interest in the museum's collection of butterflies, particularly those collected by a man named Stromeyer, Ehlers encouraged him to study the specimens.

Despite the fact that Stromeyer's collection consisted primarily of butterflies, rather than Jordan's beloved beetles, the collection appealed to him for very specific reasons; namely, the high value his zoological training placed on detail and his own interest in variation and geography. Stromeyer's specimen labels noted where and when the specimens had been obtained. Such detailed labels may have seemed unnecessary to some enthusiasts glancing over the colorful insects, but Jordan emphasized that such facts advanced the knowledge of the distribution of butterflies in Germany.[24] He cited Adolf and August Speyer's work from midcentury as evidence of the scientific use to which such a collection could be put. In paying attention not only to insect systematics but the "biological conditions" in which insects lived, the Speyers had been able to define five vertical zones for Swiss and German Lepidoptera. Similarly detailed data on the Stromeyer specimens allowed Jordan to place the different species of butterflies in a variety of forest types, pastures, and meadows. One could thus use the specimens to study not only the facts of distribution but the reasons for the occurrence of particular species in certain environments. Studying the collection allowed him to conclude, for example, that first climate and second the condition of the ground and vegetation determined butterfly distribution within the European-Siberian region.

Jordan's experience with the Stromeyer collection solidified his belief in the worth of detailed locality labels and collectors' notes. Building upon his childhood observations of the close tie between landscape and animal forms, he obtained first-hand experience of the fact that, in contrast to those who saw collections of pinned specimens as too isolated from "real" nature, a well-labeled collection could include important information from the field that allowed one to relate fauna to its "biological conditions." The work on this collection taught him the use to which a mass of natural history observations could be put if collectors recorded the right facts. And it convinced him of the importance of harnessing a network of naturalists to amass more material. For in addition to reviewing decades of natural history literature for observations and records, he devoted time to correspondence with fellow entomologists in order to amass as much information as possible.

Well aware that the resulting work completely depended on a network of ardent naturalists, he extended hearty thanks to all the gentlemen who had helped him. Even as he worked in the corridors of the museum in Göttingen, poring over journals and surrounded by the specimens amassed by butterfly collectors, the goal of providing as comprehensive a portrait of distribution as possible tied him to a network of entomologists far beyond the boundaries of the museum walls. The fruits of this time devoted to correspondence, during which symbols of science such as the microscope and microtome sat abandoned, taught Jordan that good natural history work entailed much more than time in the field or at a museum drawer. The museum desk and a pad of stationary often proved just as important in order to escape the limitations of one's own powers of observation and locale.

Having completed his thesis, "The Butterfly Fauna of Göttingen," as well as a lengthy article on the butterflies of northwest Germany for the *Zoologische Jahrbücher*, Jordan graduated summa cum laude with a PhD in 1886. After completing the single year of compulsory military service for which his status as a scholar qualified him (rather than two), Jordan passed "through the gateway to the adult world," beyond which his certificate of demobilization conferred the privileges of marriage, permanent employment, and a place at the men's table in the local Gasthaus.[25] Given his intention to marry Minna Brünnig, his childhood sweetheart (he would do so in 1891, and the couple would have two daughters, Ada and Hilda), hiking about the fields in search of Coleoptera could no longer take up much of his time. Jordan had to make a living.

By the time he graduated from university, an academic position seemed unlikely. Competition for the scarce number of positions in zoology could be fierce, and his own preference for the zoology of the museum rather than the laboratory may have turned him away from entering the fray. Jordan later recalled that during his years at university, systematics had become largely the province of amateurs, "the public Museums of Natural History being as yet in their early youth," with the professional zoologists of the universities concentrating "almost exclusively on the classes of animals which had to be preserved in liquid."[26] As we have seen, the comment belied the well-rounded and broad education in zoology he had obtained under Ehlers. Perhaps it reflected the memory of a student casting about for academic positions after graduation, for by 1893 systematic zoologists indeed complained that universities filled nearly all chairs in zoology with "morphologists."[27] Jordan had been trained to see each as dependent on the other, but such a nuanced view could disappear in fights over positions and institutional space.

Jordan thus turned to Germany's extensive secondary educational system for employment. As we have seen, the expansion of the school system that accompanied industrialization had inspired some universities to open their doors to students from the more "modern" high schools on the grounds that this would fill the high demand for teachers trained in modern languages and natural sciences. Indeed, more than two-thirds of the students in the philosophical faculty became secondary teachers.[28] Bolstered by the prospects of a secure livelihood, Jordan pursued a teaching diploma, and in 1888 he received an appointment as a master at a grammar school in Münden, southwest of Göttingen. Certainly, he need not thus abandon natural history, since many of those teaching zoology in the high schools continued work on taxonomy, faunistics, and what would soon become known as ecology.[29] The kind of questions that had fascinated Jordan since boyhood—namely, the patterns of geographical distribution—could luckily be studied without access to the labs or microscopes provided by Ehlers's Zoological Institute.

As Jordan's catalog of the beetles of Hildesheim published in the *Societas entomologica*, of Zurich, between 1886 and 1888 demonstrated,[30] there remained an ample framework for contributing facts along the lines established by Linnaean natural history outside of academia. In contrast to the physiological and morphological research that took over university research with the rise of "scientific zoology," the study of geographical distribution and the analysis of faunistic regions dominated research in the civic setting in which Jordan now lived and worked.[31] Here museums flourished, supported by both the state and private patrons, and individuals amassed their own collections of animals and plants. Immersing himself in this world, Jordan annually visited the diluvial flats of northwest Germany to collect beetles between 1888 and 1891. He captured more than nine hundred of the twenty-five hundred species he suspected occurred in the area. Once again, he noticed how this particular fauna compared with the highlands, and how the strange conditions of the soil and climate resulted in a characteristic composition of flora and, therefore, a distinct beetle fauna. Variation in the type of soil, the proximity to water (certain species could be found only at the edge of puddles), the height of a sand pit—all these factors could result in the presence of different species.[32] This study only increased Jordan's appreciation of the importance of locality and habitat. It also reinforced his belief that naming species formed just the first step to the more important endeavor of sorting out the relationships between organisms and their environment.

The fact that Münden's new schoolmaster was an ardent, academically trained naturalist soon became known among those who cared about such things in the town. In particular, A. Metzger, the professor of zoology at the Royal Prussian Academy of Forestry, welcomed him into his home, introducing Jordan to yet another context in which one might study insects. The Prussian government had lavishly financed such academies.[33] They served as the training grounds for state foresters, and natural history formed a central part of the curriculum. Not surprisingly, insects—studied under the rubric of "forest entomology"—proved of special interest since the practice of planting large areas with single species of trees had increased the problem of insect pests.[34]

This kind of natural history included the classification of animals, plants, and minerals, but focused primarily on the study of animals and plants "from a forestal point of view," particularly those insects of use or harmful to forests. Forestry schools relied on both experimental gardens and museum collections filled with specimens of forest animals and plants. Curators placed insect pests alongside healthy and damaged specimens of a tree "so that the student can see at a glance the nature of the damage, and connect it with the animal which causes it."[35] Notably, this kind of explicitly economic context for natural history was not yet strong in many other European countries, although certainly some British naturalists and administrators argued that following Germany's example would improve the efficient management of their own empire.[36]

This very distinct tradition within natural history almost captured Jordan, when, in 1892, he accepted a post as master of mathematics, physics, and natural history at the School of Agriculture at Hildesheim, sixty miles north of Münden. From here, he noted he "might possibly have ended as a professor at some High School of Forestry."[37] Had he enlisted in the ranks of forest entomology, many aspects of his working life—including both the kind of knowledge amassed and the status of that knowledge within society—would have been quite different. Forest entomologists paid a very specific kind of attention to insects, famously including more attention to life history and the interrelationship between insects and their environments. In addition, the status of "the professional" forester (defined at the time as one who applied the scientific study of natural history and physical laws to forest management)[38] was relatively secure, compared with other kinds of naturalists. After all, they could justify their existence by appealing to the importance of healthy forests to the German nation. But Jordan soon followed quite a different route, both to knowledge about the natural world and to status in society.

THE COSMOPOLITAN NATURALISTS

Some time in 1892, a thin, bearded, thirty-four-year-old ornithologist named Ernst Hartert offered Jordan, on behalf of his employer, a position as curator of insects at the Walter Rothschild Zoological Museum at Tring, in England. Jordan first met Hartert at the home in Münden of Count Hans Hermann Carl Ludwig von Berlepsch, a wealthy naturalist with whom Jordan had become good friends. Berlepsch's collection surpassed anything Jordan could have amassed during his vacations from school. As one of the wealthiest naturalists in the district, Berlepsch introduced Jordan to the enormous private resources that were channeled into collections and created sites for research outside the universities. In the hands of men and women of wealth, the honored tradition of amassing collections of local faunas—an avocation in which Jordan participated so avidly—became the inventory of foreign lands and, often, Empire.

Berlepsch, for example, had hired a network of collectors throughout the neotropics. He could also buy specimens from the many natural history dealers who had set up shop in Germany, the hunt for profit having closely tracked the enthusiasm for collecting nature's diversity. Using these resources, Berlepsch had developed one of the most important private collections of birds in Germany, and his home had become a center for both hired collectors, coming and going for their orders, and specimen setters processing specimens for display and study. Such private collections provided a tremendously fertile place for studying the geographical distribution of animals, or zoogeography, on a far grander scale than the regional studies represented by Jordan's dissertation.

The British ornithologist and secretary of the Zoological Society of London, Philip Lutley Sclater, outlined the goals of this endeavor when he used the data amassed by such naturalists to classify the world's surface into six "centers of creation" (now termed biogeographic regions: the Palaearctic, Aethiopian, Indian, Australian, Nearctic, and Neotropical). Sclater urged: "We want far more correct information concerning the families, genera, and species of created beings—their exact localities, and the geographical areas over which they extend... even local varieties must be fully worked out in order to accomplish the perfect solution of the problem."[39] A large network of collectors, natural history agents, and publications provided the means to fill in blanks on the map and refine the boundaries of known regions. Almost two decades later, the naturalist and codiscoverer of natural selection, Alfred Russel Wallace, provided a more detailed system of subregions in *The Geographical Distribution of Animals*. Subsequently much of the study of zoogeography focused on mapping out the distribution of living animals and discussing the accuracy of the realms outlined by Sclater and Wallace. Meanwhile, naturalists like Darwin, Wallace, and Moritz Wagner had changed the meaning of the faunal maps that resulted: from representing separate centers of creation, they came to show the legacies of different evolutionary pasts that could be analyzed based on the presence or absence of species.[40]

The natural history that was pursued by collection owners of Berlepsch's means relied upon the movement of capital from Europe across the globe. This capital formed the basis of a growing administrative and transportation infrastructure that opened up new regions to European investment and exploration. Given this increased ability to access faraway lands, wealthy naturalists and administrators of museums in national capitals readily channeled the tradition of compiling local fauna and flora into producing an inventory of more distant regions. As contemporaries commented, the increase in population, wealth, education, facilities for travel, and intercommunication had led to more persons working on material from all over the world.[41] Research on geographical distribution benefited enormously from both improved infrastructure for travel and the broader endeavor of geography. "It is easy to prove," wrote the zoologist Ludwig von Graff, "that the colossal addition to our list of animal forms—about 50,000 in 1832, to-day about 150,000—is due no less to the increase of means of communication and to the evolution of geography."[42]

By the time Jordan arrived at Berlepsch's home, German naturalists' participation in this project had gained extraordinary momentum from the nation's increasing imperialist ambitions. Around 1883, Bismarck had responded to both public pressure in the face of economic depression and increasing British and French protectionism by actively pursuing formal imperial expansion. He did so just in time to join the "scramble for Africa" that resulted in France, Germany, Belgium, Italy, Portugal, Spain, and Britain dividing up the continent among themselves. By the century's end, Germany had acquired Togo and the

Cameroons, the areas now known as Namibia and Tanzania, and further afield, the northeastern corner of New Guinea, the Bismarck Archipelago, and the Marshall Islands.[43]

For decades, the European movement into Africa would define international relations and much of natural history as well. Whereas prior to the 1860s, German naturalists' infiltration into Africa had relied on the support of small institutions, scientific academies, and wealthy aristocrats such as Berlepsch, the nation's colonial era inaugurated governmental patronage of natural history in earnest.[44] As director of the Zoological Collections for the Museum für Naturkunde in Berlin, Karl Möbius, for example, wrote, "A new, rich, and extremely important source opened itself for the museum with the acquisition of colonies."[45] At their naturalist society meetings back home, zoologists and botanists listened as explorers discussed topics such as the position of Africa with regard to Germany and the need to establish factories and trading stations "owing to the colonial policy at present agitating German political and commercial circles."[46]

The adventures of Jordan's new acquaintance Ernst Hartert, prior to his joining Rothschild's museum, exemplified German naturalists' increasing ties to their nation's imperial ambitions. In 1885 he had joined an expedition up the River Niger led by the explorer Eduard Robert Flegel, who canvassed for expedition funding on the grounds he would be establishing trade relations in the region. One of Hartert's tasks during the expedition had been to travel overland between Loko and Sokoto in order to discuss a potential trade agreement with the sultan and deliver presents from the kaiser. With such agenda items, it is little surprise that the British kept a close watch on the expedition's activities.[47]

Britain had long since established the model for naturalists intent on cataloging and describing the flora and fauna of newly claimed lands. Botanists at Kew Gardens and zoologists at the British Museum had called on the importance of inventorying imperial possessions as justifications of their institutions and work for decades. Building on their strong ties to geographical exploration and emphasizing their expertise in cataloging natural products, naturalists followed the establishment of colonies and spheres of influence in the nineteenth century almost as quickly as, if not with, military might.[48] By century's end, the Cambridge University ornithologist Alfred Newton could quite respectably begin some remarks at the 1887 British Association for the Advancement of Science meeting with the adage "Property has its duties as well as its rights." He continued: "Various events have given to this nation rights of property in many parts of the globe. I think we ought to justify those rights, and there is no better way of doing this than by performing the corresponding duties." By this he meant the "properly organized biological investigation" of the innumerable places in the dependencies of the British Crown not yet scientifically investigated.[49]

Although certainly not new, the emphasis on natural history in the name of the nation contrasted starkly with both traditional justifications of science for its own sake and scientists' associated self-identity as members of cosmopolitan communities rather than nations. Indeed, it bound natural history and nationalist sentiment so tightly that employing foreign naturalists to curate even a private collection in Britain might be frowned upon. Walter Rothschild's mentor at the British Museum, Albert Günther, had no doubt advised him to bear the resentment that might be inspired by sending Hartert to Germany to recruit a new curator. Günther justified his own recruitment of Continental zoologists by frankly stating, "Britain did not turn out men of the right caliber."[50] The entomologist Lord Walsingham agreed that it was better to "face the national prejudice and get competent foreigners who would be much more useful in the department, and who would require less coaching."[51]

Ultimately, Jordan, too, would argue for the importance of looking beyond the bounds of nationhood in the interests of getting good work done. For now, he must have been thankful that the effort to keep "Britain for the British," a movement urging that the scarce resources of British society be reserved for those born in the country,[52] had not extended to Rothschild's coffers. The cosmopolitan ethos was also reflected in Hartert's constant emphasis to the museum's correspondents that Rothschild gave no special privilege of place or price to English insects.[53] Rothschild's Museum, Hartert often insisted, was a "scientific, all world collection."

Walter Rothschild's act of hiring two foreign curators was not the only way the family supported cosmopolitan natural history. Naturalists' access to far-off lands often depended on capitalists having arrived there first. And the Tring Rothschilds, as the wealthiest banking family in the world, certainly had plenty of capital. The House of Rothschild's biographer writes that for a contemporary equivalent to the family's wealth and influence, "one has to imagine a merger between Merrill Lynch, Morgan Stanley, J. P. Morgan and probably Goldman Sachs too—as well, perhaps, as the International Monetary Fund, given the nineteenth-century Rothschilds' role in stabilizing the finances of numerous governments."[54]

Walter Rothschild built his collection using an infrastructure for trade and travel bankrolled by this tremendous wealth. Tring Park, the four-thousand-acre Rothschild estate, was thirty-three miles northwest from London, but in many ways existed at the center of the planning and financing of the British Empire. Founded by Nathan Mayer Rothschild in 1809, N. M. Rothschild & Sons had joined other merchant banks in providing much of the credit for world trade, acting as guarantors for bills and as intermediaries in raising long-term foreign loans.[55] They participated fully in formal empire (the exertion of official control over far-off lands through the establishment of political and economic infrastructure). Colonial Secretary Joseph Chamberlain turned to N. M. Rothschild in 1896, for example, for financial help leasing Delagoa Bay from Portugal in an effort to forestall German influence in the region.[56] The year Jordan arrived at Tring, Lord Rothschild was financing the imperial adventures of Cecil Rhodes in South Africa, though he became increasingly dissatisfied with that particular imperialist's methods.[57]

The bank also formed a central financial backbone to what is now known as Britain's "informal empire" a term for those regions in which British creditors and leaders had a crucial influence on the economic policies of ostensibly independent nations. Having helped secure loans for Brazil, for example, on the occasion of that nation's independence from Portugal in the early 1820s, the Rothschild Bank had issued £173 million in public loans to the country between 1855 and 1914, at least 20 percent marked for the construction of railways.[58] The same maps that served as a critical source of information to prospective investors in railways and other ventures guided naturalists wishing to explore newly opened regions. Once the railways appeared, often solely to move goods (such as coffee in Brazil), naturalists eagerly boarded the trains to the newly accessible regions.

With both the formal and informal support of Rothschild wealth, the vision Hartert conveyed to Jordan of a "scientific, all-world collection" could actually become reality. In traveling to Tring in 1892 to meet Rothschild and see the museum, Jordan found a cosmopolitan collection that already formed the center of a natural history network of both Continental naturalists and collectors throughout the globe. Once he made the decision to take advantage of this collection, its network, and its world, Jordan's salary would be £200 per annum for the first five years and £250 the following five. It was a sum Rothschild sometimes spent on single specimens, and one-tenth of what he would pay to mount his series of sixty-five specimens of cassowaries.[59] But the salary would place the Jordan family firmly within the respectable middle class, a status that fit quite well with the ethos and ambitions of the average professional scientist. Soon, Jordan had written to accept the job. In explaining his decision to Ernst Ehlers, he emphasized the great richness of the collections and library (though he also permitted himself the weighty aside that of course one could do without the kangaroos wandering Tring Park).[60]

In April 1893, Jordan and his family packed up their belongings and made the trip across the Channel and through the English countryside to the village of Tring. In passing through London they witnessed the busiest port on Earth, where a force of tens of thousands of dockworkers unloaded the spoils of Empire. There, "food, drink, spices, herbs, teas and coffees, animal hides, furs, feathers, ivory, gold, silver and other metals, precious stones, timber, cotton, curios, jute, hemp, objets d'art, in fact nearly everything the planet produced," were sorted, cataloged, and packed into huge warehouses behind the quays and wharves of the Thames.[61] Some of the boats carried live animals, destined for the zoological gardens of the Zoological Society of London or wealthy naturalists, such as Walter Rothschild, who had their own private menageries. The family may have passed by the markets Rothschild visited to, as rumor had it, "pay any price" for specimens and live animals. Between the unloading on the dock and the market, the products of Great Britain's empire were unpacked and repacked, sorted and processed. It would be Jordan's job to sort, catalog, and pack away into museum drawers a small part of this diversity.

THE "NICE BERTH": CURATING A ZOOLOGICAL MUSEUM

By the time Jordan accepted the post of curator of insects at the Walter Rothschild Zoological Museum (sometimes called the Zoological Museum, Tring, or simply the Tring Museum), he had achieved much in German society. In attending university, he had joined a class of elites whose social status was defined by education rather than heredity or wealth.[62] Though from humble beginnings, in obtaining the PhD he had become a member of the *Bildungsbürgetum*, or the educated, rather than the moneyed, middle class. He had obtained the coveted title of Herr Doctor in a society where one's title designated a man's place and status. Among the growing number of university graduates, he possessed one of the best degrees in the world.[63]

However, within the increasingly specialized world of the sciences, Jordan's choice of zoology did not confer rank in all places. He often recounted how a fellow student, a mineralogist, had taken the liberty to explain to him "with the fervour of youth and the insistence of an enthusiast, that the only branch of the Linnean Natural Sciences worth studying was Mineralogy, because this science was based on the exact methods of physics, chemistry and mathematics." This common belief in the methods of physics, chemistry, and mathematics as the standards by which all science must be measured would haunt zoology and the naturalist tradition throughout Jordan's life and work. At the time he defended his chosen subject passionately, but Jordan knew, as he trudged up the steps to the zoological institute at Göttingen, that his young mineralogist friend had pitied him.[64]

This youthful exchange made an impression on Jordan. In later describing the "long entomological journey" that had led him from a boyhood fascination with beetles to his position at Tring, he often placed his decision to leave his own country and accept the job working for Rothschild as firmly based in his wish to do good science. He described, for example, how on witnessing the enormous variety of forms in the Tring collections, containing "such a variety of insects as no single tropical country, however prolific, can produce," he decided to "break the journey and to collect facts in the Museum and to draw conclusions therefrom."[65] In other words, Jordan described his decision as reflecting his ambition to ultimately establish scientific generalizations about the natural world. Such generalizations would, in turn, be firmly rooted in the enormous number of facts amassed within the museum drawers. He believed that working on this collection of insects could result in useful and good science, no matter what his mineralogist friend might say.

But zoologists knew that "drawing conclusions therefrom" using such a collection often took a back seat to collecting and organizing facts, and, furthermore, that organizing facts could result in something quite different than finding the order in nature. The self-described "scientific zoologists" who established positions at German universities in the 1850s believed that a man in charge of a museum could not do really good research and insisted that professors of zoology should not be burdened by the administrative duties of collections.[66] At the British Museum, severely limited space and manpower meant the zoology department needed all available staff members to carry out the task of labeling, cataloging, and organizing the millions of specimens. And salary often depended on the number of specimens cataloged.[67] By the time Jordan first visited the grand museum in London, the number of specimens waiting to be cataloged had reached overwhelming proportions. From well over a million zoological specimens in 1880 (1,300,000), by 1895 the collection had almost doubled (to 2,245,000).[68] Other collections faced similar overloads. Curators with visions of "drawing conclusions" based on such collections could often find themselves inundated by the need simply to organize and catalog incoming specimens.

Jordan's ability to move beyond the tasks of organization and cataloging would ultimately depend on a range of factors. But the most important determinants were the interests and character of his employer. A general description of Lionel Walter Rothschild, who was to become Lord Rothschild upon the death of his father (N. M. Rothschild) in 1915, is best left to his loyal niece, the daughter of Charles Rothschild, Miriam Rothschild. She grew up in the halls of the museum and became an expert naturalist herself. Describing Jordan's arrival at Tring, she wrote:

> Walter Rothschild was then twenty-four years of age and a most eccentric figure. He was always just about to leave the Rothschild banking house in disgrace, not for any serious misdeed, but because he displayed a quite remarkable lack of business acumen coupled with curious but very extravagant tastes.... He went up to Cambridge accompanied by a large flock of kiwis, and could be seen bowling down Piccadilly behind four lively zebras. He was endowed with a most remarkable, somewhat freakish memory which retained the specific characters of large groups of animals and enabled him to pick out, instantly, any new species without recourse to the relevant literature. His collecting was also a trifle eccentric: he was for instance obsessed by size and felt impelled to collect—at whatever cost—the largest sponge, the largest tortoise, the largest ape and the like for his museum. He himself was a large man with an enormous head, standing six feet three.... In addition he suffered from a peculiar impediment in his speech together with a certain inability to control his voice, so that he alternately spoke with an embarrassing stammer or in a loud bellow. Although he was essentially gregarious and jolly, with boundless good nature, this physical disability, coupled with the knowledge of his father's constant, bitter disapproval of everything he did or liked, had made him shy and uncouth in society. In a sense he never grew up, remaining all his life the truant, rather irresponsible schoolboy, escaping with his butterfly net and pill boxes into the fields, and in later life secretly buying the largest boa constrictor in the world when his curators at the Museum were urging the purchase of some less spectacular but more important addition to the collection. Yet Lord Rothschild was both so very peculiar and so very kindly, and also so richly endowed with those nameless qualities which combine to make a natural gentleman, that all those who knew him well liked him and found him delightful, and those with whom he came into daily contact grew, as time went on, increasingly fond of him. Everyone was always astonishingly ready to rally round and take responsibility for the latest scrape, whether it entailed coping with cassowaries or chorus girls, a final income tax demand or a hungry iguana. Moreover,

working with him was always easy, even if at times it must have appeared to his curators like a partnership with a rogue elephant![69]

Miriam Rothschild's account captures well the complex nature of Hartert's and Jordan's positions at Tring, for over the years they would become as much mentors as employees to Walter. They were not Rothschild's first mentors. By the age of thirteen, Rothschild paid regular visits to the British Museum, where the keeper of zoology, Albert Günther, took him under his wing.[70] After noticing that Lord Rothschild did not, like his wife Emma, indulgently nourish their son Walter's interest in natural history, Günther resolved not to encourage the boy's effort to form a large collection.[71] But Walter soon found other teachers. Jordan later described how, when Rothschild went to university, the Cambridge ornithologist Alfred Newton transformed the young "amateur collector who derived pleasure from collecting into a naturalist whose collections had the object of increasing our knowledge of nature."[72] Though Rothschild had begun his collection by purchasing specimens "as and when they were offered," Newton inspired him to hire the collector Henry Palmer to go to the Chatham Islands, off New Zealand, in 1890 and then the Sandwich Islands (Hawaii) from 1890 to 1893.

The resulting masses of specimens from such ventures soon overflowed Rothschild's rented sheds at Tring. Meanwhile, his parents, concerned at their son's growing fondness for chorus girls, decided that a proper museum would be a far more seemly hobby, and so, when Rothschild graduated and took up responsibilities at the family bank in London, his parents built him a natural history museum for his twenty-first birthday.[73] In addition, in August 1892 Walter opened a two-story exhibition gallery for the public, with mounted specimens of mammals, birds, fish, amphibians, reptiles, and insects. Hiring Hartert had been the first phase of this exciting venture; bringing Jordan on board had been the second. The official title of the museum became the Walter Rothschild Zoological Museum, although Walter Rothschild always called it "My Museum."

Certainly, at times, the latter could be the most appropriate title. Even as they gratefully acknowledged his enthusiasm and loyalty, Jordan and Hartert had to constantly temper Rothschild's passion for specimens, books, and live animals in the interest of redirecting his wealth into more "scientific"—and less expensive—channels. Rothschild often made his "own little arrangements" when visiting dealers in London, though Hartert, at least, absolved himself of all responsibility for such "city-bargains."[74] Although Hartert and Jordan often reprimanded collectors for sending unsolicited specimens, they sometimes obtained replies that directly quoted Rothschild, who had already agreed to pay for all specimens sent.[75]

Rothschild's expensive purchases plagued his curators' efforts to maintain a reasonable budget that would keep the Rothschild family's purse strings loose. Hartert and Jordan knew that the entire enterprise of turning Tring into a scientific collection depended on the willingness of Walter's father to give the museum access to the family's financial reserves, and relations were already strained. Jordan had no doubt heard of the infamous 1892 "Felder affair" before coming to Tring. Dr. von Felder, burgomaster of Vienna, had publicly announced that he had negotiated a sale of his insect collection to Walter for 50,000 guldens (£15,000).[76] Rothschild, who at the time was given an annual allowance of £1,300, had, on demand from his irate father, tried to call off the sale, but the deal eventually went through (*Science Gossip* reported that in the end Felder received £5,000 for the great collection).[77] Lord Rothschild understood neither his son's obsessive interest in natural history nor the disposition that led to such scrapes, and by 1893 "with a ruthlessness born of profound and baffled disappointment and regret, abandoned Walter and turned with relief to his second son Charles."[78]

Walter's mother, Emma, took a sincere interest in her eldest son's natural history endeavors, so long as they did not bode ill for his health. Indeed, Miriam Rothschild thought that Rothschild's curators took the place of her son's German governess and tutor in Emma's mind. In Jordan's case, this impression no doubt arose from the fact that, in contrast to Hartert's "good-humoured handling of the rogue-elephant aspect of Walter's drive and personality," Jordan tended to treat him with "tact, good humor, and affection"—as an overgrown schoolboy whom a few "mildly patronizing" reproofs could keep in check.[79]

Jordan would later blame his employer's eccentricities on his upbringing, particularly the mother's overprotectiveness and the father's impossibly high expectations.[80] "Admiration for the intelligent boy and early flattery were not missing," he explained after Walter's death in 1936. This coddling, Jordan wrote, had accustomed Walter "to regard himself as the center of the world and to expect the fulfillment of his boyish wishes as a natural corollary of his important position. Shy by nature, as he grew up he became unduly self-centered and averse to asking advice."[81] This aversion would have tremendous repercussions for the future of "My Museum" and would try Hartert and Jordan's patience continuously over the years. Still, Rothschild's reticence came accompanied by a deference to his curators that must have tempered their frustrations. Ernst Mayr, visiting the museum to prepare for an expedition sponsored by Rothschild in the late 1920s, recalled how Rothschild would never intrude upon his curators' work, preferring to "wait by the door until noticed and invited to enter."[82] And his loyalty knew no bounds, whatever his other faults. Asked how he got on with Jordan, Rothschild replied, "We have terrific arguments, but the fellow is always right!"[83]

Whether Rothschild's newly hired ornithologist and entomologist had "terrific arguments" seems lost to the historical record. While there is little evidence that Hartert and Jordan formed a strongly personal friendship, they pursued broadly similar research programs, and when building Rothschild's collection that counted for a great deal. Certainly Jordan must have been grateful for Hartert's willingness to support his, "the junior curator's," own entomological research interests. Soon after Jordan arrived, Hartert began informing Tring's network of collectors that "we are especially fond of Anthribidae, of which Dr. Jordan makes a special study."[84] The only recorded unpleasant, yet telling, exchange in their relationship occurred when the University of Marburg in Hessen conferred an honorary doctorate on Hartert in 1904. Jordan's own title of PhD meant much to him, and he seems to have imbibed at least a certain amount of the arrogance noted as a common failing among graduates due to the social position academic achievement conferred in German society.[85] Upon Hartert's return from the degree ceremony, Jordan "flatly refused to accord him the title of Doctor, which everyone else hastened to do." Jordan thus refused to accord his fellow curator equal academic status. But Hartert, the son of a German army officer, managed to get in his own snubs. Miriam Rothschild recalled that he was "faintly contemptuous" of Jordan's lower social status, which is perhaps why Jordan resented the apparent removal of his one claim to superiority. Otherwise, Jordan and Hartert worked together well enough, although their competing claims for social status kept their wives from becoming friends. As Miriam Rothschild recalled, "Mrs. Hartert was openly contemptuous of Mrs. Jordan."[86]

The Tring natural history circle grew when, soon after arriving at the museum, Jordan was joined at his microscope by sixteen-year-old Charles Rothschild, Walter's younger brother. Miriam Rothschild recalled that although "Hartert remained essentially Walter's man, Jordan became a devoted and lifelong friend of his younger brother Charles."[87] Charles, too, was fascinated by insects, and chose the Siphonaptera, or fleas, as his favorite group. It was an unusual choice, since for much of the century beetles, butterflies, and moths had monopolized entomologists' attention. Although, as years passed, he assiduously took up his duties at the bank, Charles continued to work at his entomology in the evenings. Jordan drew the illustrations for his publications (including five hundred new species of fleas) and they would compose dozens of joint papers before Charles Rothschild's tragic death in 1923.

Eventually, the obligations conferred by his friendship with Charles would determine how Jordan spent his final decades at the museum. For he would eventually set the butterflies, moths, and beetles aside to carry on the study of fleas in Charles's honor. But in the early days, the obligations of memory and friendship were far off and those of a salary predominated. Since Walter Rothschild had to work during the day—or, as Jordan put it, "at least be present" at the bank in London[88]—Jordan sometimes spent fourteen hours or more in the museum. He rarely worked alone. A large collection of butterflies, moths, and beetles, with new specimens coming in daily required a small group of "setters," hired to pin specimens in their boxes, so that the height and manner of setting would be consistent throughout the museum's drawers. Jordan later recalled how sometimes the museum hired "several ladies in Tring and neighbourhood" to help, "so that at one time there were more than half a dozen outside helpers to cope with the many collections that came in."[89] Meanwhile, caretaker Alfred Minall and his assistant Fred Young saw to cleaning and heating the premises and acted as taxidermists; artist Frederick W. Frohawk painted illustrations for Rothschild's books and journal; and soon after Jordan arrived, Arthur Goodson joined the staff as assistant to the curators to help them label and sort specimens.

Before he could make the most of this workforce, Jordan's first task upon arriving in Tring was to organize the countless insect specimens piling up in boxes around the museum. In two years Jordan processed the more than three hundred thousand specimens of beetles of sixty thousand species—and described four hundred new species in the process.[90] Meanwhile, more specimens arrived daily. For during Tring's heyday in the 1890s, natural history agents, private collectors, friends, and colleagues constantly delivered specimens supplied by hundreds of collectors moving along the trade routes of the British Empire and beyond.

When new shipments of specimens arrived at Tring, Rothschild would unpack the boxes and roar to his curators, "Schauen sie, Schauen sie . . . COME AND SEE WHAT WE'VE GOT!"[91] Jordan's immediate task after dutifully marveling at the new shipments was to sort the specimens into their ordered spaces between the hundreds of thousands of beetle specimens already distributed into their appropriate cabinets. In other words, Jordan's job, both as a curator and as a naturalist, was to find some sense of order within all the diversity arriving in packages from afar. He had to do so in the midst of carrying out, with Hartert, the day-to-day administration of a museum that was becoming increasingly popular with both the public and other naturalists.

The curators opened the museum to the public after 3.00 p.m. four days a week, plus a few hours in the morning on Fridays. Hartert, at least, complained of the time the average of 136 visitors a day cost.[92] Jordan, perhaps nostalgic for his days as a schoolteacher, seemed more tolerant of such interruptions, evidenced in the gratitude expressed by the headmaster of Stowe School, Buckingham, after Jordan gave his pupils a tour of the museum in the 1920s.[93] At least a few of the boys must have looked up to the tall man in charge of the enormous collection of insects, as children do, with the thought that they, too, would be an entomologist in a museum one day.

Of course, by the time his young visitors walked the halls of the museum, and certainly by the time they would reach their twenties and be casting about for careers, the world was quite a different place than when Jordan fell in love with beetles, obtained an education in zoology, and became a museum curator. Meanwhile, though, life at the Walter Rothschild Zoological Museum proved a "nice berth" indeed. The captain of the ship may at times have seemed unmanageable and in danger of placing the whole voyage in danger. But the sea seemed calm enough and, thanks to a robust naturalist tradition, the rest of the crew, from collectors in the field to the specimen setters at home, stood by ready to assist.

MOBILIZING THE NATURALIST TRADITION

Soon after arriving at Tring, Jordan took over that part of the museum correspondence that dealt with insects, writing to collectors, natural history agents, and booksellers. It was time-consuming work, during which the specimens of insects lay neglected and the cabinet drawers stayed closed. But, having depended on the cooperation of many entomologists for his own work on butterflies and beetles, Jordan knew the importance of establishing cordial relations for Rothschild's museum with as many members of the diverse world of natural history as possible. He knew that the success of the collection depended on his and Hartert's ability to cultivate this network and inspire cooperation. Good natural history, at least that based in collections, was a social endeavor, not a solitary one. And building the collection and putting it to use thus depended upon maintaining relationships through which Tring could acquire and process specimens and distribute and share information.

He began by enquiring about prices, ordering specimens, arranging payment, and organizing exchanges from the dozens of collectors and natural history agents that made a living from the naturalist tradition, including, to name just a few, Hermann Rolle, of Berlin, Dr. J. Bohts, of Paraguay, Gustav Schneider, of Brazil, Gerrard, Fruhstorfer and H. Deyrolle, of Paris, and Otto Staudinger, of Dresden. For example, he confirmed with O. E. Janson, of London, that Rothschild took beetles at nine pennies per specimen; and he informed Messrs. H. Deyrolle and Douckier of Paris that their specimens did not justify their high prices. He reminded W. Niepelt, of Freiburg, that the museum would not accept all of the specimens in a dispatched collection of butterflies and beetles; and he ordered books from R. Friedländer, of Berlin. For much of this correspondence, the museum used the Rothschild banking house stationery and postal network, relying on the same system of communication that made the bank so efficient. The infrastructure of commerce proved of excellent use, it turned out, when it came to amassing specimens and books of natural history as well.

By far the bulk of the correspondence was directed to fellow collection owners and collectors in the field, two distinct groups often separated by divisions of class yet united in their common dependence on specimen exchange. Between 1890 and 1908, before his parents reined in his finances, Rothschild was rumored to have "employed" more than four hundred collectors (although Hartert insisted that "the idea of our keeping such a number of collectors at one time is absurd").[94] Indeed, Jordan carefully pointed out to inquiring collectors that all of Tring's correspondents traveled and collected at their own risk and expense. Although the museum sometimes advanced small sums for the payment of initial costs, none of its collectors received salaries.[95]

Guided by their wish to build a distinct kind of collection, the curators certainly had their favorite collectors. These included W. J. Ansorge, William Doherty, Rollo Beck, Albert Eichhorn, Alfred Everett, C. Hose, S. M. Klages, Heinrich Kühn, and A. T. Meek.[96] They also had a blacklist. "Has Mr. Fruhstorfer paid you a visit?" Jordan once asked a colleague. "He is probably in search of new customers, we shall not take again a set of his Lepidoptera."[97]

The curators valued good collectors and knew the debt they owed to those who risked their lives to send Rothschild's treasures. The human capital at the root of the entire endeavor is chronicled throughout the memorials of Tring's journal, *Novitates Zoologicae*. Occasional obituaries of collectors reflected the hardship entailed in building a collection like Rothschild's. Despite the dangers, Rothschild pressed his collectors to work unexplored areas such as the mountainous regions of New Guinea. In "On the Birds of the Orinoco Region," Hartert explained that the collectors, Mr. George K. Cherrie and Mrs. Stella Cherrie, had not only suffered severely from fever but had "many losses and annoyances from some of the endless revolutions for which Venezuela is so notorious." Another collector on the river had lost his collection of birds, plants, and minerals through an accident in the rapids, "and very narrowly escaped starvation."[98] Kühn lost a collection from Wetter due to shipwreck, and fever made the expedition an exercise in misery. Both Eichorn and Meek had to turn back much earlier than planned from their expeditions into the mountains of New Guinea—Eichorn due to an illness that deprived him of the use of his right arm,[99] and Meek due to the death of some of his "boys" (the word used for native guides and bearers of all ages) from beri-beri.[100] Well aware of the dangers collectors faced in the field, Jordan marked his admiration for his favorite collectors, such as Doherty, by naming new species after them. The act is noteworthy since he usually avoided the practice on the grounds it placed taxonomists under suspicion of creating new species to name after their friends, rather than to selflessly advance science.

As the natural history collection at Tring grew, Hartert's and Jordan's curatorial duties also included hosting visiting specialists who wished to consult the collection. Zoologists arrived from London, while foreign visitors took up even more time. Olof Christopher Aurivillius, of Sweden, visited Tring shortly after Jordan's arrival, as did Ignacio Bolivar, of Spain.[101] Although the scientific collection of butterflies was closed to the public, Hartert directed those interested in examining Rothschild's insects to Jordan, who would show them the available collections. If naturalists could not visit, Jordan and Hartert dealt with their requests and demands from afar by mail. Though such letter writing consumed enormous amounts of time, they had good reasons to respond as efficiently and kindly as possible to such requests. Rothschild, Hartert, and Jordan emphasized in all their correspondence that they wanted a worldwide collection upon which to base their classificatory and zoogeographical conclusions, and for this they needed a worldwide network of correspondents willing to send both specimens and information.

Within a short time, Jordan wrote to and received letters from entomologists from all over Europe and beyond. Letters from Guillaume Severin, of Brussels, Géza Horváth, of Budapest, Hermann Julius Kolbe, of Berlin, Ignacio Bolivar, of Madrid, and scores of other entomologists fill the museum correspondence, the vast majority thanking Jordan for sending specimens or notifying him that they had sent specimens for determination. As Jordan became an acknowledged expert on the Anthribidae and then other groups, he obliged hundreds of requests to determine specimens from both collectors and museums.

Over the years, many famous entomologists and ornithologists would become associated with Tring, a reflection of the open access Rothschild and his curators provided to fellow naturalists of all nations. Richard Meinertzhagen, Admiral Hubert Lynes, John James Lewis Bonhote, C. I. Forsyth Major, Martin Jacoby, J. Faust, W. F. Kirby, H. Grose Smith, William Warren, Oldfield Thomas, Osbert Salvin, Albert Günther, Ernest Olivier, and many others worked on the collection and published in the museum's journal, *Novitates Zoologicae*. The journal also included occasional papers from the editors' German colleagues, including F. W. Riggenbach, Carl Edward Hellmayr, of the Munich Museum, Anton Reichenow, of Berlin's Humboldt Museum, Oscar Neumann, and others. Tring often recruited these experts to work through groups on which they were specialists. Rothschild allowed contributors to keep a specimen in return for their descriptions, an obvious incentive for specialists to maintain cordial relations with the museum and its curators.

For this highly skilled work they also contracted out to experts from a range of economic backgrounds, paying William Warren, for example, to study and arrange the Geometridae. Martin Jacoby, by trade a professional violinist, requested that, if he agreed to identify and name the Phytophagous Coleoptera for Rothschild, he receive some remuneration. Working through such a troublesome group was slow work, Jacoby explained, "owing to the enormous mass of described species and the great variability of the insects of this family." He would "determine" Rothschild's insects (or name the old and describe the new) for "1/species, which is usually charged." Jacoby, the only specialist on the group in England, was paid £27 for 373 species.[102]

Curating the Tring Museum, then, involved directing a factory of sorts—a factory of natural history made up of various assistants, hundreds of collectors in the field, a range of specialists, and dozens of natural history agents throughout Europe and abroad. At the center of this network for the next three decades, Rothschild, Jordan, and Hartert—"the Tring Triumverate"—worked away at the collection, naming, cataloging, and organizing the specimens, directing work by specialists and assistants, classifying the groups in which they specialized, negotiating with collectors and dealers, and maintaining an international network of correspondents. They were ordering life, indeed, at a range of levels.

This diverse network depended in turn on much more than individuals' financial or intellectual interests in natural history. By the end of the nineteenth century, the presence of railroads, steamships, canal systems, roads, and ports created reliable means of travel, shipping, and correspondence that naturalists used to accumulate specimens. They could do so in part because of the relative stability of the Age of Empire in which they lived. Though this world ultimately depended on dangerously brittle diplomatic alliances, for a time a three-pronged division of powers between the British Empire, the Franco-Russian Alliance of 1894, and the Triple Alliance of Germany, Austria-Hungary, and Italy in 1897 permitted a fragile peace.[103]

As those at Tring took advantage of this stability to direct a web of naturalists throughout the world, N. M. Rothschild and fellow capitalists directed a financial network that moved millions of pounds along the pathways of Empire, both formal and informal. Sometimes, Rothschild's naturalists tracked the paths of capital and Empire so closely that the endeavors seemed inextricably bound to each other. Walter Rothschild joined with the Royal Society, for example, in funding an expedition to accompany the governor of Aden's "political trip through the interior of Southern Arabia" in 1899.[104] Hartert wrote about birds from the Congo Basin after receiving a collection from W. Bonny of the "notorious Rear-Column of Stanley's Emin Pasha Relief Expedition."[105]

Supported by social and economic stability and a network of correspondents, collectors, and workers in Tring, Rothschild, Hartert, and Jordan eventually published more than seventeen hundred scientific papers and described more than five thousand new species of animals based on the resulting collection. The amount of money required to amass that collection was considerable, and when one adds the financial framework upon which access in general to far-off lands depended, the monetary commitment becomes formidable. Of course, the life and means of a middle-class naturalist such as Karl Jordan contrasted sharply with the minikingdom of natural history that Walter Rothschild could create. Rothschild's museum served as an extraordinary example of the opulent lifestyle of the Edwardian rich. The public gallery was graced by beautifully constructed cabinets and tall cases that displayed his possessions. The library—three stories high—had railings the same style as the Eiffel Tower in Paris. Rothschild himself embraced luxury, collecting art and expensive books and paying meticulous attention to his beautiful clothes. Jordan, by contrast, became known for his austere, even severe existence. Indeed, he often seemed eager to separate himself from the extraordinary wealth upon which the natural history work done at Tring ultimately depended. When later recounting his first years at Tring, he emphasized the difficult circumstances under which he began work on the collection:

The cottage where the work in systematics was carried out was full of cabinets, books, and boxes, and there were only two desks with good light available for the three scientists; the junior Curator had to find a place at a corridor window when he wished to use a compound lens or the microscope.[106]

Memories of this kind joined Jordan, even as an employee of a Rothschild, with tales of museum curators enduring difficult conditions, driven only by the search for knowledge of nature. Rothschild's old mentor, Albert Günther, recounted his own arrival at the British Museum by describing his dark, damp, subterranean and "rheumaticky quarters." He then added, "But what did I care for that? There were several thousand bottles of unnamed snakes to be examined and cataloged, the most congenial work that I could have wished."[107] Jordan modeled himself on the Günthers of the world, not the Rothschilds. While he pointed out in his memorial to Rothschild that his employer had not been "one of those hardy travelers who can stand any amount of discomfort,"[108] he prefaced accounts of his own travels with the claim that "a natural scientist does not need much comfort."[109] He considered it a point of pride that, when traveling in South-West Africa at the age of seventy, "the simple living and the shaking one gets on the often very vile tracks seem to agree with me."[110]

Jordan's famously stoic lifestyle exemplified the values of a rising professional class that would ultimately transform late nineteenth- and twentieth-century society. As a member of this new, academically trained class, the standards by which Jordan wished to be judged differed from those of the landed and capitalist elite who ruled Victorian and Edwardian society. Wealth and the display of wealth did not matter in the world to which Jordan paid allegiance. Rather, it was specialized, educated expertise in the service of some professional calling that commanded respect.[111] As these transitions played out and as the old survived alongside the new, Rothschild exemplified the wealth upon which a certain kind of natural history depended by the end of the nineteenth century. Jordan, by contrast, served as an excellent example of the opportunities that had opened up for members of the lower and middle class amid changing economic times—opportunities that would sometimes work in concert with established traditions and at other times would inspire discord.

For a few decades, the Tring Museum proved an astonishingly productive mix of these two worlds. Each man depended absolutely on the other to produce reliable knowledge about the natural world. Rothschild drew on his curators' academic training and heightened status in the new hierarchy of the professional class to ensure a role for "My Museum" in both science and a changing community of naturalists. Hartert and Jordan relied on Rothschild's enthusiasm and wealth to build the collection in certain ways and spend their time doing certain kinds of work, even while Jordan, at least, tried to separate himself from the ostentatious luxury of Rothschild's class. But to what extent was that work inextricably bound to the financial resources and even the interests and values of that class? Certainly, it may have been to Tring's advantage during its heyday years to be able to move within both classes—that defined by wealth and that defined by professional expertise. Whether the kind of work carried out at the museum could ultimately survive the social, economic, and cultural changes of the century that lay ahead is a question Jordan would constantly face over the following six decades.

TWO

Reforming Entomology

During what would become known as the heyday of the Tring Museum, the 1890s, specimens of butterflies, moths, and birds arrived at the museum from agents, collectors, friends, and colleagues. Jordan's task as a naturalist and his job as a curator was to both build the collection and find some sense of order within the myriad of biological forms being unpacked in the corridors of the museum. Doing so could not occur in an intellectual or social vacuum. The British entomologist J. W. Tutt once described the entomological world that Jordan must navigate in order to build Rothschild's collection as a "strange human mixture," its units "bound together by a common interest in the handiwork of Nature."[1]

When Jordan arrived in England, this world was engaged in heated debate over the role of natural history collections and the goals of the naturalist. Faced by boxes of specimens from the far reaches of Empire and beyond, Jordan had to take a stance regarding everything from the rather mundane issue of how high to set insects on their pins to whether collections of dead specimens could contribute anything to science. Early on, he became associated with a small set of reforming entomologists, including some of the most revolutionary proposals regarding how to name species since Linnaeus first outlined the binomial system of nomenclature in the eighteenth century. These reforms entailed, at their foundation, an attempt to adjust naturalists' naming practices to the fact that they were now attempting to order biological forms that evolved. Putting these reforms into practice required much more than simply the ordering of insect life, however. They also depended on reordering those who collected, sold, and/or studied insects. In other words, Jordan and his fellow reformers built on the naturalist tradition even as they transformed it.

THE "STRANGE MIXTURE" OF ENTOMOLOGISTS

A short notice in the British natural history magazine *Natural Science* in May 1893 announced: "Dr. Jordan has arrived at Tring from Hildesheim, and will take charge of the Entomological collections belonging to the Honorable Walter Rothschild."[2] This notice formed the first in a series of introductions, whether in print or in polite exchanges at society meetings, of Jordan to the British naturalist community. Such introductions, and the assessments of Jordan's status that inevitably followed, reflected the social aspects of doing natural history. J. W. Tutt once wrote that his fellow entomologists were adept at "picking to pieces the scientific character of a mutual acquaintance."[3] And indeed, Jordan soon found that he would have to navigate more than a large collection of beetles silently displaying the chaos of their diversity. He was joining a new community of entomological fellows, one embroiled in debates over the goals, methods, and priorities of natural history and taxonomy. In his new post in Britain, Jordan had to learn what Tutt described as "the internal workings of that strange human mixture which makes up 'the entomological world.'"[4]

Certain ideals existed regarding how this "strange human mixture" should interact. As one frontispiece of the journal the *Entomologist* read, "By mutual confidence and mutual aid, Great deeds are done and great discoveries made." But this ideal of cooperation and exchange—particularly where specimens were concerned—had to be balanced by judgments of who deserved to be included in the community, who deserved confidence, and who deserved aid. Certainly some of the London entomologists gave the new Tring curator no special privileges during his first trips to their collections. Jordan later recalled that when he first became interested in the beetle family Anthribidae, he described some species inaccurately because, during visits to the British Museum, "Kirby told me that a visitor was not permitted to remove the lid from a drawer and so I had to look at them through the glass of the lid."[5] Having learned the great importance of cooperation during his work on German butterflies, Jordan knew that to be a successful curator, the Kirbys of Britain must learn to trust him.

As Jordan set out to both gain access to British collections and prove himself a valuable part of the British entomological world, his hailing from Germany, a nation in heated competition with Britain for industrial and imperial superiority, could be a blessing and a curse. Some reforming naturalists looked to German zoology as a model of good, scientific work. Indeed, German zoologists occasionally inspired resentment due to their tendency, and indeed ability, to ignore the work of those across the Channel. "They all seem to be rather careless over there with regard to results published in this country," complained one entomologist to Jordan some years after his arrival.⁶ The more conservative British naturalists, warily eying the elaborate phylogenetic trees of naturalists such as Ernst Haeckel, found their German colleagues too inclined toward speculation. The stereotype of the "philosophical" German reached deep into popular culture. "All these Germans imbibe philosophy from the cradle," observed a *Punch* cartoon featuring a Teutonic barber.⁷ But while some worried that German zoology delighted too much in imaginative flights of fancy, others thought the presence of a German zoologist at Tring could only improve British natural history, on the grounds a little "philosophy" would bring their compatriots into the modern scientific age.

As for German zoologists' detailed emphasis on morphology using microscopes, not everyone believed that even counted as "zoology."⁸ Indeed, stark differences between practices back home and in his new community of workers extended even to whether or not particular species should be classified as butterflies or as moths. British lepidopterists placed one confusing species, *Pseudopontia paradoxa*, with the moths, while Continental entomologists doing detailed morphological work (including Jordan) placed it, correctly, with the butterflies.⁹ Inevitably, the arrival at Tring of a German naturalist willing to change time-honored classifications at the turn of a microscope made some entomologists nervous.

The fact that Jordan worked for the Rothschilds also constituted a challenge to Jordan's effort to establish a respectable place in British natural history. Miriam Rothschild blamed zoologists' lack of appreciation for both her uncle's legacy and his museum on anti-Semitism. If so, it was the sort that remained limited to whispered asides in the corridors of museums and society meetings, rather than in correspondence or natural history journals. Rothschild's younger brother, Charles Rothschild, one of the few flea specialists in the country, hinted at such chauvinism when he once complained that "Fleas are increasing anti-Semitism, as the few other students of the Order think I'm too keen a competitor!"¹⁰ Certainly naturalists could use insidious stereotypes as easily as other members of the public. In 1897, E. R. Bankes offered to broker the sale of some specimen cases for Lord Walsingham on the grounds that the seller "is such a dreadful 'Jew' that he would be sure to ask you an outrageous price for them."¹¹

Still, it seems that when judging scientific character other divisions concerned naturalists more. Jordan would reflect in his nineties that upon arriving in Britain he had "become aware of some jealousy and personal antagonism among working Entomologists, official and amateur, which warned me to be very cautious in order to avoid public controversy."[12] The tension between "official" naturalists at places such as the British Museum, and "amateur" naturalists such as Rothschild had a long history, as generations of often poorly paid curators tried to wrest control of natural history from the command of upper-class patronage. As curator of insects at Rothschild's museum—neither official nor amateur but working with both—Jordan found himself having to navigate long-standing tensions.

The fact that Rothschild had a troublesome habit of throwing his weight around did not help matters. When the British Museum herpetologist G. A. Boulenger, for example, described one of Rothschild's tortoises living in the zoological gardens without his permission, Rothschild not only protested against such "extremely rude and ungentlemanly" behavior but threatened to use his influence to divert everything he could from London to Frankfurt. He would, he insisted, show men "of Mr. Boulenger's type that they make a great mistake in being too bumptious." The British Museum's Albert Günther, in efforts to "keep the peace," had occasionally reprimanded Rothschild for being so high-handed toward his fellow zoologists. He tried to intervene, for example, when Rothschild threatened to boycott the journal *Proceedings of the Zoological Society of London* in favor of the *Ibis* if the editor of the *Proceedings* insisted on checking his identifications.[13] By the time Jordan arrived at Tring, Rothschild had mutinied from the *Transactions of the Entomological Society of London* altogether on the grounds that the editor delayed his papers. In this case, Hartert supported his employer's revolt. "The Honbl. Walter Rothschild asks me to inform you," Hartert had written to the editor of the *Transactions*, "that after so much delay he does not wish to publish the returned paper in your periodical, and that he will not publish anything again in the Entomological Society's Transactions, but send his papers somewhere else, for example, the "Iris" edited, I believe, by Dr. Staudinger in Dresden."[14] The letter proved a sly attempt at disinterested naiveté; Hartert knew quite well who edited the *Iris*. The same day he wrote to Staudinger himself, attributing what he called the "strange difficulties in publication" to entomologists' petty jealousy of Rothschild's considerable collections.[15] How silly and stupid they all appeared, Hartert exclaimed, compared with a man like his boss, "of such warm interest and understanding, no less for entomology than for the whole of zoology!"[16]

Indeed, naturalists had much to envy. Miriam Rothschild attributed Alfred Newton's estrangement from his ardent pupil to the fact Newton resented Rothschild's ability to send collectors to unexplored regions at will.[17] Entomologists expected Rothschild to be the highest bidder at natural history auctions as a matter of course. One entomological satirist described one such auction as follows: "The men in yonder corner? They're the professional element. Buy on commission for some of our rich collectors. That one probably has a commission from one of the Rothschilds."[18] Jordan's correspondence occasionally mentioned naturalists' jealousy of Rothschild and, by extension, his curators. When in 1896 Lord Walsingham's curator, John Hartley Durrant (Walsingham called Durrant his secretary), requested Jordan's help finding specimens in the Tring collection, Jordan cited naturalists' jealousy of Rothschild as a justification for insisting Durrant promise not to back date the resulting publications, a practice sometimes used to ensure that one's names for new species received priority. Jordan feared that Tring would be implicated in Durrant's occasional use of this practice. "You know that there are some people who are more than jealous of us," he wrote, "and will not fail to take the opportunity to hint in the above direction."[19]

Both Jordan and Hartert spent a lot of time trying to temper such envy, constantly countering rumors that Rothschild routinely spent exorbitant sums on his specimens. But newspapermen loved telling tales of Rothschild's latest supposed exploits. One notice of his book on cassowaries described how the work depended on "an enormous outlay which no mere man of science could have borne. It entailed, to begin with, the purchasing of hundreds of live cassowaries, which sometimes cost as much as £150 each."[20] After reading a similar story about Rothschild's insect collection in the *St James's Gazette*, one naturalist requested "the name of the Ecuador Butterfly which has been purchased by Mr. Rothschild for 1000 pounds."[21] In 1910, Hartert admonished the editors of *Popular Science Siftings* for an article entitled "A 200 Million Pound Butterfly Industry" for their "gross exaggerations," particularly the "dangerous" claim that a handsome profit could be gained by collectors. "Nor is it true," Hartert insisted, "that the Hon. Walter Rothschild has ever paid £200/-/- or anything near it for a single moth or butterfly," the maximum price being £2 at the utmost.[22] Sixteen years after arriving at Tring, Hartert was still carefully correcting such tales. "That Mr. Rothschild wanted to give £1000 for the green parrot with the yellow patch is, of course, one of the many lies that are told about his collecting," he informed one correspondent.[23] But both the public and fellow naturalists found the stories of the museum's purchasing power easy to believe, given the Rothschilds' standing as the wealthiest family in the world.

Fortunately, a naturalist like Jordan, keen on avoiding antagonism with other naturalists, had a powerful ally in the very nature of the naturalist tradition.

Scientific natural history was defined at the time as the specimen-based description and classification of organisms. Thus, naturalists were bound together by the need for specimens. Given that doing good entomology depended as much on exchanging specimens and expertise as on access to Rothschild's allowance, and well aware of the central place of the British Museum, establishing cordial relations with the London naturalists proved particularly important. Jordan soon called on Albert Günther and met those working for the Department of Zoology and in the Insect Room: C. J. Gahan, R. I. Pocock, F. A. Heron, E. E. Austen, G. F. Hampson, W. F. Kirby, C. O. Waterhouse, and A. G. Butler. He traveled by train to visit London at least every two weeks, a habit that would ward off at least one contemporary complaint that entomological "provincials" did not mix together as the metropolitan ones did.[24]

Jordan even engaged in a few strategic uses of the pinnacle of naturalist diplomacy; naming new species after colleagues. He did so despite the fact that museum taxonomists were sometimes criticized for making species purely for the sake of naming them after their best friends. Alfonso L. Herrera, of Mexico, who often attacked "name-mongers," once complained that "the aim of the modern explorer, as gathered from his own writings, is to amass the largest number possible of birds, butterflies, eggs, plants, fishes, etc., etc., in the hope of discovering as many new sub-species as possible to bear the name of himself, his family relations, his friends, and his native servants."[25] Jordan exhibited a similar abhorrence of collectors who wished specimens be named after themselves "and their wife!"[26] But he also knew the power of a judicious coining of names in acknowledging—and inspiring—useful help from both collectors in the field and museum workers back home. As part of his developing partnership with London entomologists, he named a species in honor of Charles Joseph Gahan "to whose help in comparing our species with those in the British Museum collection I owe so very much."[27] Writing from Petrópolis, Brazil, F. G. Foetterle was so encouraged by Jordan's dedication of a subspecies to him in 1906, that he was spurred to further contributions to the "palace of science."[28]

Such politic use of naming and Jordan's diplomatic routine of constantly traveling to London soon paid dividends. Within two years, the entomologist David Sharp, a staunch opponent of private collections, wrote to Hartert that, although he planned to take the British Museum's Anthribidae in hand, if "Dr. Jordan likes to do them first—I shall be quite as pleased."[29] The recognition of Jordan's status as a fellow entomologist was formalized when, in 1894, the Entomological Society of London elected him a fellow. No doubt the election also represented an acknowledgement of Rothschild's resources as much as

Jordan's expertise. The society members soon requested that Jordan bring exhibits of certain beetles, if Mr. Rothschild had no objection.[30]

The fact that both Jordan and Hartert inhabited powerful positions as the gatekeepers to Rothschild's collection aided their ability to establish good relations with naturalists who needed specimens for their research. Museum curators controlled access to specimens, and these feathered, furred, and scaled objects were required for doing scientific natural history by the end of the nineteenth century—a fact that, as we will see, caused much consternation to some. Entomologists at the British Museum had an interest in maintaining good relations with wealthy collectors such as Rothschild, even if ultimately they hoped to break the connections between wealth and the ability to do research.

Strapped for resources, the British Museum had long since harnessed the enthusiasm of those who did not need remuneration. A whole team of ornithologists, most of them wealthy collection owners, had compiled the museum's famous *Catalogue of Birds*, for example. The museum's curators and administrators recognized that, as one entomologist explained in 1900, "the field-naturalist and the amateur are in a far better position for doing valuable work than is the average professional scientific man—tied to a museum workroom or a lecturer's table."[31] The British Museum entomologists saw the Tring curators and their collection as useful resources even as they sometimes competed for specimens. Even naturalists with whom Rothschild had come into conflict, such as Sclater and Boulenger, continued to correspond with him and exchange specimens. The draw of good specimens often trumped pride. Besides, maintaining good relations with wealthy owners of private collections made it more likely that these specimens would eventually end up in London via a bequest when the owners died.

Even more than the museum staff, collectors in the field who intended to make a living from selling specimens depended on museum curators for access to the extensive resources of wealthy collection owners such as Rothschild. A collector named Dodd acknowledged his dependence upon the good will of Walsingham's curator Durrant when protesting that he had not criticized Durrant's dealings with collectors. "Pray consider," Dodd insisted, whether anyone knowing Durrant's important position under Lord Walsingham "would attempt to cast reflections upon you?"[32] As we will see, how a curator used this power influenced his ability to tailor incoming specimens to particular research needs. While Walsingham's collectors often complained that they never heard whether their consignments had been of any interest, Jordan's correspondents received grateful notes, surely helpful in inspiring them to pay more attention to his detailed requests. This ability to influence collectors would become absolutely key to the type of research Tring pursued in the following decades.

Simply citing Rothschild's name was enough to influence some correspondents. Hartert often commenced letters with "the Honorable Walter Rothschild requests me to say to you," and he once reminded a collector to "Please note that this letter is copied and that on the smallest impertinence or irregularity from your side it will be laid before the Hon. Walter Rothschild."[33] The *Hon.* certainly conferred part of the authority Hartert relied on to convince the wayward collector. But it is also important to point out that many respected Rothschild as an expert naturalist in his own right. Although he would later be described (and often dismissed) as an amateur, for much of the history of the naturalist tradition little had differentiated amateurs and professionals beyond the latters' eking a (usually precarious) livelihood from the work. Paid naturalists such as the workers at the British Museum simply did not have the status later associated with professionals. In natural history, the term *professional* often implied lower status, since to accept money for any but a few professions "was to consign oneself to a social limbo."[34] Indeed, the community often used the term *professional naturalist* for those who made their money collecting specimens in the field or as dealers or natural history agents back home.[35] (Dealers lived even lower on the ladder than professional collectors. Of one pair of dealers, Günther wrote, "I believe although dealers, both are sufficiently honest.")[36] In this world Walter Rothschild had as much qualification to be considered an expert entomologist as any of the "official" naturalists with paid positions in London.

Of course, the criteria by which expert status was conferred were changing. Some naturalists, raised in a spirit of political and educational reform and chaffing under the power of wealthy naturalists who controlled the societies and journals, no longer considered ardent commitment as an adequate criteria of authority. By the time Rothschild opened his museum, scientific administrators from Thomas Henry Huxley to Richard Owen had been working for decades to wrest control of science away from traditional title and fortune. In this new arrangement, science would be controlled by those with professional merit, and professional merit would be judged by the possession of academic training, specialized expertise, and salaried positions.[37] Although Rothschild had a degree in zoology from Bonn, Albert Günther noted while campaigning unsuccessfully for Rothschild's election as a fellow of the Royal Society in 1902 that he feared not many of the fellows appreciated Rothschild's work.[38] "Among scientists," Günther's biographer explained, "a Rothschild was at a disadvantage, for few believed that a wealthy man could be anything other than an amateur."[39]

As Jordan's position demonstrated, wealthy collection owners themselves increasingly deferred to a rising class of experts to process their material in the latest "scientific" manner. Lord Walsingham admitted that although he possessed an excellent collection and had a good eye for species, "when it comes to classification by minute differences of structure, which differences have to be compared again and again with infinite care and trouble so that in no case can they be overlooked without falling into errors, I am simply nowhere." Here, his assistant Durrant came in, although Walsingham sometimes hesitated to follow him "to the more or less logical conclusions which he puts forward for my acceptance."[40] Within an increasingly professionalized, academically trained scientific community, Rothschild often depended on his curators for intellectual prestige and legitimacy, rather than the other way around. And in a world in which standards of science, status, and trust were changing, Tring's possession of two routes to legitimacy—wealth in the person of Rothschild and the new professionalism in the person of Jordan—aided their efforts to negotiate the diverse "strange human mixture" of the entomological world and create an extraordinary collection. If one adds Hartert's status as an experienced collector, the "Tring Triumvirate" represented a formidable team.

HOW TO DO ENTOMOLOGY

Entomologists had long disputed precisely what criteria should be used to pick "to pieces the scientific character of a mutual acquaintance," as Tutt once put it.[41] Such debates often appeared in the guise of discussions regarding the proper aims and methods of studying insects. When Jordan arrived in Britain, the questions of how and why to do entomology had recently inspired a particularly heated controversy among entomologists in London. For decades, and like their fellow naturalists at the Zoological Society of London, and the British Ornithologists' Union, contributors to the journals of the Entomological Society of London focused on the accumulation of specimen-based facts, both of taxonomy and geographical distribution. Those contributing to these broad research programs counted as "scientific" entomologists. This definition of scientific entomology and the associated work of describing thousands of new forms was founded on a methodological ethos firmly entrenched within British science. Within such a context, naturalists' societies existed to collect facts, and as a result their journals generally upheld a strictly descriptive format, with comments on the speculative theories of men such as Darwin reserved for annual presidential addresses. Indeed, many of those in charge of such societies felt it their duty to admonish any "philosophizing" by members, delivering quick rebuttals to any naturalist who attempted explanations of the all-important "facts."[42] Supported by this methodological framework, collectors, naturalists, and explorers inundated museum halls with specimens from the outskirts of empire, and the careful naming and description of these forms dominated naturalists' journals.

By the century's end, however, some British entomologists insisted that their status as a science depended on entomologists doing more than purely descriptive work. They must, reformers urged, decisively and explicitly engage with the theoretical debates of the day. The controversy that resulted from such calls sets the stage for some of Jordan's preoccupations and concerns as he navigated this new community, for the calls for new methods and goals would have a profound influence on the role of collections, museum curators, and taxonomy in general.

The reformers' campaign was inspired in large part by their effort to deal with the problems raised by evolution theory. Just as in Germany, some argued that the theory must provide more than just an explanatory framework for data; it must transform the very goals and methods of natural history. Such claims inspired heated debate at the Entomological Society of London just prior to Jordan's arrival, and the scars of the fight still lingered. The controversy began when F. A. Dixey, demonstrator at the University Museum in Oxford, read a paper in which he proposed a phylogeny of certain butterflies based on wing markings. He used experiment and observation to trace the origin of colors and then speculated on whether characters were ancestral or adapted. Furthermore, as a protégé of the ardent Darwinian E. B. Poulton, Dixey used the Oxford University Museum of Natural History's collections of insects to defend natural selection, explicitly basing his studies on German biologist August Weismann's 1882 edition of *Studies on the Theory of Descent*. Sides quickly formed in the responses to Dixey's work. H. J. Elwes sharply reprimanded Dixey for his excursion into evolution theory, and warned members that such investigations would mire entomologists "in a maze of unprofitable speculation upon points about which we could at present come to no certain conclusions."[43] By contrast, Tutt praised Dixey's experiments as a "most unusual treat" in the pages of his journal the *Entomologist's Record and Journal of Variation*, begun precisely because the Entomological Society of London refrained so strictly from evolutionary questions. He no doubt saw Dixey's work as the perfect antidote to those entomologists who aimed only at the possession of "a series of a certain species in rather finer condition than those of their entomological neighbours, and yet who never want to know the 'go' of it."[44]

Elwes's attempt to maintain the traditional boundaries of the society's journal soon inspired a group of fellows, led by Poulton, to challenge his subsequent nomination as president on the grounds that his election would prioritize systematics to the detriment of more "biological" work. Indeed, Poulton and his friends threatened to retire from active work in the society "if Biological Entomology were, as we deemed it, slighted" again.[45] When the fellows elected Elwes in spite of their threat, Elwes fired back at the "biologists" with a presidential address on the proper methods of British naturalists. He insisted that the actual facts available to bring the "great and important" laws of development, variation, and distribution of insects remained "quite insufficient to bring such speculations to a definite end." Furthermore, he felt the number of persons of sufficient talents to "steer a straight course through the numerous difficulties, contradictions, and doubts, which constantly surround such enquiries is very limited." Finally, he urged that to insist that one must try and discover explanatory laws in order to join the "highest ranks of science" would place a dangerous check on the humble work of those naturalists "who find their greatest pleasure in collecting and arranging the material which must form the only solid foundation for such work as has been done by Darwin, Wallace, Bates, Weismann, and others."[46]

Entomologists such as Elwes, wary that premature speculation would distract naturalists from the necessary stage of collecting facts, certainly had the argument of numbers on their side. In 1891, Francis DuCane Godman had pointed out that perhaps two million species of insects existed in the world, of which about two hundred thousand had been described. This left at least one million and eight hundred thousand unnamed and undescribed, of which probably one hundred thousand already sat unprocessed in collections. And the specimens kept coming in, overwhelming the small number of entomologists available to describe them. Godman pointed out that "unless our numbers largely increase the arrears of work will become more and more unmanageable." Like Elwes, Godman firmly believed that "the extent of the subject of Entomology is so vast that nothing but a systematic and continuous effort to amass collections, work them out, and preserve them, can place us in a position to proceed safely with the larger questions which follow the initial step of naming species."[47]

In 1895, Raphael Meldola offered the other side of the argument when he succeeded Elwes as society president. He entitled his presidential address "The

Speculative Method in Entomology" and used his opportunity at the entomological pulpit to emphasize that although science necessarily begins with observation or experiment—that is, the description of facts—"it is perhaps necessary to assert that no mere collection of facts can constitute a science. We begin to be scientific when we compare and coordinate our facts with a view to arriving at generalizations on which to base hypotheses or to make guesses at principles underlying the facts." Meldola readily admitted the pitfalls of generalizing. He recalled once remarking to Darwin how difficult it was "to get Nature to give a definite answer to a simple question." Darwin, "with a flash of humour," had replied: "'She will tell you a direct lie if she can.'" But Meldola insisted that the "bugbear" of speculation would be harmless if looked boldly in the face. He, at least, expressed relief that some workers refused "to submit to the plea of inability because all the existing species of Lepidoptera have not been collected and named."[48]

Despite their different responses to what the entomologist must do in the face of an incomplete catalog, both sides in this controversy believed the scientific status of entomology to be at stake in the discussion. One noted how theory guided research in other sciences; the other feared that overspeculation would discredit entomological work. Such fights over presidential chairs and the contents of journals represented profound challenges to traditional definitions of what it meant to do important work in natural history. In contrast to those who considered "systematic arrangement and naming of specimens" as the "principal occupation of scientific entomology," Tutt insisted that entomologists must study both origins and life history.[49] "We want to know not only what is," he urged, "but what are the probable causes which have made entomological facts what they are."[50] In his view, the "scientific entomologist" must read the essays of such men as Weismann and Herbert Spencer, and have a firm grounding in the debates over evolutionary theory.[51] Furthermore, Tutt drew a distinction between systematists, those "so-called scientists" who "are generally museum species-namers," and biologists who knew the life histories and stages of insects. Even excellent museum entomologists could not be biologists, he insisted, due to the simple fact that they had no access to living insects. "Let our museum people name specimens and describe them," he declared. "This is their natural work, because it is the only one for which they have the slightest training; but for goodness' sake let our biologists classify."[52]

Other entomologists came to the defense of taxonomists working in museums. "The work of discriminating species is a necessary one," insisted G. H. Carpenter, "and the generalizations of the philosophical naturalist must largely depend on the technical labour of the systematist."[53] Another urged that "such men as Cope, Allen, Merriam, Coues, Scudder, Agassiz, Gill, and a score of others in America, and as many other bright lights in England, are systematists. Do let us concede that they are naturalists and scientific men also."[54] Even Meldola conceded that museums had been central to the development of the "philosophical" theories he wished to see included within the purview of entomologists' work, particularly the explanation of mimicry by natural selection. Henry Walter Bates himself, Meldola noted, had developed his theory of mimicry "while looking over his specimens when he had reached London." But Meldola had to confess he sometimes thought that perhaps the hard, chitinous skeletons of insects had been expressly designed for the retardation of entomological science; "they lended themselves so conveniently for preservation as cabinet specimens!"[55]

By 1893, then, as Jordan joined Rothschild's museum, some British entomologists were questioning the dominance of collection-based work in their society journals. Defenders of systematic naturalists even referred to them with sympathetic sarcasm as "this despised person."[56] The reasons for the disdain proved complex, but rebellion against collection-based natural history that was strongly dependent on wealth inspired at least some of the critiques. Taxonomy was based on museums and private collections, and those spaces required an enormous amount of resources, creating a hierarchy in which those with money had the most influence. Men trained in T. H. Huxley's world were primed to see an endeavor in which scientific achievement depended on wealth as abhorrent. Tutt's journal included anonymous, sarcastic descriptions of visits to the saleroom, "where these beautiful insects become objects of barter," and men preferred to buy their insects rather than collect them in the field.[57]

As a rising class of professionals tried to remake the scientific world, stereotypes abounded of the once-honored fact collectors, including the quite serious charge that they carried out the work of schoolboys, something that real scientists outgrew. The year Rothschild opened his museum, the *Daily News* described collecting as the activity of boys, a symptom of the pleasure arising from having what other collectors did not have, and the enjoyment of the chase. "Boys will go collecting things," the article read, "It is difficult to say why. Indeed, it is not easy to explain why anybody collects anything.... Butterflies, beetles, and moths we could do without."[58] At least one observer advocated that schoolboys' interest in natural history be rechanneled into areas of science that would "not aid the whole-sale extinction of all the rarer creatures, and in which success cannot possibly be accomplished by money."[59] This equation of collecting with childhood, wealth, and the charms of possession could be quite an insidious charge for a scientific-minded curator working for one of the wealthiest naturalists in Britain, particularly a Rothschild rumored to obtain his treasures at all cost.

Various proposals for reform in the face of criticism existed among those who believed in the continued importance of both collections and descriptive work. One author of a note entitled "Species-making and Species-taking" bemoaned the "stupid and unprincipled action of species-mongers of the baser sort" and wished, despite wariness regarding "too great extension of professionalism or officialdom in science," for "something in the nature of a diploma without which no one should be permitted to practice, *i.e.* to publish."[60] Notably, this author and others made a distinction between the "species-making" of those conferring names from base motives of personal prestige and competent and "honest" workers. But for those who, like Jordan, thought conferring names on animals an important step in the study of the natural world (and who were paid to do so), what exactly did it mean to be a competent and honest worker and what kind of work did they produce? And how could Jordan establish himself as a member of this "competent" fraternity, while fulfilling the demands of his employer to process and name his incoming specimens?

Jordan was obviously very aware that the constant stream of specimens coming in and the dozens of new species descriptions going out of Rothschild's museum epitomized the stereotypes of museum work. Within the first ten volumes of the museum's journal, *Novitates Zoologicae,* 56 new species or subspecies of mammals, 345 birds, 14 reptiles and amphibians, and 4 fish appeared. Jordan's own contribution to the first volume consisted of seven papers on Coleoptera in which he described 400 new species and 39 new genera. With good reason, he became quite sensitive to the impression given by the long lists of new species emanating from the museum halls. He could not change some aspects of this work, such as the dependence of the museum on Rothschild's wealth. Specimens cost money, as did good cabinets and the all-important library. Nor could he alter the basic directive under which Rothschild had hired him; namely, to name and order incoming specimens. But some criticisms of the museum wing of the naturalist tradition could be taken firmly in hand.

Some of the adjustments Jordan made may seem quite small. For example, when giving the name of a species, he abandoned the common practice of adding the name of the first describer. This apparently minor move, representing a few letters on a manuscript, takes on larger meaning when one remembers that Darwin blamed "a very wrong spirit" in natural history on the tradition of including the author of each new name alongside that name; for example, *Strangalia armata*, Herbst, after the German naturalist Johann Herbst. Darwin feared that such a practice implied "some merit was due to a man for merely naming and defining a species." "Species-mongers," he wrote, often acted "as if they had actually made the species, and it was their own property." Darwin thought the custom of adding the name of the first describer to the species name only exacerbated this tendency. "Why," he asked, "should naturalists append their own names to new species when mineralogists and chemists do not do so to new substances?" Like Jordan's mentor Ernst Ehlers, Darwin thought that if a naturalist "works out *minutely* and anatomically any one species, or systematically a whole group, credit is due, but I must think the mere defining a species is nothing."[61]

Jordan's reasons for abandoning the practice, given such criticisms, are telling. "As I am the 'author' of about 300 species of *Anthribidae*," he wrote, "and shall probably become the 'author' of a still larger number, I should have to bring my own name before the reader again and again, and thus acquire that kind of cheap immortality to which critics of classificatory work have nowadays so often alluded as being the chief aim of the publication of mere descriptions of species." By not including the author's name in the species name, Jordan insisted that "this cutting criticism loses the point."[62] The editor of *Natural Science* noticed Jordan's words on this subject with approval, concluding, "Example is always better than precept, and we hope that this one will find many followers."[63]

Given some of the proposals being vetted in the entomological literature, Jordan knew that taxonomists must consider more profound reforms than this simple yet meaningful omission of their own names from the titles of new species. Tutt, for example, was proposing that only those studying live organisms counted as scientists and biologists. Certainly Jordan was sympathetic to Tutt's concerns. He, too, often urged that entomologists must not view their specimens as "mere bits of chitin with beautiful or interesting colouring and structure, but as 'units of life functioning each in its particular sphere of activity.'"[64] But as a curator in charge of the insect collection at the Walter Rothschild Zoological Museum, Jordan could hardly defend the claim that species work based in a museum should not count as science. He rooted his own approach to reform in an assumption that his day-to-day task of sorting, naming, and describing Rothschild's collection was a valuable endeavor; that, somehow, the demands of studying insects as "units of life" could in fact be combined with the task of examining thousands of dead specimens. Furthermore, Jordan's particular approach to the question of "how to do entomology" illustrates his early concern with the place of taxonomy within biology, a concern that would become a driving force in determining how he spent his time at his museum desk over the following sixty years.

THE "MAKING" OF SPECIES

Like all museum curators, Jordan's first step toward finding some order in the specimens coming in from all over the world was to distinguish the units, or species, to be ordered. Then he must determine whether or not a certain species had been named previously. If not, he must coin a new name and publish a description so that the information would be available to other naturalists. Of course, as Darwin had found during his eight years' work on a monograph on barnacles, the variation both within and between species confounded naturalists' efforts to determine, or "diagnose," species. "Everybody who is a little acquainted with the diagnostic works on Zoology or Botany," Jordan wrote in 1896, "will know sufficiently that a continuous question of contest amongst us species-makers is whether a given form of animal or plant is a 'distinct' species or not."[65] (This was Jordan's first adoption of the term *species-makers*.) A form that looked slightly different might simply be a "variety," and thus not warrant its own binomial name. Or it might indeed be a separate species, in which case a binomial name would be required.

Naturalists recognized that a great amount of confusion existed in the natural history literature due to some workers describing as varieties forms that another might describe as a species. Some entomologists, for example, described slightly different forms as species on the grounds that the new name brought attention to the observed difference, while other entomologists ignored slight variations under the assumption they were unimportant or in order to avoid the multiplication of names. In any case, many found the apparent arbitrariness of such decisions an easy target of ridicule.

In practice, naturalists had long recognized that more specimens could influence one's decision regarding whether a form should be described as a new species or a variety. The German zoologist Carl Claus described the accepted rule that guided the discrimination of species and varieties in his 1885 *Elementary Textbook of Zoology*. "If forms which are widely different can be connected by a continuous series of intermediate forms, they are held to be varieties of the same species. But if such intermediate forms are absent, they are held to be distinct species, even when the differences between them (so long as they are constant) are less." Claus conceded that "under such circumstances we can understand that in the absence of a positive test, the individual judgment and the natural tact of the observer decides between species and variety."[66] It was in view of "such circumstances" that Darwin had noted how often "one form is ranked as a variety of another, not because the intermediate links have actually been found, but because analogy leads the observer to suppose either that they do now somewhere exist, or may formerly have existed." This judgment by analogy, Darwin wrote, opened a "wide door for the entry of doubt and conjecture" since inevitably naturalists had slightly different experiences with the groups in question. Darwin confessed that as a result he had been struck by "how entirely vague and arbitrary is the distinction between species and varieties."[67]

The problem inspired the editor of the *Entomologist* to include a long quote from *On the Origin of Species* on its frontispiece: "What a multitude of forms exist, which some experienced naturalists rank as varieties, others as geographical races or subspecies, and others as distinct, though closely allied species!" Darwin had famously refused, given such arbitrariness, to accept his contemporaries' definitions of species as specially created by God and varieties as a product of environmental influence. He had then offered a theoretical explanation for all these difficulties (and a host of other natural history puzzles) by speculating that varieties represented incipient species, which could pass from "a state in which it differs very slightly from its parent to one in which it differs more" through the action of natural selection. Naturalists' struggles to differentiate between varieties and species resulted from, and were thus explained by, descent with modification.

At first glance, Darwin's designation of varieties as incipient species may seem to have placed both the careful collection of varieties and the experienced knowledge of systematists at the center of the study of the origin of species. And where else to study these varieties, than in collections and museums staffed by naturalists processing the thousands of specimens coming in from the far reaches of the British Empire? Darwin famously asserted that his theory would cause a revolution in natural history, after which systematists would pursue their labours as at present, without being "incessantly haunted by the shadowy doubt whether this or that form be in essence a species."[68] But he did not thereby believe his conclusions degraded the work of the systematist. Indeed, he implicitly upheld the place of expert naturalists as arbiters of the natural order when he wrote, "in determining whether a form should be ranked as a species or a variety, the opinion of naturalists having sound judgment and wide experience seems the only guide to follow."[69] But for some, the Darwinian statement that a species was simply what a good naturalist claimed it to be only proved one more reason to discount any claims to scientific knowledge that natural history and systematics may once have held. In other words, it furthered the impression that taxonomists "made" species rather than discovered them.

Certainly, in convincing naturalists that evolution explained the diversity around them, Darwin completely transformed the conceptual framework in which many naturalists viewed species. From specially created forms with limited variation, species had become the end result of a process of change, while varieties had become the starting point from which species developed. Soon, a practical need to distinguish varieties and species and the philosophical interest in studying these transitions melded in the work of some naturalists. Faced with a collection of extraordinarily variable shells before which a naturalist's most immediate reaction might be a sense of hopeless confusion, John Gulick insisted that "to one who is studying the origin of species this profusion of intergrading forms is a sight of the highest interest." Though the systematist "who insists that every form shall be exactly classified" found such a collection, full of variations, a stumbling block, Gulick insisted on variation as the chief interest of the collection, a stance he explicitly tied to the study of evolution.[70]

When Jordan arrived in England, an increasing number of entomologists, intent on taking up evolution theory as a guide to research, also insisted that the study of variation must be central to entomological work. In his rogue journal the *Entomologist's Record and Journal of Variation*, Tutt campaigned relentlessly for entomologists to shift their attention from describing species to the study of variation. He took every opportunity to acknowledge work on geographical and local variation. After noting exhibitions of variation using long series of specimens at society meetings, Tutt announced that he thought "the above report will show how strong a hold 'Variation' is obtaining on the minds and sympathies of entomologists."[71] More famously, William Bateson also urged entomologists to devote attention to the study of variation. "If facts of the old kind will not help, let us seek facts of a new kind," he wrote in 1895. Once naturalists refocused their attention accordingly,

> we should no longer say "*if* Variation takes place in such a way," or "*if* such a variation were possible"; we should on the contrary be able to say "since Variation *does*, or at least *may* take place in such a way," and "since such and such a Variation *is* possible," and we should be expected to quote a case or cases of such occurrence as an observed fact.[72]

Tutt, Gulick, and Bateson each tied calls for the study of variation to their wish that naturalists focus on problems arising from Darwin's theory. Jordan, Hartert, and Rothschild agreed, and actively joined in the attempts to reform systematics by an emphasis on variation. A brief "Note from the Editors" that opened the first volume of *Novitates Zoologicae* stated their methods and "the grounds on which they departed from many naturalists." The most controversial of these departures entailed their use of a third name, or trinomial, to designate geographical varieties, or subspecies, a practice that they explicitly tied to the need to adjust both the rules of nomenclature and museum work to the new theoretical framework provided by evolution.

Jordan later cited the German entomologist Eugenius Johann Christoph Esper as the first to distinguish between species, subspecies, and other kinds of varieties in 1791. As Esper's work lay ignored, the ornithologist Christian Ludwig Brehm used subspecies in 1831 as a way to designate different geographical forms of a single species.[73] In 1844, the ornithologist Hermann Schlegel proposed a trinomial naming system in which a third name would designate those forms entitled to the designation of "subspecies," or "local races." By the 1880s various German and American ornithologists were campaigning for "trinomialist" reforms, and in 1884 the American ornithologist Elliott Coues even visited Britain in order to explain the practice to the British.[74] Trinomialists argued that the provision of a third name for geographical varieties helpfully broke down each species into smaller, geographically defined units, illustrating their relationship to one another and their distribution in space. Linnaeus, they argued, had rendered order with his binomial naming system. The use of trinomials would render detail and order according to geography.

Jordan's long-held interest in geographical variation primed him to take an interest in such innovations. As a child he had marveled at the diversity of the landscape. Under the influence of reading Karl Möbius and others he had learned to think of these landscapes in terms of habitats (or biotopes), distinguishable by soil, vegetation, and climatic conditions, and inhabited by definite and well-characterized animal communities. This had been the reason he had found the Stromeyer collection, with its carefully labeled specimens, so fascinating. In the calls for a trinomial nomenclature, Jordan discovered a campaign intent on ensuring that naming and descriptive work served a scientific purpose and for which being scientific entailed an incorporation of the insights of evolution theory. Henry Seebohm had defended the use of trinomials, for example, on the grounds that naturalists must incorporate their oft-stated belief in evolution as a working hypothesis within taxonomy.[75] Jordan particularly praised the work of entomologist Otto Staudinger, who in 1871 applied the term *sp. Darw.* to those forms that were distinct but whose status as species he doubted. Staudinger, Jordan later explained, had thus been the first to deliberately introduce evolution into taxonomy.[76] Notably, trinomialists' attempt to reform the practice of those working on collections by incorporating evolution theory combined traditional institutions and new scientific reforms, something Jordan had been taught to value under the tutelage of Ernst Ehlers.

Rothschild, Hartert, and Jordan laid out the principles upon which their own use of trinomials would proceed in the first volume of *Novitates*. Since the basis of "scientific systematic work" was "the knowledge of the species and their geographical distribution," even slightly different forms must be distinguished from one another, provided such differences are constant. They outlined how they would fulfill this principle in practice as follows: If very closely allied forms were found to be connected by intermediate specimens, they would not be admitted as distinct species but would be degraded to the status of subspecies. They defended their brevity by noting that "most of these points of view and theories have so often been discussed, or are of such eminent practical usefulness, or else are merely postulates of logical reflection, that we think it unnecessary to dilate upon them."[77]

The Tring naturalists' optimistic announcement that naturalists would surely see both the practical and logical reasons for their nomenclatural practices proved premature. Although American ornithologists outlined a set of rules for using trinomials in the first American Ornithologists' Union Code of 1886, and the German Ornithological Society agreed on their use in 1884, most British naturalists opposed the nomenclatural innovations. The controversy that took place among British ornithologists over trinomial nomenclature, in large part inspired by Ernst Hartert's massive coining of new trinomial names, has often been characterized as having been between evolutionists and creationists, particularly once trinomialists gained ground and could caricature their vanquished rivals. But opposition to the new nomenclature did not arise from creationist sentiment. When the American ornithologist Elliott Coues visited England in 1884 to debate the matter, most British naturalists peppered him with anxious concern regarding the practicality of the system, rather than its theoretical assumptions. The curators of large collections such as the British Museum mounted the strongest opposition. British Museum ornithologist Richard Bowdler Sharpe, for example, believed that "the burden imposed upon the Zoologists who follow this method for the naming of their specimens will become too heavy, and the system will fall by its own weight. That races or sub-species of birds exist in nature, no one can deny, but, to my mind, a binomial title answers every purpose."[78]

The practical constraints of ordering life proved, in this instance, a conservative force as some naturalists attempted to reform the tradition's focus. Based on their experience with the wide diversity of priorities and means within the naturalist tradition, many feared that trinomials would be applied prematurely, before the required study of geographical variation had been completed. Albert Günther, who expressed a willingness to adopt the trinomial system "in all those cases in which the geographical range of certain forms is clearly ascertained," was obviously skeptical of ornithologists' ability to extensively apply the system in practice, for example, given the "great diversity of value" they attached to the distinctive characters of various birds. Philip Lutley Sclater, secretary of the Zoological Society of London and editor of the British Ornithologists' Union's journal the *Ibis*, feared that "if too much stress were laid upon the value of trinomialism we should open the flood-gates to an avalanche of new names by naturalists who have not taken enough trouble to investigate the matter under consideration." Even Henry Seebohm, one of the most ardent proponents of trinomialism in Britain, admitted that ideal statements regarding the usefulness of trinomials "credits ornithologists with an amount of discretion which their past history does not justify, and totally ignores the inordinate desire to introduce new names."[79]

As one of the few converts to the trinomial system in Britain, Jordan's fellow curator Ernst Hartert waged a battle for trinomials in the pages of the *Ibis* for the next two decades. After Hartert published *Vögel der paläarktischen Fauna*, in which he demoted numerous species described by fellow ornithologists to the level of subspecies, Philip Lutley Sclater responded: "As you know well we of 'The Ibis' are of the old school and are not fond of 'trinomials,' but we will endeavour to treat you fairly, while we hold to our own views!" Sclater had reason to be diplomatic, for in the same letter he asked Hartert to bring a bird that would be figured at Rothschild's expense in the *Ibis* to the next British Ornithologists' Club meeting.[80] Ornithologists could be far more cutting in private correspondence. Alfred Newton confided to a friend that he thought "all this new-fashioned stuff and nonsense about trinomials and nomenclature generally is begotten by pride (or self-conceit) upon illiteracy, and a very pretty progeny is the consequence!"[81] Faced by such powerful opposition, at least one zoologist expressed gratification that Rothschild, "the distinguished author and patron of zoological science," was prepared to lead the campaign for a trinomial nomenclature.[82]

As an entomologist, Jordan had an even more difficult audience than the one Hartert encountered. No British entomologist defended the system at the 1884 meeting, and each speaker who had experience with insects voiced concerns rather than support for trinomials. Their material was too chaotic and insects too varied, entomologists cried. Worse, many feared that entomologists' notorious reputation for creating new names at the slightest excuse would be exacerbated by the adoption of trinomial names. Lord Walsingham, for example, expressed concern that entomology ran the risk of "drifting into the inconvenient multiplication of names." David Sharp thought the "adoption of a system of names for forms lower than species, would lead to complete chaos." Henry John Elwes similarly warned that "the system of naming varieties was liable to great abuse, especially in entomology, where the number of species is already so great."[83] Others feared that this extension of binomial names would begin a trend toward "quadrinomials" or worse, obliterating the advantages obtained by the Linnaean binomial. As one critic later lamented: "If we are to have trinomial names, why not multinomial, and so back to the old pre-Linnaean times when species had no definite names, but were designated by phrases, very diffuse and indefinite."[84]

A few years after Tring issued its methodological manifesto, the editor of *Natural Science* generously conceded that if zoologists could make their usage of trinomials "invariable and consistent," then by all means they should use them.[85] Every reader knew what a tall order that *if* represented. Both American ornithologists and the Tring naturalists used trinomials only for geographical varieties. But critics pointed out, and proponents certainly acknowledged, that determining whether a certain variety represented a "geographical variety" rather than some other kind of variation required numerous specimens over a large geographical area. Sharpe highlighted this point when he conceded that trinomials brought attention to the presence of allied geographical forms. But he believed ornithologists of Old World birds knew far too little regarding geographic distribution compared with their American colleagues, who had "a clear idea of the natural geographical divisions of their continent."[86]

Such criticisms captured the strong dependence of accurately applying trinomials on the existence of certain kinds of collections. Collections amassed according to previous priorities were often useless for the accurate determination of geographical variation. To use trinomials according to the principles outlined in the first volume of *Novitates*, one needed long series of well-labeled specimens from various places—dozens if not hundreds of individuals of the same form. In other words, now that reforming naturalists focused their attention on geographical variation, they needed a new kind of collection in order to ensure that "species-makers" did not simply become "variety-makers." While the Tring naturalists were by no means the first to develop a collection aimed at sorting out varieties in the interest of studying geographical variation, Rothschild's museum was certainly one of the first collections organized entirely with this aim in mind. And given the inertia that could build up in museums with massive amounts of material to process, a workload that tended to overwhelm any effort to rearrange or change direction, this was indeed something.

A NEW TYPE OF COLLECTION

Given the contents (hundreds of descriptions of new species) and indeed, the name of the museum's journal, observers sometimes had a difficult time differentiating between the "principled" research coming out of Tring and traditional taxonomy. A reviewer in *Natural Science* provided a somewhat backhanded acknowledgment of the triumvirate's manifesto when he described *Novitates Zoologicae* as having "been started mainly for the purpose of describing new species, and, we presume, of thereby adding to the type-specimens in the Rothschild Museum. The species-mongering is, however, not to be done rashly, or at the mere caprice of writers; it is to be subject to certain principles which are laid down by the editors in an introductory note."[87] Hartert responded indignantly that although there would be many new things described, "you may rest assured that it is not only 'species-mongering' that will fill our journal."[88]

The Tring naturalists' stance on trinomial nomenclature reflected their ability to strike out on a relatively independent and resource-intensive path, even as they were tied via specimen-exchange, society meetings, and correspondence, to the broader naturalist community. Walter Rothschild had both the means to create the kind of collection required to apply trinomials in practice and the ability and willingness to proceed in defiance of ruling British naturalists. (Indeed, Miriam Rothschild attributed her uncle's decision to found *Novitates Zoologicae* to the refusal by Philip Lutley Sclater, at the time editor of both the *Ibis* and the *Proceedings of the Zoological Society of London*, to publish his trinomial names.)[89] Rothschild's foreign curators, though conscientious of the utility of good relations with British naturalists, could also afford to rebel if needed. Although both Hartert and Jordan worked in Britain, they remained loyal to a cosmopolitan community of naturalists that answered to a call for increasingly international standards. With their own journal, established on distinct principles, and a new, rapidly growing collection to catalog, process, and describe as they saw fit, they could not only define for themselves what counted as "more than species-mongering," but they could actually deliver on the ideal.

In doing so, Jordan and his colleagues at Tring carved out a middle road between those who dismissed the knowledge claims of those in collections and those who insisted collections must be at the center of natural history. The collections must remain, but the work must be revised by paying close attention to both the implications of evolution for research and the "higher goals" conferred on "species making" by the theory of evolution. They firmly believed that, when Darwin published his statements regarding the role of variation in natural selection, collections useful in one framework had suddenly become relatively worthless. Certain that species did not change, for example, naturalists from previous generations had often picked out the most "typical" specimen of each species (if more than one specimen had arrived) and disposed of "duplicate" specimens to save space. The old zoology keeper of the British Museum, John Edward Gray, had taken great pride in conscientiously donating any unnecessary duplicates to more poorly stocked, provincial museums on the grounds such specimens required too much care in an already overburdened institution.[90] Darwin, by contrast, had seen the slight variations present in such specimens as central to the process of diversification. In other words, the Darwinian framework had made the concept of a "duplicate" specimen spurious.

In his effort to champion natural selection, Alfred Russel Wallace had described the new kind of collection required by naturalists intent on studying evolution. Since the foundations of any theory must be absolutely secure it had become "necessary to show, by a wide and comprehensive array of facts, that animals and plants *do* perpetually vary." Wallace cited the American ornithologist (and trinomialist) Joel Asaph Allen's long suites of specimens as an example of such a collection—one that allowed "accurate comparisons and measurements demonstrating this large amount of variability." Wallace particularly noted how a long series of specimens of the same species countered the claim that slight variations occurred too rarely for natural selection to work. Wallace similarly praised Gulick's studies on land snails in Hawaii. Gulick had found between 700 and 800 geographical varieties of 175 species in just one forest region of Oahu. Notably, Wallace could give few examples to support his claim that a large amount of variation existed in insects. Entomologists had not devoted themselves to the investigation of variation, he lamented, although Henry Walter Bates had written of *Heliconius numata* that "this species is so variable that it is difficult to find two examples exactly alike."[91] Indeed, Wallace thought many supposedly distinct species of insects probably represented varieties, a confusion he attributed directly to entomologists' lack of attention to variation.

Ironically, given Wallace's complaints, some naturalists had been insisting on the importance of collecting dozens of specimens of the same species for decades. Only then, it was argued, could one truly determine what forms warranted status as full species. The staunch antievolutionist Louis Agassiz, for example, had insisted on the importance of amassing large series of specimens, rather than dispensing with so-called duplicates in his *Essay on Classification* of 1857.[92] Otherwise one might describe an isolated specimen as a new species while a long series of the form actually showed it intergraded with an already described species. The form would then have to be relegated to a variety, according to every good naturalist's practice of describing species. By 1892, Frederick DuCane Godman thought that, although "in former times a pair or two of a species was considered enough to represent it in a collection," entomologists now understood "that 40 or 50 or even 100 specimens are necessary to show the stability or instability of a species, its range, and all the many points connected with a satisfactory comprehension of its limits."[93] But despite the obvious practical importance of long series for clarifying species boundaries, evolutionary or not, Jordan found upon arriving at Tring that one "eminent lepidopterist" considered the "only safe course to take with 'intermediates,' i.e. specimens that didn't fit his concept of species, was to put them in the waste paper basket."[94] As Gulick explained to George Romanes in 1888: "If the naturalist has all the forms in his hand he can create species by throwing intergrade forms into the fire."[95]

In contrast to Darwin and the Tring naturalists, Agassiz believed that the existence of so many doubtful species arose solely from "the insatiable desire of describing new species from insufficient data."[96] He insisted that more data would remove the apparently blurry lines between varieties and species that evolutionists claimed as evidence for speciation. And indeed, faced with Agassiz's long series of specimens at the Museum of Comparative Zoology, at least one of his students noted that their daily museum work demonstrated "the permanence of species" and "showed Darwinism to be absurd."[97] In other words, not all proponents of the study of varieties, long series, and even the use of trinomials, were evolutionists. The German ornithologist Otto Kleinschmidt, to whom Hartert attributed the first-published illustrations of series of specimens, contributed important studies on geographical variation (and used trinomials) but staunchly insisted that varieties did not represent incipient species.[98] Similarly, the French entomologist Charles Oberthür, whom Jordan cited as one of the first entomologists to collect long series of specimens, continued to "define the species as a created entity."[99] Nowhere was the diversity of the naturalist tradition more evident than in this completely different interpretation of the same data.

Jordan found allies in entomologists such as Oberthür, because they saw the importance of collecting the full range of variation in nature, even if they explained the implications of that variation differently. And Tring needed such allies. Amid the debates over how and where to do natural history, even ardent Darwinians dismissed Rothschild's collection of long series of specimens as misguided if not entirely useless. Despite the fact that studying Revd. Henry Baker Tristram's long series of specimens of the same species of birds had been important in his own conversion to Darwin's theory, Alfred Newton protested staunchly against Rothschild's collection practices, writing that he "could not agree with you in thinking that Zoology is best advanced by collectors of the kind you employ. No doubt they answer admirably the purpose of stocking a Museum; but they unstock the world—and that is a terrible consideration."[100]

Newton already thought the natural history magazines "blood-stained" due to naturalists' focus on the collection of dead specimens.[101] He must have been dismayed to hear that Rothschild, writing to his collector in the Galápagos, demanded that "Birds at least 50 of a kind" be secured, with particular attention to "the slightest difference in bill or size."[102] Newton took the high road of conservationist sentiment, but others had less principled reasons for ignoring the calls for long series of specimens. Those willing to describe new species based on a single specimen had a distinct advantage over their more cautious fellows. Lord Walsingham saw his own discoveries scooped by other, less scrupulous naturalists, for example, while he waited for more specimens from his collectors.[103]

Those willing to wait for long series in order to determine the limits of variation and making "good" species had two methods available: one could either breed the insect and preserve the broods or one could collect many specimens in the field. Jordan, Hartert, and Rothschild put their energies—and Rothschild's wealth—into obtaining large series of specimens of each form from the field. Thus, Alfred Russel Wallace preached to the converted when in 1905 he urged on Jordan the necessity that a collector "must go to some place where he can get quantity both in species and specimens."[104] Jordan recalled years later that Rothschild formed his collection with the specific intent of rectifying the unsatisfactory state of affairs in which "a large number of described species were known from only a specimen or two, and many of these were probably due merely to imperfect knowledge of the latitude of variation within some other species." This situation had "induced Lord Rothschild to insist on acquiring large series, maintaining that the number of specimens available for research could never be too large."[105] One entomologist later recalled how, in marked contrast to John Edward Gray's diligent disposal of duplicates, Walter Rothschild would open drawers filled with hundreds of specimens of the same species, and announce proudly, "I have no duplicates in my collection!"[106]

The pages of Novitates Zoologicae are peppered with astonished references to the results of this new way of collecting and looking at specimens. In an 1894 paper on Sphingidae, Rothschild described the remarkable variation found in a series of sixty-four specimens of a single species sent from one locality in Queensland, Australia: "no two are alike; in some the transverse bands are wanting, in others broken up into spots, and lastly in some these bands are nearly double the usual width."[107] As Wallace had noted, long series of specimens provided quite a different portrait of the amount of variation in nature than that given by collections purged of "duplicates." And long series of specimens arranged according to the geographical regions from which they came showed something more. A series of specimens with detailed locality labels demonstrated the prevalence of geographical variation. Cross a river valley and a known species differed ever so slightly. Scale a mountain range and one might encounter a bird or butterfly that could be a different species or perhaps just a variety of the winged inhabitants of the land from which one came. In 1869, Henry Walter Bates had lamented the fact that, while entomologists recognized the role of the study of distribution in evolutionary theory since "certain local forms are modifications of their sister-forms," most collections contained few geographically arranged suites of specimens.[108] The Tring naturalists commenced an unprecedented effort to provide such collections, as convincing evidence of the importance of both recognizing subspecies as incipient species and using a trinomial nomenclature.

Exactly how many specimens the Tring naturalists required for a good series depended on the organism being studied and the questions at issue. In the 1890s, Hartert recommended that collectors send up to twelve specimens of birds (twenty of hummingbirds), eight of butterflies and moth species, and as many as possible of *Papilionidae*. In cases where there were questions about subspecies to settle, they would "take much more than 12, up to 20, 24 or so."[109] Furthermore, the Tring naturalists insisted that such series must not be collected randomly. "It is also not enough to have fifty or more individuals of a species from a certain place," Jordan once wrote, "but it is most important to have the insect in numbers from every district where it occurs."[110] They wanted collectors who surveyed entire landscapes, so that relationships between geography and variation could be discerned.

Even as they urged long series of specimens as the basis of more accurate conclusions regarding species, the Tring naturalists constantly highlighted the provisional nature of their conclusions. Hartert warned that his decisions regarding specimens arriving from the Celebes "must be subject to alterations, or even be erroneous for a great part," until a more exhaustive collection could be made of the islands.[111] Rothschild and Jordan peppered their endless pages of new forms with constant caveats that new specimens might alter their conclusions. In other words, they conceded that many of their species had been "made" pending more material. This entailed quite a different sense of naturalists' tenuous grasp on species, however, than that of critics who dismissed the whole endeavor of taxonomy as unscientific. As we shall see, in urging the importance of as many specimens as possible, the Tring naturalists entered into a long campaign to retrain an entire natural history network in order to ensure that those specimens that did arrive would aid, rather than obscure, judgments regarding species.

RETRAINING THE NATURAL HISTORY NETWORK

A few years after the debate over the proper content of entomological journals at the Entomological Society of London, J. W. Tutt described how "students now are breaking up rapidly into systematists, who still continue to describe genera and species, and biological entomologists, who combine their biological studies with a study of the relationships of the insects to one another and to their environment."[112] Through amassing long series of specimens for the purpose of studying geographical variation, Rothschild, Hartert, and Jordan united systematics with Tutt's "biological studies": a long series of specimens collected over a wide geographical area explicitly related insects to their environment. The Tring naturalists' stance was that in doing so one could then adequately study the relationships of insects to one another. "Biological entomology," then, could indeed be done within the walls of the museum.

Of course, relating specimens in a museum drawer to their area of provenance required that the specimens' original locality be carefully recorded. Yet, as Jordan had found during his work on the Stromeyer collection, many lepidopterists failed to provide detailed labels for their specimens. This lack of detail rendered the specimens useless for both the purposes of studying geographical variation (some of Rothschild's older specimen labels simply said

"Africa") and the project upon which Tring had staked its claim as a scientific institution.

To build a collection in which one could analyze geographical variation, Rothschild and his curators had to reform and standardize both the collection of specimens in the field and the processing of specimens by natural history agents, other museum curators, and private collection owners so that specimens arrived tailored to their needs. The four surviving out-letter notebooks, from 1892 to 1898, show Hartert and Jordan constantly advising collectors, natural history agents, friends, and colleagues of their rules and procedures for collecting, arranging, and processing specimens. Hartert, for example, constantly reprimanded collectors who sent a few specimens of the biggest and most striking birds and butterflies with orders that they pay "special attention to all the dull-coloured small and insignificant looking birds as well as the bright-coloured ones."[113] The most common and constant exhortation to collectors was to include detailed locality information, a basic prerequisite for the Tring naturalists' wish to study the relationship between geography and variation.[114] "The place where species are found is very important," Jordan explained to J. Schröder in 1893.[115] Hartert struck the same note: "One thing is most important, that is proper labeling of the specimens. Every specimen must have a label with the exact locality and date and if possible sex."[116] He admonished one collector who attached unsatisfactory labels that "in future I advise you to be as careful as possible with the localities, or you will not receive anymore work."[117] They expected labels to be completed in the field in order to ensure accuracy, faulting even "the immortal Wallace" for producing some rather dubious locality labels as a result of filling them in later.[118]

What exactly the Tring naturalists meant by "exact locality" took some explaining. By requesting twelve specimens from every locality, they did not want 144 butterflies from different stretches of beach on the same island, but 12 butterflies each from localities having a "somewhat different fauna."[119] Hartert advised one collector that "It would be desirable that every small and big island of the above named groups would be visited, and every bird should be collected on every island, for they may seem to be similar to those of the other lands, and yet be different perhaps new species, as you know."[120] Hartert could give as detailed directions as he wished, but ultimately the collector in the field had to judge what constituted a "somewhat different fauna." Such decisions depended on the collector's skill as a naturalist, an important factor in why the Tring curators developed favorites among Rothschild's dozens of collectors.

In composing hundreds of letters to those in the field, the curators tried to tailor incoming specimens to Tring's distinct research aims. Ernst Mayr, having witnessed at first hand Hartert's and Jordan's shrewd dealing with collectors and the organization of expeditions (Mayr collected for Rothschild in the 1920s), would attribute Tring's success in building a first-rate collection to this constant stream of letters to collectors and colleagues.[121] The letter writing of course took time away from moments that might otherwise have been spent processing incoming specimens and describing new species. But both Hartert and Jordan knew the letters had to be written, as a part of their constant effort to make the species described "good." For they had to deal with a tradition and a network that did not always work by their standards. Ten years after beginning their campaign, Rothschild and Jordan lamented that, judging from the materials offered to Tring, "there are still Entomologists, dealers, and their collectors in the Tropics who are unaware of the great importance of correct labeling."[122] Lord Walsingham's curator, Durrant, had similar troubles when trying to convince collectors to regard their work from a scientific point of view. "Your prices," one collector replied indignantly, "would scarcely repay me for the extra work required in carrying out your conditions." Besides, the collector insisted, he could not very well neglect good paying insects for the long series Walsingham wanted at so low a price.[123]

Rothschild and Jordan conceded in 1906 that it was perhaps unfair to complain of past workers. "If we find the localities given in the works of the earlier writers often deplorably vague and frequently erroneous, we must remember that the majority of the specimens were collected by people who did not take an actual interest in Natural History, but brought the specimens home as curiosities from foreign countries." But they insisted that the opportunity for excuses had now passed; "Every serious student of some branch of systematics is aware that specimens without exact locality are of little value to the scientist." They conceded that progress had been marked; labels now arrived printed with exact locality, date or season of capture, altitude, name of collector, and "even some biological fact." But the general collecting public, dependent on dealers for additions to their collections, had not yet followed in rejecting with disgust the specimens offered with such vague localities as East Africa, Australia, or the Amazon.[124] Forty years after Tring began its campaign to retrain its collectors and colleagues, some naturalists continued to ignore their new "scientific" directions for collecting. The persistence of poor labeling prompted Jordan to propose with exasperation in 1937 that "collectors who do not label their collections before selling specimens should be punished in some way. Scientifically speaking, they are criminals."[125]

The fact that natural history was also pursued for profit both aided and abetted the Tring naturalists' efforts to enforce their priorities on collectors. In the United States alone, more than one hundred natural history dealers operated by the end of the nineteenth century.[126] Dealers' advertisements peppered the back pages of natural history journals. One dealer courted the business of those collecting in the field by announcing: "Collections containing good species purchased for PROMPT CASH, and the most liberal prices paid, W. W. possessing an exclusive and wealthy clientele."[127] The presence of these natural history dealers complicated Jordan and Hartert's efforts to influence the methods of specimen collecting. In their direct communications with collectors, they made it quite clear that satisfactory collectors would receive orders for further consignments, while collectors who deviated from their rules risked being cut off from Rothschild's purse. But natural history agents, who often possessed less discriminating clients, did not always fall in line with Tring's new-fangled methods. Jordan's and Hartert's insistence on accurate, original labels made by the collector in the field inspired the following outburst from one New York dealer: "We certainly cannot understand that you complain about removal of labels from the last shipment." They had had no such complaints from their other correspondents or patrons.[128] Jordan later recalled that when Tring asked one Continental insect dealer in the 1890s to give a precise locality for some specimens of Lepidoptera, "he was very angry that 'New Guinea' was not considered good enough for Tring."[129] Agents also sometimes made it difficult for the Tring curators to obtain direct access to the collectors in the field, a basic requirement of Tring's efforts to reform collecting methods. One of the largest natural history agents, Janson & Sons, reminded Rothschild that it was not customary to give out the names of their collectors to buyers, following the practice "for obvious reasons" of agents and dealers in all businesses.[130] This particular business custom conflicted directly with the Tring curators' wish to hold collectors accountable for the quality of their work, and in doing so direct the methods of those in the field.

That natural history was seen, at least by some, as a business and therefore bound by rules of supply and demand meant that the new scientific value placed on so-called duplicates could prove costly if dealers decided to start charging more for large series. As a result, Hartert actively downplayed the value of duplicates when negotiating prices. In stark contrast to Rothschild's proud claim that he possessed no duplicates, Hartert wrote to one correspondent that "there are so many thousands of birds, which we are in want of from other countries, that it would be foolish to waste money on duplicates. Besides, the duplicates take up <u>room</u>, and you know how short we are of room."[131] Indeed, Hartert expressly contradicted the scientific arguments for series—that is, that no duplicates existed in nature—when he wrote to one collector: "We take, naturally, only a limited number of specimens of every species.... There is nobody who cares for many hundreds of one kind, but everybody wants as many species as he can get."[132]

However, the Tring naturalists did not thereby lose the opportunity to study the "many hundreds of one kind," for they insisted that the museum get the first pick of any collection. Those specimens not chosen would then be returned after Rothschild's curators "scientifically worked" the collection.[133] They also insisted that they be allowed to first study specimens not purchased "in case a series is required to settle different questions." They could thus achieve their scientific standards without paying for the entire consignment. Even if they bought entire collections, Hartert downplayed the monetary value of these more scientific collections. When the collector Edward Gerrard argued about a price paid for a long series, Hartert replied that "the Hon. Walter Rothschild wished him to say" that the 115 butterflies taken by the museum from the collection were "almost all of the commonest types and for their greater part merely taken for the sake of locality."[134] One shilling per specimen was, under these circumstances, a very fair price. A shrewd collector could have countered, of course, that locality conferred *all* the value on such a collection. And it was not long before some collectors became wise to the shift. "Collectors are just learning," wrote one observer, "that the study of these varieties and the affixing of locality labels mean money," since a few hours' labor adding locality labels to even a small collection could mean a difference of £100 or more.[135]

On the other hand, financial incentives failed to impress some influential British entomologists who had their own money. Yet these men, too, had to be convinced if the entire collecting network was to be retrained. That much of the entomological and ornithological community had not altered their collection practices by the turn of the century is evident from the constant pleas throughout Hartert, Rothschild, and Jordan's papers for long series of well-labeled specimens rather than single samples. Even H. J. Elwes expressed little sympathy with Rothschild's complaint that some specimens arrived without detailed localities. "There may be a few more odds and ends without locality," Elwes admitted when Rothschild purchased his collection, "but not enough to be worth talking about." When asked for more detail than "South India," which was marked on some of the birds, Elwes replied they came from various places on the west coast: "I believe you might safely put Malabar on them," he suggested.¹³⁶ Clearly, Elwes had a very different idea of what contributed to a collection's value. He boasted that his collection did not contain a large number of common European birds nor a lot of specimens of one species; "Mine will not average more than 2 or 3 specimens of each species and all the rubbish has been turned out."¹³⁷ Still, Rothschild bought the collection on the grounds it contained many fascinating rarities.

In the interests of educating the natural history network about their requirements, Hartert and Jordan knew how important it was to show visitors around the collection. In 1898, they had a welcome opportunity to explain their methods when the members of the International Congress of Zoology meeting in Cambridge made a day trip to see the museum. One reporter described how Rothschild's curators had explained at length how Rothschild's museum differed from other local museums in "providing facilities for practical scientific work." The visitors had finished their journey "amid cheers and waving of salutations to Mr. Rothschild and his two curators, both of whom had done their utmost to display the treasures of the Museum, and to make visitors acquainted with the methods of research there carried on."¹³⁸ Demonstrating those methods was even more important than showing Rothschild's rarities, for the Tring naturalists could thus illustrate the principles that guided their development of Rothschild's collection. The long series of specimens impressed most, Jordan recalled; "It was a frequent topic of conversation with visitors."¹³⁹ In their writings, Rothschild and Jordan explained their constant exhortations for better locality labels and long series as founded on the hope that in time smaller collections would be labeled better; thus, when they ultimately came to public institutes, the material would be worth preserving.¹⁴⁰

Though surrounded by the pinned and stuffed specimens that had formed the primary data of scientific natural history for centuries, the Tring naturalists were intent on tailoring these specimens to new questions and methods. After all, Darwin himself had insisted that to properly examine variation in nature, "a long catalogue of dry facts should be given."¹⁴¹ The Tring naturalists joined the campaign to revise the facts of importance to help ensure that even as the Walter Rothschild Zoological Museum produced "long catalogues of dry facts," the endeavor of naming species was indeed aiding the attempt to find the natural order. Through mapping that order at the fine detail of geographical variation, the museum could then contribute to the goal of explaining the origin of living diversity by ensuring the facts amassed actually helped answer evolutionary questions. But whether this "revision of facts," commenced with the first issue of *Novitates Zoologicae* in 1894, would be sufficient to bring natural history museums and systematics into the twentieth century safely was certainly an open question. It was a question to which Jordan turned again and again, over the ensuing decades. For where systematics fit within the modern order of both the sciences and society more broadly was a pressing question, if naturalists were to continue their quest of finding order in nature.

THREE

Ordering Beetles, Butterflies, and Moths

Jordan often told the story of how, sometime in the 1890s, a professor of zoology at Cambridge University tried to discourage his young friend Charles Rothschild from studying the systematics of insects. Charles had wanted to make a collection of fleas as part of his coursework and would not back down when the professor tried to channel his energies elsewhere. Jordan recounted how ultimately the professor, "unable to overcome the love of a born Entomologist for Entomological investigations, and not being big enough to look beyond the fog of prejudice and ignorance, dismissed the subject in high dudgeon, crying: 'Oh, well, if you *like* sticking insects between two bits of glass!'"[1]

In later years, Jordan cited this episode as evidence of the fact that many academic zoologists viewed systematics as beneath the aspiration of a "scientific mind." Jordan did not need to pay attention to such attitudes: Cambridge zoologists were not reviewing his submissions to *Novitates Zoologicae*; and his salary did not depend on their good opinion. But the ethos of professional science goes beyond practicality. As an academically trained, professional zoologist, Jordan could not ignore the reputation of the scientific tradition in which he worked. His identity as a scientist, and the very meaning of his hours at the microscope, were at stake. He also knew that the disdain of misguided college professors might inspire potential recruits to abandon the naturalist tradition altogether.

Jordan firmly believed that fending off stereotypes and criticisms of systematics must include a reply to the question of *why* "stick insects between two bits of glass." Hartert, Rothschild, and Jordan would each launch ardent defenses of Tring's methods. The form of Jordan's defense was influenced by his position as a museum curator, his concepts of the relationship between systematics and biology, and contemporary debates over the relationship between facts and theories. Most importantly, his writings about *why* entomologists stick insects between two bits of glass led him directly to the question of *how* systematic entomologists could "make species" good. For the strength of the justifications he outlined for all those hours at the microscope, naming, describing, and ordering insects, depended on entomologists' willingness and ability to adopt certain methods in their work. And ultimately, this returned him to the effort to reform an entire natural history network's way of ordering both nature and itself.

"THE GREAT DESIDERATUM"

By the time the first issue of *Novitates Zoologicae* appeared, taxonomists had offered various justifications for naming, describing, and classifying organisms. Some naturalists saw their primary task to be the complete documentation of all nature. The call to inventory imperial possessions gave further impetus to this cataloging project. Certainly, naturalists recognized the dangers entailed in attempting comprehensive catalogs of the Earth's animals and plants. The zoologist Leonhard Stejneger pointed out toward the end of the century that the scale of one famous catalog taken on by the German Zoological Society meant it would be ages before the task was completed (he particularly wondered how many volumes would be required for the insects). Even if the sixty-four specialists enlisted to complete the catalog finished their job, Stejneger wondered whether the first volumes would not be completely antiquated before the last volume appeared.[2] Yet catalog projects such as those of the German Zoological Society and the British Museum reflected a methodological ethos firmly defended by zoologists H. J. Elwes, Albert Günther, and others. Furthermore, as we saw during the debate regarding the role of speculation in entomology, some commentators insisted that only a careful accumulation of specimens and the "completion of the catalog" could provide a firm basis upon which to theorize.

As Jordan, Hartert, and others often noted, naturalists' use of the Linnaean system meant that they were in fact finding the order in nature (transformed by Darwin into a search for the lines of descent) through the very act of choosing an organism's name within the catalog. But "biological" entomologists, intent on reforming their discipline, had little time for nuance. Ignoring the robust tradition of a nonevolutionary, and often accurate, search for the natural order, Harry Eltringham described "the older naturalists, secure in the comfortable belief in the fixity of species," as having "occupied themselves with the compilation of voluminous catalogues of all the forms then known to them, the result being a mere list of names."[3] The new entomologist, by contrast, produced biological classifications that traced the descent of each species.

Caught up by the rhetorical power of such dichotomies, some entomologists began distinguishing between those who applied evolution to research and everyone else studying insects. In 1894, W. E. Sharp described how, although Darwin's work had little influence on the entomologists of the 1850s and 1860s since they were "too busy describing new species," eventually entomologists began to see how evolution might raise entomology "from a study to a science." Previous generations had supposed, he explained, "that when every insect form in the world had been described and catalogued, and the whole of the order finally and unanimously arranged in methodological series of divisions, then their warfare would be accomplished and their occupation gone." Modern entomologists, by contrast, knew that even with a full list of existing species "our real labours would be but begun." Sharp made the incentives to reform explicit: entomologists now worked according to new ideals that would "redeem Entomology from the charge of triviality or lack of adequate intellectual interest," as well as the justifiable "mild contempt from their public."[4] In other words, Sharp and others believed that the cataloging project had become dangerous to the scientific status of the discipline.

Of course, those "too busy describing new species" had their defenders. Alfred Russel Wallace argued that in a Darwinian world the humble work of the "species-monger" continued to have great fascination and importance, for they carried out the useful work of creating order out of chaos.[5] Firmly invested in the descriptive work demanded by large collections, Albert Günther pointed out that although many systematists accepted evolution, "the needs of minds differ" and not everyone needed "some universal concept into which their daily work may be fitted."[6] But Darwin, who had certainly put in his own time tediously sorting species, provided a less generous analysis in his observation that few generalisers existed among systematists. He once wrote to Joseph Dalton Hooker, "I really suspect there is something absolutely opposed to each other and hostile in the two frames of mind required for systematizing and reasoning on large collections of facts." To Henry Walter Bates he was more blunt, "I have long thought that *too much* systematic work [and] description somehow blunts the faculties."[7] By the end of the century, young, self-proclaimed "biologists" intent on carving out a place for themselves within science tended to side with Darwin's characterization of the systematic mind. The dichotomy had certainly been implicit during the debates over entomological method between Raphael Meldolda, H. J. Elwes, and E. B. Poulton at the 1890s meetings of the Entomological Society of London. When Poulton fought for command over the insect collection in Oxford, he took advantage of the shift in what counted as valuable within science by insisting he would not "encourage the growth of a great centre for the exclusive study of Systematic Entomology and the description of new species at Oxford," but that he would devote the collections and his lectures to "philosophical biology."[8]

Having been summarily booted from the halls of science, some naturalists bemoaned the ingratitude of "biologists" who poured "such vials of wrath" upon the describers of new species, who, as one wrote, "if they did nothing else, at any rate provided the theorists with the foundations of their airy structures."[9] But clearly, in the face of the "vials of wrath" (real or imagined), those working in large collections of specimens at the end of the nineteenth century had to ask themselves precisely what kind of work should be done and why, given changing scientific priorities and methods. They were not the first to reflect on the point of the thousands of descriptions of new species filling naturalists' journals. Darwin's staunch critic Louis Agassiz had often warned that although the discovery of new species had been the great object when naturalists knew less of animals and plants, "this has been carried far, and is now almost the lowest kind of scientific work." To temper naturalists' mania for new species, Agassiz had urged them to publish new names only when analyzing genera or families since only then could one pay adequate attention to variation and geographic distribution.[10]

Decades later, conscientious entomologists intent on doing good systematics were still admonishing their fellows that the day for endless descriptions of new species had passed. "The number of brief notes and papers in which 'new species' of insects are described is appalling," the editors of *Natural Science* complained. For most such species it had surely been "absolutely impossible for the authors who write on many of the groups to have read a tithe of the papers previously published on the subject."[11] "Species-mongers" who described thirty-two alleged new species within three pages added to the problems of the "overwhelmed systematist" and, more importantly, retarded science.

Synonymy (the existence in the literature of more than one name for the same animal) was seen as one of the worst problems created by "species-mongers." By the time Jordan took up his tasks as curator of insects, the problem of synonymy had reached "hideous" proportions, according to the Mexican zoologist Alfonso Herrera, who determined that thus far 8,327 names had been used for 1,485 distinct birds (this meant 6,482 of these names were meaningless). As the naturalist Thomas R. R. Stebbing pointed out, this chaos of duplicate names meant that "before any comprehensive work on zoology can now be produced, the naturalist finds his studies in a manner divorced from nature. There is so much to read that little time is left for observing. He is in a sort of bondage to an antecedent literature which cannot be ignored." The fact that naturalists published descriptions in many languages, in small and large society journals, expedition reports, pamphlets and expensive monographs, made it difficult for the "conscientious zoologist" to ascertain whether his proposed species had been previously described.[12] Unfortunately, the new focus on varieties had exacerbated the problem. J. W. Tutt bemoaned the fact that "scattered over the entomological literature are endless descriptions of local forms, varieties and aberrations of lepidoptera, confusing an already chaotic literature."[13]

Given the confusion—a legacy of diverse naturalists working according to various means, priorities, and methods—some entomologists insisted that what the discipline required most was not descriptions of new species, but good taxonomic "revisions" that mined all these obscure publications and cleared up the existing chaos created by the tradition. In other words, the past needed to be ordered as much as the specimens crossing one's desk in the present.

Doing a taxonomic revision thus became one way for taxonomists to both counter the jibe "species-monger" and actually fight against such characterizations by reforming and improving the data available. Indeed, the editors of *Natural Science* minced no words in 1896; those who produced revisions and monographs "are bridge-builders and road-makers for science, the others only raise obstacles in her path." "The great desideratum at the present time," they wrote, "is not the multiplication of species whether good or bad, but the re-description, in the light of modern science, of the species that have already been named... and the arrangement of all species under their proper and finally determined names in the genera to which they are thus shown to belong."[14] "What is really wanted nowadays," they insisted, "is careful and detailed work, accompanied by a thorough revision of all the imperfect and inaccurate statements launched upon the world."[15] Such work would take the mass of overwhelming facts in species descriptions and synthesize them into a useable form, no matter what particular use one might then wish to make of them, since a revision's primary aim was to accurately identify previously described species and "delimitate" species and their respective varieties, rather than give detailed descriptions of new species. Although naturalists would often disagree as to the details regarding how they should be done (for example, on what grounds certain names should be chosen as final), all the entomologists involved in the debates of the 1890s agreed on the utility of sorting out the chaos of names through careful revisions.

REVISING THE SWALLOWTAILS

As the curator of a large collection of insects, Jordan could have spent his entire career describing new species. He certainly could have described more than the three thousand or so he had named by the time he penned his last name. But Jordan's training and inclination inspired him to side with those who believed monographs and revisions constituted the most important and useful work in systematics. A revision represented an attempt to discern order amid chaos in various realms. In reviewing the existing literature, one sorted the entomological wheat from the chaff to discern which statements about insect relationships (in other words, which Linnaean names) should be allowed to persist. In deciding on certain names for new species one continued the effort to find the natural order in nature, since the names represented conclusions about that order. And all of this ordering was accompanied by the constant effort to sort the incoming specimens that might adjust prior decisions.

It was also sometimes accompanied by a need to reorder one's own expectations about what kind of specimens would be arriving. After he completed his first assignment of bringing "some order" to Rothschild's huge beetle collection, Jordan had hoped to await incoming beetle specimens with more leisure, process them more carefully, and "draw conclusions therefrom" at a gentler pace.[16] He would indeed be able to devote more of his time to drawing conclusions in the next few years. But any incoming beetle specimens mostly lay to one side, at least during the museum's official working hours. For although Rothschild fully supported his curators' efforts to plan the collection along scientific lines, he ultimately controlled what kinds of specimens would be purchased. Having set out on a project of completing a revision of the butterflies of the islands east of Wallace's Line (a biogeographic boundary between Borneo and Sulawesi), he had decided Jordan must help him. Miriam Rothschild describes how, conscience-ridden for switching his curator from beetles to butterflies, her uncle decided to simply instruct his collectors abroad to stop sending specimens of Coleoptera. As the number of beetle specimens in the boxes arriving at the museum plummeted, "K.J. soon realized what was afoot, but made no bones about abandoning the beetles and forthwith began work upon the Swallowtails.... At any rate he was now given a desk with an excellent light."[17]

Eventually Jordan himself warned collectors that beetles were of interest only to him and that no special effort should be made to get them.[18] Five years later, as the work on Lepidoptera continued and Jordan realized he could not possibly attend to a worldwide collection of both orders, Rothschild sold his entire beetle collection—half a million specimens—to various buyers for a total of £2,300.[19] Initially Jordan had kept the Anthribidae and Longicorn beetles, his particular favorites, but eventually gave up the idea of tending to all but the Anthribidae and sent the Longicorn beetles to Charles Oberthür in France.[20] In his nineties, Jordan would refer to this episode as when he "was taken off the Coleoptera and became a sort of private secretary to Walter R. who wanted to publish Revisions of Lepidoptera."[21] From then on Jordan reserved the daylight necessary for detailed descriptive work on butterflies. (The museum possessed no electric lighting until the erection of a new entomological building during the first decade of the twentieth century.) As a result, Jordan could describe far fewer new species of beetles than he found in the packages arriving from collectors in the field, natural history agents, or colleagues at other museums. In 1906, he returned beetles sent by Poulton for determination, for example, by explaining, "I am sorry to have kept the beetles so long: but I can devote so little time to Coleoptera that I have not yet done as much with them as I should have liked."[22] The Oxford "biological entomologist" probably did not begrudge the time Jordan spent on butterflies (Jordan's work would prove critical to Poulton's work on mimicry), but the coleopterists must have been dismayed.

Meanwhile, under Jordan's guidance the original project of revising a whole region's Lepidoptera quickly became more targeted as the problems posed by both Rothschild's growing collection and the errors of prior naturalists threatened to overwhelm the whole endeavor. In the end, as Rothschild explained, he decided to "give a series of monographic revisions, dealing with a few families both of *Rhopalocera* and *Heterocera* from the whole world."[23] They would revise the Papilionidae, or swallowtail butterflies, first.

Before turning his attention to this popular group, Jordan had known few exotic Lepidoptera.[24] But no doubt he recognized the advantages of accommodating himself to Rothschild's plan. First, a good revision of any group would provide an excellent response to criticisms of the Tring Museum as a center of "species-mongering," principled or not. Second, Jordan had learned the use to which a good collection could be put during his work on the Stromeyer collection, and Rothschild not only had an excellent collection of butterflies but the resources to make it even better. Finally, as we will see, butterflies proved more amenable to certain microscopical methods he had learned under the tutelage of Ernst Ehlers. For while Rothschild may have dictated the kind of organisms piling up in drawers on his desk, he did not demand that Jordan study those specimens in any particular way. Indeed, Jordan later recalled with gratitude how Rothschild gave him permission to do dissections "although at the time this was still strictly forbidden in museums."[25]

Jordan plunged into the work with dedication. "I studied the butterflies with a magnifying glass and microscope, in order to find morphological differences," he recalled; "on weekends I submitted the results to Walter Rothschild."[26] The culmination of this collaboration appeared in the 1895 issue of *Novitates Zoologicae* as "A Revision of the Papilios of the Eastern Hemisphere, exclusive of Africa." Rothschild's name appeared as sole author, with a notice in the introduction that "throughout the work will be found a number of notes by Dr. Jordan, dealing with our methods of investigation, such new or unfamiliar facts as have presented themselves to us, and, lastly, detailing the scope of the present article."[27] In fact, Jordan wrote the entire "Introductory Notes," since cited as one of his most important commentaries on the study of geographical variation and subspecies.[28] This apparent co-option of authorship was common enough at the time. Lord Walsingham "freely asserted" that his "secretary," John Hartley Durrant, "did the major part of the work on the many pretentious publications, bearing Lord Walsingham's name."[29] Tring's subsequent Lepidoptera revisions listed Jordan as coauthor, but readers remained confused by the nature of their collaboration. The American entomologist William Jacob Holland, for example, had to ask: "In designating the species in my manuscript whom shall I credit with the authorship? Are the new species described in this paper to be credited to Rothschild & Jordan, or to Jordan & Rothschild, or to Jordan?"[30] Not surprisingly, given his need to distance himself from charges of species-mongering, Jordan seemed unconcerned by the confusion.

By contrast, he was obviously very concerned by the continued opposition to Tring's nomenclatural reforms. Expanding on the blunt statement of the editors that had opened *Novitates Zoologicae*, Jordan began his introduction to Rothschild's first "Revision" with a lengthy review of the history and practices of zoological nomenclature. He did so on the grounds that the results of his and Rothschild's investigations often differed from those of other entomologists. Therefore it became, he wrote, "necessary to give a short account of the method of our researches, the means upon which they are based, and our views with regard to nomenclature and variation of the Papilios." He began with their objectives: a revision, he explained, entailed the identification of previously described Eastern Papilios and the delimitation of the species (a monograph, by contrast, would provide "detailed descriptions of the insects dealt with"). Then, he precisely defined the criteria he and Rothschild used in order to "delimitate" the species. "We consider," he explained, "... all those Papilios as varietal forms of the same species which are connected with one another, in one or both sexes, by intergradations; and treat those forms as specifically distinct, however closely allied they may be, which no chain of intergraduate specimens combines." Following the lead of American ornithologists, they defined subspecies, in contrast to other varieties, as well-characterized local or geographical forms, giving these forms alone a trinomial name. Notably, they abandoned the use of the term *variety* altogether due to systematists' long history of using the term indiscriminately for every kind of variation.

The stance Jordan outlined in the "Revision" on the study of geographical variation, trinomials, and the boundaries between species drew on precise ideals regarding the ultimate goal of systematics. Delimiting the species meant placing them in their correct genera and so on, an activity that inevitably entailed statements about relationship, and thus the path of evolution. Yet, as Jordan insisted in his detailed defense of their emphasis on geographical variation, "it is impossible to understand the relationship of closely allied species, without a knowledge of the varieties." Thus, "when one neglects the latter, one neglects also the most striking facts which can serve to explain the origin of species."[31] Jordan insisted, in other words, that this largely descriptive-based ordering of the natural world provided a foundation for accurate generalizations regarding how that order had come about.

Jordan's stance on trinomial nomenclature and naming varieties relied in turn on numerous ideals of how methods could be changed and natural history reformed given debates over how and why systematics should be done. By contrast, the stance he outlined toward the problem of synonymy, a primary target of a "Revision," arose from a highly pragmatic recognition of, first, the reality of how naturalists worked and, second, the limitations of both the diversity and historical legacy of the naturalist tradition. Given the existence of various "synonymous" names for the same species, much discussion had centered on how to decide which name should be given "priority." A traditional rule that, in the case of two names having been given to a particular beast, the first one to have been published is the one that stands proved in practice the source of heated debate. Leading authorities differed regarding how far back the "law of priority" should go, whether long usage should trump the discovery of some long-forgotten name, and what should count as a valid description. French entomologist Charles Oberthür argued, for example, that only descriptions accompanied by a figure be accepted. Keenly aware that to support such a practice would exacerbate charges that success in "species making" relied on wealth, the Tring naturalists warned that such a rule would require "thousands of good and true workers" to give up the study of entomology since it would thenceforth "become the privilege of a number of very rich men" who could afford the cost of publishing illustrations.[32] Clearly some kind of limit had to be placed on who could describe new species, however. Proponents of a strict priority rule insisted that it would ensure that naturalists did the tedious book work required to avoid synonymy. "By insisting on this essential test of honest intentions," one defender of priority maintained, "we retain the taxonomic and phylogenetic work within the circle of a class of men who are competent to it, and cease to hold out rewards to picture-makers and cataloguers."[33]

Opponents, by contrast, worried that the strict interpretation of the priority rule was inspiring the inconvenient "resuscitation" of old and forgotten names. Others criticized the rule on the grounds that names must be at the mercy of scientific progress. Tutt, for example, adamantly opposed the idea that "hazy indications" might replace those founded on "exact science" simply because they had been published first. "How in the name of common sense," he thundered, "can Hübner's names replace Chapman's, worked out on the minute structure of ova and larvae of which no previous author seems to have had the slightest knowledge?"[34] Intent on a merciless elimination of inadequate entomological practices, Tutt argued that names must reflect the most current state of knowledge, and the man who had done the *best* work should be given priority, whether he had been first to name the organism or not.

Ultimately, Jordan's awareness of the tentative nature of scientific knowledge, mixed with a large dose of pragmatism regarding how naturalists actually worked, determined his and Rothschild's stance on synonymy. Despite their obvious sympathy with Tutt's idealistic vision of naturalists adjusting names according to the latest scientific knowledge, they sided with a strict law of priority as the only realistic response to the enormous diversity in naturalists' aims and methods. In defending their stance in a later 1903 revision, Jordan insisted that to allow a change of name with every advance in science would lead to chaos due to both differing opinions as to what constituted an "advance" and the constantly shifting nature of knowledge. Thus, Jordan and Rothschild refused to coin new names for forms deemed wrongly diagnosed, described, or named, since "there is no line to draw between good and bad definitions, sufficient and insufficient descriptions; and every description is incomplete."[35]

Guided by this strict rule of priority, Jordan pored over the natural history literature and laboriously compared descriptions to type-specimens (the original specimens used to describe new species) in order to avoid erroneously describing forms as new. To clarify (and expunge) the synonymy, he and Rothschild provided long lists of the numerous names conferred on some species. For one subspecies, for which they decided on the name *Papilio memnon agenor* L., the list of thirty different names under which it had been described as either a species, subspecies, or aberration (sometimes naturalists had even described the males and females as different species) ran to almost two full pages.

In a moment of exasperation inspired by the endless library research required to establish the correct name for each species, Rothschild and Jordan once confessed that the energy spent on "book-research" in order to disentangle entomological nomenclature reminded "one too much of the famous fight against windmills."[36] Jordan would return to this image of taxonomist-as-Don Quixote more than once over the next six decades. Meanwhile, frustration with the tradition that had created such an imaginary battle escaped in the form of constant exhortations to their fellow entomologists to do better in the future.

Even as they wrestled with the tradition's cumulative errors, Rothschild and Jordan built their meticulous revision upon the huge amount of effort expended by their predecessors to describe and order swallowtail butterflies. The works of Rudolf Felder and Wallace provided starting points, as did the publications of F. Moore, W. L. Distant, Georg Semper, J. H. Leech, Otto Staudinger, and H. J. Elwes. As they mined these works for information on swallowtail butterflies, the legacy of the naturalist tradition's varied ways and means loomed over their efforts. Jordan confessed that he had initially even feared that despite having the resources of a Rothschild, the mistakes engendered by the wide range in entomologists' methods would prevent them from completing the work.

But they soon found a way through the chaos, beyond the deceptive windmills, in the realization that the mistakes of their predecessors often arose from a failure to follow certain methods of research. This failure could be surmounted, they insisted, by adopting better methods and amassing both a good collection and a well-stocked library. For example, previous entomologists working on the group had not carefully compared the descriptions of older writers with the type specimens, a practice that led to misidentifications. They had also worked with small amounts of material, completely inadequate for establishing the limits of variation. As Jordan wrote, "many of the errors could be avoided by the help of long series of specimens, a good library, and the Felderian types."[37] By examining Felder's type specimens, for example, they found that his habit of characterizing new species from a few specimens led him to mistake individual or geographical differences for specific characters.

All of the resources Rothschild and Jordan used to correct such mistakes cost money, of course. (*Science Gossip* reported that Felder received £5,000 for his type specimens.)[38] But although they could not necessarily escape the charge that success in taxonomy came to those with money, Rothschild's and Jordan's emphasis on geographical variation in the "Revision" certainly addressed another of the most fundamental criticisms of museum work, namely that museum taxonomists proceeded in isolation from a knowledge of insects in nature. Jordan reflected in his memorial to Rothschild in *Novitates* forty years later, that the "Revision of the Papilios of the Eastern Hemisphere" represented

> the first time in Lepidopterology [that] geographical races described by various authors as distinct species were almost consistently (as much as the material at hand warranted) reduced to the rank of subspecies and the classification thereby much clarified.[39]

In thus relating variation explicitly to the changing nature of landscapes, Jordan and Rothschild brought information on living insects from the field into the drawers of the museum and into their classifications. Defending their division of species on the island of Java into subspecies, for example, they described how their predecessors, Staudinger and Fickert, "forget that the fauna of the mountains is different from that of the lower districts, and that many mountain insects are local forms of the species of the plains and hills."[40] Jordan explicitly tied the study of variation to a more focused attention on animals' environments when he noted how the number of aberrant specimens gradually increases "under the influence of the altered '*biocoenosis*,' as it has been termed by Möbius."[41] Thus, Rothschild's long series of specimens and their close attention to geographical variation, type specimens, and extensive library research embraced some of the basic elements of the Linnaean tradition of cataloging species even as they transformed it from within.

This transformation absolutely depended on the nature of Rothschild's collection. For example, fulfilling the principles outlined on the first page of the first volume of *Novitates*, and in Jordan's introduction, in practice relied first and foremost on their possession of long series of specimens. Based on such series, where authors had paired up males and females as different species based on a few specimens, Rothschild and Jordan found that the supposed "different 'males' of these 'species' run all into one another." One could not possibly separate them if faced by a long series of specimens, they insisted, "unless one picks out the typical specimens and burns the intermediate ones, or describes every third specimen as a distinct species."[42] Constantly impressed by the influence of new specimens on their conclusions, they repeatedly emphasized the need for more specimens, even though the "Tring Museum contains of most species greater numbers of specimens than entomologists usually keep in their collections."

This awareness of taxonomists' dependence on the material available escaped in a steady stream of asides and caveats regarding the tentative nature of their determinations. "Many of the new Papilios to be discovered in the future," they warned, "will doubtless turn out to connect some of the insects which now appear to us specifically distinct."[43] They spoke from experience, well aware that their ability to demote many of their predecessors' species to subspecies depended upon long series of specimens to capture a more accurate portrait of the variation in populations in the field. Meanwhile in demoting species to subspecies wherever possible in the interest of revising entomologists' knowledge about certain groups, they were *unmaking* species even as they filled *Novitates Zoologicae* with hundreds of "sp. nov."

MAKING SYSTEMATICS SCIENTIFIC

Rothschild and Jordan would follow the 1895 "Revision of the Papilios of the Eastern Hemisphere" with a 1903 "Revision of the Sphingidae" and a 1906 "Revision of the American Papilios." Good revisions such as these provided one way of maintaining and improving the best parts of the naturalist tradition while answering critics who ridiculed museum workers. Still, many who could not differentiate between careful revisionary work and "species-mongering" continued to ridicule the whole endeavor. Even those who should have known better dismissed the detailed study of geographical variation so central to Rothschild and Jordan's revision. The zoologist R. Lydekker described in a 1906 memorial of Sir William Flower, director of the British Museum of Natural History, how Flower's "sympathies were with the wider and more philosophical aspects of zoology." Lydekker took pains to point out that Flower had thus taken no part in describing new species or subspecies, or redefining genera, which, he noted by the way, "have scarcely any more right to be regarded as real philosophical science than has stamp-collecting."[44] Statements such as this, which explicitly dismissed an emphasis on subspecies as outside the purview of the true scientist, inspired Jordan to became increasingly explicit regarding the relation between describing subspecies and the "wider and more philosophical aspects of zoology."

One of the most damning criticisms of the description of species, encapsulated in the jibe "species-maker," was that naturalists' decisions regarding what counted as species and varieties were arbitrary. Some argued, for example, that Darwin's revolutionary redefinition of varieties as incipient species meant species themselves must be figments of taxonomic imagination. Under this view, the entire cataloging project could seem a quixotic attempt to find a static order amid constant change. As a trained zoologist who valued his scientific credentials highly and with strong interests in keeping the work of museums within the purview of the sciences, Jordan took these accusations of arbitrariness to heart. On those occasions that he set aside the careful description of particular species and tried his hand at phylogeny, Jordan's guiding refrain was the constant effort to avoid arbitrary, and therefore unscientific, decisions. In a paper on the antennae of Lepidoptera, for example, he insisted that entomologists must end the common practice of selecting characteristics for classification based on a priori opinions regarding their importance. Such arbitrary selection of characters, Jordan explained, created the endless debates for which systematists were renowned, since obviously entomologists would differ regarding which characters to choose. To end these debates, and by implication divest the discipline of its low status, entomologists must examine *all* characters. Only after such extensive examinations could the naturalist then conclude that forms were closely related insofar as they agreed in the most number of characters, combined with educated judgments on the part of the classifier regarding the probable phyletic development of each character.[45]

Jordan's defenses of the use of trinomials were similarly grounded in the need to reduce the apparently arbitrary nature of much taxonomic practice. When A. Pagenstecher, director of the natural history museum in Hamburg, criticized Rothschild's and Jordan's use of trinomial names for geographical varieties, the jointly authored rebuttal echoed arguments Jordan used throughout his own work. They defended trinomial nomenclature on the grounds that the practice avoided arbitrary and subjective judgments based on superficial knowledge, "a vice not to be suffered" in scientific research. In describing any difference that exists for the respective district, for example, the systematist must leave aside personal opinions regarding the importance of that difference. And in order to prevent this record of difference from being forgotten or overlooked, "and hence lost to science," each geographical variety must receive a name, "however minute the distinction may be." "There is nothing arbitrary here," they emphasized, since the describer could always prove whether or not a difference existed.[46]

But how to separate the individual views of an author from the "real" boundaries in nature? For the purely practical purposes of the first "Revision," Jordan had searched for a standard criteria for deciding whether to name forms as species or varieties. He explained in the introduction that, after dividing the six hundred named forms into preliminary genera, and then tentative species, they had then studied comparatively every form of each "species" in respect to the extent of variation "with a special view to find a practical rule which might lead us to delimitate the species scientifically, not arbitrarily, in the case of each Papilio."[47] Jordan consistently couched this search firmly in the context of a search for more scientific methods. He knew, he wrote, that an endeavor in which "it is usually said that the specific distinctness or non-distinctness of a Papilio or other animal very often depends on the individual views of an author, and that there is no general parting line between species and varietal forms," hardly seemed scientific. A rule that allowed one to proceed "scientifically" would provide an objectivity lacking in systematic work and remove the basis for the criticism that carefully described species reflected boundaries in the naturalist's mind rather than in nature.

Jordan's plan for finding a nonarbitrary rule for determining the limits between forms arose from both his training at the University of Göttingen and his personal experience with the prevalence of geographical variation. During his time as a student of Ehlers, he had learned the importance of detailed, meticulous microscopical and morphological work to the accurate assessment of variation. As we have seen, this foundation had already allied him with a reform-minded segment of entomologists in Britain. It demanded that in order to find nonarbitrary means for selecting characters by which to determine and classify forms, he avoid relying on the most conspicuous characters should inconspicuous, but more reliable, characters be available. But sacrificing specimens to dissection was not standard practice among lepidopterists. Indeed, one naturalist lamented the prevalent injunction against such methods as due to a "museum conscience" that viewed the violation of a rare or perfect specimen by scalpel and scissors a sin not to be forgiven. Jordan's training had given him a new, "anatomical conscience" associated with biologists such as T. H. Huxley that saw shutting up a rare specimen in a glass bottle as "miserly stupidity."[48]

Generally, this division of attitudes regarding the sanctity of specimens had become institutionalized as distinct traditions: academic zoologists imbibed an "anatomical conscience" and museums and private collectors a "museum conscience." But though Rothschild himself "had never taken kindly to the microscope and microtome" (despite his academic studies at Bonn and Cambridge), he let Jordan do as he wished with the specimens. And so, as Rothschild began going through the drawers of Papilio butterflies to prepare the first revision, comparing their patterns and ordering them according to form, color, and pattern, Jordan had commenced taking the butterflies apart and carrying them to his microscope for dissection.[49] Thus, as a university-trained museum curator with a directive to order and describe the possessions of a wealthy collector, Jordan brought both traditions to bear on the specimens before him.

As he searched for helpful characters, Jordan soon began spending a disproportionate amount of time examining the genital armature. He did so for a very specific reason; namely, the role the male genital armature (the penis and its attachments) had already played in improving species determinations. Entomologists had occasionally found that upon closer examination of the genital armature, seemingly identical forms proved to be different species, or vice versa. Jordan had been particularly impressed by the fact that among all those who had tried to classify the Papilios, only Erich Hasse had not been fooled by mimetic forms due to his detailed study of the genital armature. An insect mimicking another group of insects still possessed the genital armature similar to its closest relative, rather than its model. Similarly, a seasonal dimorphic in wing pattern still possessed the genital armature of its fellows.

As he checked his and Rothschild's conclusions based on more conspicuous characters against the genital armature to sort through deceptive appearances, Jordan also considered the question of whether the genital armature might prove useful in deciding whether a new form represented a new species or a variety. As an evolutionist, he assumed these tiny parts of insect anatomy would vary like any character, but given the role genital armature must play in encouraging or preventing fertility, he wondered whether they might vary relatively less. Others had already assumed as much. As Elwes and Edwards put it in their revision of the Hesperiidae: "In the separation of species from species nothing can be more efficient than structural alterations which may render sexual intercourse between diverging groups a physical impossibility."[50]

Amid the constant effort to decide whether a form represented a species or a variety, some microscope-inclined entomologists had recently taken the morphology of the genital armature as a clear, infallible guide. F. Buchanan White wrote that he had so far found the genital armature to be distinct in every species, and unvarying in its form within species.[51] But critics of the practice complained that while the study of the genitalia "may amuse an idle hour" and "the drawings of them are very pretty," they proved of little value in determining closely related species.[52] At a more practical level, the entomologist W. Bloomfield protested: "If this is to be the only way to distinguish the species, I think we may as well do away with the differentiation of species altogether."[53]

Some entomologists took a middle road between these two camps. Augustus Grote, of Bremen, used differences in male genitalia to separate closely related forms, but warned that while he had found some closely related species differed markedly in their genital armature, other easily distinguishable species did not. One must study all parts of the insect, he concluded. W. H. Edwards also expressed wariness, noting that drying distorted the armature. "It is not to be supposed that they are cast in moulds like so many iron pots," he warned, "and knowing that every other organ varies, we have the right to believe that the genitalia vary also. How much, is the question."[54] Jordan had set out to answer this question in order to determine the utility of these characters for his taxonomic decisions. Meanwhile, he used the detailed locality labels on each specimen to constantly attend to any ties between geography and observed variation.

Poring over Rothschild's specimens at his desk in the museum, Jordan soon discovered a significant difference in how variation in the genitalia correlated to other kinds of variation. When he found geographical variation in the wing pattern, he also found geographical variation in the genitalia. Nongeographical variation did not, by contrast, correlate in any way to variation in the tiny reproductive structures examined under his microscope. He quickly noted that this discovery could provide a "non-arbitrary" way to distinguish between the different kinds of varieties during his day-to-day task of sorting through Rothschild's specimens. For based on these correlations, one could assume that if, after finding some variety, one determined the genitalia varied as well, in all likelihood one had a case of geographical variation and a justification for coining a trinomial name to alert other naturalists to the fact. When, by contrast, he found variation in the genitalia following no definite geographic pattern, he concluded that the specimens did not represent a geographical variation. A trinomial would, in such a case, not be needed.

Jordan clearly considered his findings a triumph of both the detailed morphological methods in which he had been trained and the careful attention to geographical variation for which he and other reforming entomologists had been campaigning. Most importantly, the observed correlation seemed an excellent guide for work on varieties based on scientific, rather than arbitrary, means. But to many intent on making entomology a science, being "scientific" entailed much more than simply increasing the exactness of research or finding guiding rules for species work. It also meant explicitly engaging in theoretical debates concerning the origin of species. Jordan himself had been trained to be wary of too sharp a shift in the methodological ethos of those working to describe and classify the living world. Ehlers had instilled a methodological caution that inspired his loyal student to constantly return to the microscope before being seduced by either the appeal of a conspicuous character or some generalization regarding the origin of species. Strong allies backed such methodological conservatism. As the famous physicist Ernst Mach wrote in a general physics textbook that Jordan, as an accredited teacher of physics, no doubt knew, "We err when we expect more enlightenment from an hypothesis than from the facts themselves."[55]

Methodological caution in one's own work did not, of course, preclude a naturalist from reading the speculations of others. Soon after his arrival at Tring, Jordan was ordering books such as Wallace's *Natural Selection* from the museum's network of booksellers. He read George Romanes's *Darwin and after Darwin* and knew August Weismann's works and those of British naturalists who had taken up the crusade for "philosophical natural history"—certainly the least that could be expected of a scientifically trained entomologist who supposedly "imbibed philosophy from the cradle." Jordan would disappoint such expectations on more than one occasion, particularly when pressed—half a century later—by younger colleagues to publish any thoughts a lifetime's work had inspired on the origin of species. But he quite happily read the speculations of others, inevitably comparing their generalizations to the facts before him in the museum drawers. Perhaps in these extensive discussions of the origin of species, he might even find clues to a practical, nonarbitrary guide to determining species' boundaries. And so, as he carried out the meticulous work on specimens for the swallowtail revision, Jordan compared what he found with the statements of those engaging more explicitly with the implications and problems of the theory of evolution.

These more theoretically inclined authors wrestled with questions unresolved by Darwin's theory of evolution by natural selection—questions ranging from the nature of inheritance to the efficacy of natural selection and the environment's influence on variation. A particularly vociferous debate during the years of the Tring revisions concerned the nature of variation and its role in evolution. Two general possibilities had been outlined, each with distinct implications for how species could be differentiated in the museum. Either evolution arose via the accumulation of small-scale, continuous variation or it resulted from more large-scale, discontinuous saltations. Darwin had constantly insisted that *Natura non facit saltum*, yet he could not explain how such small variations might persist in a population, much less lead to speciation. Initially he thought that geographical isolation might protect small variations, but he eventually abandoned this answer, unable to explain how variation would avoid being swamped in small, isolated populations. (Darwin accepted the nineteenth-century consensus on heredity that held traits from both parents "blended" in the young; any slight deviation from the norm would thus be blended out of existence.) In the face of such difficulties, some wondered whether in fact relatively larger variations could provide the jump start required for speciation. Notably, if variation occurred by large enough jumps, or mutations, geographical isolation need not occur as a prerequisite for speciation. The discontinuous nature of the change would ensure that a barrier to breeding quickly appeared. This view left little role in speciation for the slight, "continuous" shifts in color, size, and form characteristic of geographical varieties; indeed, it left little role for the study of geographical variation at all.

Given Jordan's close experience with the ties of diversity to landscape, such a conclusion struck him as odd, as it did many naturalists. His years studying the ties between organisms and their environments, from the landscapes of the Weser valley to the patterns displayed by Rothschild's long series of specimens, convinced him that the changing landscape must be taken into account in theories regarding the origins of nature's diversity. He had observed correlations between certain species and both climate and vegetation during his dissertation work on the butterflies of Göttingen. He knew the beetles of northwestern Germany could be arranged in particular faunas depending on the landscape, or "conditions of life." The fact that many "varieties" tracked certain geographical areas had in turn inspired trinomialists' efforts to reform nomenclature. Convinced that the patterns shown by long series of specimens were important, Jordan, Rothschild, and Hartert had highlighted the importance of geographical variation from the very first issue of *Novitates Zoologicae*. All their time consuming work advising collectors to include localities, admonishing colleagues to amass large series, and, indeed, the payments being sent out in return for specimens, assumed the study of geographical variation was important.

After spending two years studying the enormous variation in one group in order to complete Rothschild's first revision, Jordan vaguely acknowledged the weighty debate over the nature of variation when he announced in his introduction that they had found many remarkable facts concerning variation and geographical distribution. He wrote, however, with characteristic caution, that they thought it wise to deal with the Papilios of the globe before offering a more general account of their conclusions. At that point, he explained, they would "be able to illustrate more fully the relations between continuous and discontinuous variation on one side and the characters of subspecies and closely allied species on the other."[56] It would be another ten years before they felt they had amassed enough material from the Americas for such a project. Meanwhile, in marked contrast to his own caution, some biologists not only decided that evolution occurred *per saltum*, but explicitly urged that naturalists shift attention away from the types of data amassed by museums such as Tring.

The view that evolution relied on discontinuous variation had its most vociferous champion in William Bateson. His work represented one answer to an increasing stream of complaints that the arguments over the agencies at work in evolution had been "too much in regard to possibilities, too little in regard to observed facts of variation."[57] A fellow of the Entomological Society of London, Bateson had listened, like Jordan, to the constant appeals for more careful attention to variation and the correlation of available facts. Bateson attempted to fill both gaps with his 1895 book *Materials for the Study of Variation, Treated with Special Regard to Discontinuity in the Origin of Species*. Described at the time as "the most recent important contribution to the study of evolution in this country,"[58] Bateson's book seemed to provide the comprehensive organization of facts called for by naturalists who were overwhelmed by the information amassed on natural variation. David Sharp happily noted that insects had received a fair share of attention in the book, "a departure," he noted, "from the usual custom of 'biologists.'" Sharp noticed, however, that Bateson included under the term variation a wider range of facts than those to which entomologists usually applied the word. Specifically, Bateson had focused on "monstrosities" (another term for discontinuous variations) rather than smaller, continuous variation.[59]

Sharp blamed a lack of space for Bateson's lacunas and hoped that a future volume would include the phenomena of continuous variation that his fellow entomologists knew so well. But Wallace read Bateson's book with a less forgiving eye. Indeed, he accused Bateson of selecting only certain facts in order to downplay the extent of small, continuous variations in nature. Those who studied the close relationship between living organisms and their inorganic environment, Wallace wrote cuttingly, would see that Bateson's work dealt "to a large extent with words rather than with the actual facts of nature." "Species-makers," for example, knew "too well" that Bateson's claim that species are most commonly distinguished by qualitative differences rather than the slight differences "is not so!"[60] Wallace pointed out that Bateson had ignored thousands of examples of continuous variation printed in the pages of naturalists' journals. Only through ignoring such data, Wallace insisted, could Bateson conclude that "The Discontinuity of Species results from the Discontinuity of Variation."

As Wallace's review made clear, Bateson dismissed the facts upon which Tring had both composed its revisions and staked its status as a scientific institution. Worse, frustrated judgments of systematists peppered Bateson's writings as he wrestled to apply their data to new questions. "Beginning as naturalists," he complained, "they end as collectors, despairing of the problem, turning for relief to the tangible business of classification, accounting themselves happy if they can keep their species apart, caring little how they became so, and rarely telling us how they may be brought together." Faced by museum collections useless to the study of variation (and certainly the Tring naturalists would agree that existing collections were of little aid), Bateson turned in relief to the experimental work of animal breeders, who could produce stocks of known parentage and track variation over the course of generations. He argued that "The only way in which we may hope to get at the truth is by the organization of systematic experiments in breeding, a class of research that calls perhaps for more patience and more resources than any other form of biological inquiry."[61] In the meantime, he conceded that the collection of series of specimens was useful, but within a decade he announced more bluntly that the investigation of "Heredity by experimental methods, offers the sole chance of progress with the fundamental problems of Evolution."[62]

Given such arguments regarding how the study of variation should be done and where, Jordan decided to join the fray, whether the worldwide revision of swallowtail butterflies was finished or not. For he knew that the facts lying in the museum drawers flatly contradicted Bateson's and other theorists' statements regarding both the nature of species and the process of evolution. Using the Rothschild collection, he could demonstrate the importance of the "species-making" that had produced such facts, the crucial role of detailed morphological work in guiding one through the complexity of nature, and how the careful study of subspecies was inextricably bound to philosophical natural history.

CROSSING OVER TO BIOLOGY

Jordan's contribution to the debate over the nature of variation appeared in the 1896 issue of *Novitates Zoologicae*, under the title "On Mechanical Selection and Other Problems." Well aware of the recent controversies regarding how to do entomology, he began by outlining a middle road between those naturalists who treated animated nature "from the point of view of a philosopher" and those who worked "with the individual specimens." While he journeyed onto theoretical ground, the cabinets full of specimens would serve as constant reminders that "possibilities arrived at by general reasoning are often impossibilities in nature, in so far as what is *a priori* possible or even probable may be found not to occur in nature."[63] He emphasized, in other words, that the hard-won experience of a naturalist working through a collection stood behind all his generalizations, strategically tying his excursions into philosophical natural history tightly to the specimens before him.

Ultimately, Jordan's excursion into the debates over the nature of variation can be seen as one answer to increasing calls by naturalists such as J. T. Cunningham for naturalists to determine whether "the principles of the theorists are in harmony with the details of the empiricists."[64] Jordan's contribution was also an ardent defense of the institutions and methods of the "empiricists," so long as they amassed the right facts.

To begin his discussion, Jordan argued that no matter what mechanism philosophical naturalists championed, their concept *of* species often led them astray because they did not take into account the data available *on* species (Jordan's italics). Yet this data was available in a well-planned, scientific collection and summarized in good taxonomic revisions. Jordan particularly faulted theorists' statements regarding the difference between varieties and species, a common topic of discussion during debates over how varieties became species. For example, George Romanes, while insisting that species arise from the selection of variations in the sexual organs ("physiological selection") and varieties arise from the selection of nonsexual differences, relied on an assumption that varieties differed from each other in nonessential anatomical characters, while species differed from each other in their sexual organs. (Notably, Romanes's version of speciation could occur without geographical isolation since variation in genitalia could lead to speciation within the same area.) But Jordan countered that no naturalist who had examined varieties under the microscope could countenance so inaccurate a distinction between species and varieties. Species differed in their genitalia, but so, too, did varieties. Similarly, the German naturalist Theodor Eimer began his theorizing regarding speciation from the premise that species were physiologically isolated, while varieties were not. Jordan replied that in fact varieties *could* be physiologically isolated, and thus such a distinction did not exist in nature. Finally, Alfred Russel Wallace stated that species possessed adaptive differences, while varieties did not. Jordan insisted, again, that such a difference between the nature of species and varieties could not be upheld when comparing actual specimens.[65] All of these traps, Jordan argued, could be avoided by taking account of the information *on* species.

The implications of accepting any of the above definitions of the term *species*, each of which did not, as Jordan pointed out, exclude every kind of variety, struck at the heart of Jordan's concerns regarding the status of systematics. Romanes, Eimer, and Wallace each premised their theories regarding the origin of species on definitions *of* species. Those definitions broke down once the systematist attempted to use the definitions to distinguish between varieties and species in nature. As a result, using such definitions in practice, Jordan wrote, "leads naturally to the conclusion that there is no real distinction between species and varieties, and that it is purely conventional whether we call a form species or variety, an opinion by no means rarely met with even amongst us species-makers."[66] Jordan had already expressed deep concern with the seemingly arbitrary methods of systematists who relied on conspicuous characters. He feared the above theorists' claims regarding species would similarly further the impression that naturalists "made" species, rather than described real entities in nature.

Although the particular concept of species that authors used in taxonomic work often remained highly implicit, Jordan's effort to counter the inaccurate statements of some biologists led him to offer the working definition of species he had developed by studying Rothschild's collection. The definition encapsulated both the requirements of his job as a museum curator (sorting specimens), and the demands of his scientific training under Ernst Ehlers. As a museum taxonomist, for example, he included morphology as one of the primary guides to determining species. But as an evolutionist intent on applying the logic of a historical science to the facts before him, he insisted that morphology could not be the only arbiter of one's decisions. First, if one took morphological difference as the criteria of species, then every individual would be a separate species, since every individual is morphologically different in some way. Second, he pointed out that since "we all pretend to be evolutionists . . . accepting the theory of the transmutation of the species in the course of time as the base of scientific work in Natural History," one must assume that if all the specimens of history could be collected, "the gaps between the various forms would all be filled up by intergradations." Thus, a solely morphological definition would mean all life forms would be "one variable species."[67]

Jordan insisted that anyone who truly accepted evolution must, in accepting that species diverge, accept that species are those branches of an evolutionary tree that cannot reconverge. Therefore it should be possible to differentiate species from those forms at a lower degree of development that could reconverge. The chief criterion of what defines a species, then, must be the impossibility of fusion. Jordan then offered his definition of species as

> a group of individuals which is differentiated from all other contemporary groups by one or more characters, and of which the descendants which are fully qualified for propagation form again under all conditions of life one or more groups of individuals differentiated from the descendants of all other groups by one or more character.

He pointed out that this definition countered the claim, used by systematists and biologists alike, and to many one of the most damning criticisms of museum work, that no real distinction between species and varieties exists. Species could not reconverge. Varieties, by contrast, while they might differ morphologically, could still interbreed successfully.[68]

Defining species as those forms that can interbreed successfully was not a new practice or concept. Later called the "biological species concept," such a rule had provided a guide to working naturalists for centuries. In the eighteenth century, naturalist Cristoph Girtanner, for example, insisted that "all animals or plants which generate fertile young with each other belong to a physical species."[69] When debates broke out over whether particular forms represented varieties or species, naturalists used whether or not the forms interbred as the ultimate test. "Assuming the ordinary view to be correct as to what constitutes a *species*," wrote Louis B. Prout in 1892, "the evidence wanted is—Do the two forms pair in a state of nature?"[70] Inevitably, this definition meant determinations based on dead specimens had to be tentative, pending more information. W. L. Distant noted in his *A Monograph of Oriental Cicadidae*, for example, that "in treating other entomologists' species as 'varieties,' I am of course, in the absence of breeding experiments, expressing my own views alone."[71]

Jordan was not the first to argue, then, that a definition based on breeding provided an objective criterion through which to ultimately distinguish species from varieties. But some naturalists feared that a definition based on interbreeding could not be used in practice, especially given the enormous influx of new forms. As Jordan explained years later, even well-endowed Tring had never had the space, time, nor resources for breeding insects, as "the collections were so large that the staff was fully occupied."[72] Could a large collection of dead specimens be useful, given a definition based on what the insect did in life? Jordan answered firmly in the affirmative, *if* naturalists amassed collections in certain ways. The information coming out of museums *on* species, in other words, had to be good. Thus, his paper "On Mechanical Selection," which began as a lecture to philosophical naturalists on checking their assumptions against the facts, quickly turned into a lecture to working systematists on his criteria of both good method and useful collections.

Using his work on the Tring butterflies as an example, Jordan argued that naturalists could indeed use a definition of species based on interbreeding to accurately determine species in the museum, if they took advantage of two additional aids. First, they must meticulously compare the characters of forms known to be specifically distinct based on breeding experiments with those closely related but for which no breeding experiments had been done.[73] Jordan explained how this extension from knowledge gained by breeding species worked in practice in a paper on fleas in 1908:

> From a careful comparative study of a very large series of specimens of such species, we arrive at a knowledge of the approximate limits of variation. The range of variation not being the same in the various organs, the most important point to ascertain is which organs are variable and which are comparatively constant in such undoubtedly distinct species. The characters here observed to be specific are a guide and a basis of classification in the case of allied forms about the life history of which nothing is at present known.[74]

This determination-by-analogy required long series of both bred and collected specimens and careful comparative attention to a range of characters.

Jordan admitted that the careful systematists' dependence on analogy plunged him right back to the problem of enough material. And he conceded that based on his definition of species, a systematist could never prove with certainty from the specimens alone "whether the distinguishing morphological characters they exhibit are of specific value or not." But he emphasized that

> we species-makers do, in fact, not pretend, at least many of us do not, that in every case the form which we pronounce to be a species really is a species; we work, or ought to work, with the mental reservation that the specific distinctness of our *species novae* deduced from morphological differences will be corroborated by biology (in the widest sense).[75]

However, this agnostic stance regarding practice was quite different from the belief, prevalent among nonsystematists and some systematists, that determinations of species reflected no reality in nature or that it was purely conventional whether a naturalist described a form as a variety or a species. Jordan hoped that, guided by long series of specimens and close attention to morphological variation, the decisions of systematists regarding species and varieties could be included within the realm of science. Like the physicist Ernst Mach, Jordan held that although knowledge is always provisional, through right method one could establish increasingly certain knowledge.[76]

So far, in writing about systematics, Jordan had kept well within the accepted, fact-gathering framework that had dominated natural history journals for decades. Descriptive work on specimens could (indeed, must) continue in good conscience, in the interest of right method. He insisted, in other words, that the collections could still provide useful information and that systematists could come to scientific, rather than arbitrary, decisions regarding species so long as they worked with meticulous attention to detail and variation. Good entomology, then, could still be about collecting facts so long as entomologists paid attention to the *right* facts and amassed them in the right way. But Jordan also had a second requirement for coming to accurate decisions given an absence of breeding experiments for each and every form: he insisted that, in order to arrive at the correct determination of species and varieties, the systematist must also take into account the way in which divergence comes about. And in urging systematists to take into account the way in which divergence occurred, he was asking them to accept certain conclusions regarding speciation during a time in which naturalists hotly debated both alternative evolutionary mechanisms and the role of theory in natural history.

Jordan took a firm stance on these debates when, as part of his insistence that naturalists take into account *how* divergence occurs, he proceeded to outline his discovery that only geographical variation correlated with variation in the genitalia. Therefore the divergence of species must, he concluded, depend upon the isolation of geographical varieties (by a mountain range or a body of water, for example). Such isolation permitted geographical varieties to diverge still further, until ultimately they could not reconverge. Thus, the patterns he had discovered during his search for a scientific guide to determining species convinced him that, of the various kinds of varieties, only geographical varieties represented incipient species. This conclusion had then provided a guide to separating the value of the many types of variation observed in Rothschild's long series of specimens. In his and Rothschild's next revision, published in 1903, he summarized the resulting stance as follows: "*Geographical variation leads to a multiplication of the species; non-geographical variation at the highest to polymorphism.*"[77]

This firm statement was a resounding contradiction of Bateson's strident claims. Well aware that the geographical isolationists Moritz Wagner and J. T. Gulick had been called to task for holding that geographical isolation could alone produce speciation, Jordan conceded that some additional transmutative factor—selectionist, orthogenetic, or Lamarckian—must drive variation in certain directions. But no matter what mechanism won the day, he insisted that geographical isolation must play the primary role in the development of boundaries to breeding. In other words, all of naturalists' close attention to geographical varieties—and the use of trinomials to bring attention to those varieties—made *scientific sense*. "The study of geographical races, or subspecies, or incipient species," he concluded firmly, "is a study of the origin of species."[78]

From then on, Jordan's "On Mechanical Selection" provided a convenient footnote whenever Rothschild and Jordan defended Tring's methods of nomenclature. Jordan had proved, they insisted, that geographical varieties represent various steps in the evolution of daughter species. Indeed, at one point they stridently claimed that "whoever studies the distinctions of geographical varieties closely and extensively will smile at the conception of the origin of species *per saltum*."[79]

British entomology's team of reformers were impressed. In favorably reviewing "On Mechanical Selection" for *Natural Science*, George H. Carpenter noted how, although Jordan seemed content to class himself with the systematists, he had clearly demonstrated the "biological importance the careful study of species may have" by reminding systematists of the "many unsolved problems that still lie behind the dry labour of the species-maker."[80] Tutt, a hard one to please, insisted that "every scientific entomologist" must read Jordan's paper.[81] In 1901, Tutt included Jordan in a list of lepidopterists (including August Weismann, Alfred Russel Wallace, William Bateson, John Henry Comstock, Edward Bagnall Poulton, and others) who had built on the possibilities that Charles Darwin had "opened out to biological science" by dispensing with the idea that "the be-all and end-all of entomological science was the easy determination of known, and the detection of unknown, species."[82] That same year, Louis B. Prout listed Jordan among those "too well known to need more than a passing word" who had "revolutionized our later nineteenth century literature."[83] Those intent on increasing the status of entomology by shifting methods and priorities obviously found in Jordan a useful ally. G. C. Champion soon wrote to say how glad he was to see Jordan at the Entomological Society of London meetings since "the influence of such a worker as yourself is much wanted and is not of a character too abundant at the Ent. Soc."[84] Poulton, too, hoped Jordan would be present when he read his presidential address "What Is a Species?" "Elwes and others will be there and we shall have a good meeting I expect," he urged. And by the way, could Jordan bring some of Rothschild's specimens of *lormieri* to the meeting?[85]

Yet although Jordan carefully couched his discussion regarding the role of geographical variation and isolation in terms of the facts he had discovered in Rothschild's collection, the conclusion he put forth regarding the role of geographical isolation in evolution inevitably took him beyond the boundaries of some colleagues' definitions of good science. Years later, Jordan recalled how Albert Günther took him aside during one of his occasional visits to Tring soon after "On Mechanical Selection" appeared. The sixty-six-year-old British Museum zoologist admonished him "in no uncertain terms" that he had "gone over to biology, and made it clear he did not approve."[86] Like many naturalists, Günther thought the answer to the problem of distinguishing species from varieties would be achieved only through the careful collection of taxonomic facts, rather than premature philosophical debates. In contrast to those who believed Darwin's greatest contribution had been inspiring the search for general laws in natural history, Günther believed the increasingly vociferous debates over the mechanism of evolution violated the tenets of good science.[87]

Bound to the museum yet trained to look beyond it, Jordan had conscientiously expressed loyalty to both the so-called theorists and the empiricists in his 1896 paper. During much of the discussion regarding evolution, he wrote, "over the consideration of 'characters,' it has been lost sight of that speaking of a specific character and the variability of characters means, in fact, speaking of *abstracta*, while our work must be based upon *concreta*, upon the individual specimens."[88] Detailed work on individual specimens had demonstrated, for example, that geographical variation was important. Yet those theorizing about the origin of species constantly ignored such work. Furthermore, Jordan feared that some of the statements being made regarding speciation might in fact distract naturalists from paying attention to the very facts necessary for understanding the origin of species. The result, of course, would be theories that indeed did not match the data of the empiricists.

The work of Theodor Eimer, a professor of zoology in Tübingen, served as an excellent example of how Jordan's concerns played out in practice. In the 1889 book *Artbildung und Verwandschaft bei den Schmetterlingen*, Eimer had argued that evolution occurred via an internal force, or orthogenesis, rather than natural selection. To demonstrate this force in action, he developed a phylogenetic classification of Papilios based on wing markings. Eimer had then used his classification to illustrate a "law of markings" that in turn was explained by his theory of evolution. The classification, Eimer announced, demonstrated how longitudinal stripes always transform into spots, cross-strippings, and uniform coloring.[89] Eimer's work was widely read and some entomologists were obviously won over. Impressed by Eimer's work, J. W. Spengel, of Giessen, for example, sent Jordan a list of specimens needed to complete his arrangement of a series of butterflies according to Eimer's classification.[90]

No doubt concerned by this clear evidence of Eimer's influence on how collections might be amassed and arranged, Jordan composed a cutting reply when Eimer protested against the fact Rothschild and Jordan had ignored his work for the purposes of their revision. Eimer's results, Jordan wrote, were riddled with basic errors that clearly showed he had not extensively studied the literature available on the insects about which he had written. Nor had he examined specimens in fine enough detail. Examination of the sexual organs of a large series, for example, would have prevented his numerous errors in specific determinations, particularly regarding polymorphic species. Most importantly, if Eimer had studied the literature of natural history, he would not have claimed that species do not vary geographically, for to do so was in direct contradiction to the facts amassed within museum collections and published in natural history journals. Naturalists had proven, Jordan insisted, that various forms Eimer claimed as separate species in fact possessed transitional forms over a geographical range.

Indeed, Jordan took the opportunity of responding to Eimer to note how astonished he was that naturalists gave "the absence of such transitional individuals" again and again as the main argument for the "origin of species *per saltum*." Eimer and others could easily learn of the presence of transitional forms, and similar facts of importance, "by looking over the writings of entomologists." These errors of fact, Jordan insisted, led directly to errors of theory. Because he had not distinguished between geographical and nongeographical variation, for example, Eimer was able to claim that species originate within the same locality and that therefore the importance of geographical isolation for the separation of species was diminished. In response, Jordan again insisted, citing his paper on mechanical selection as evidence, that "Geographically separate races are entirely different from aberrations, seasonal forms, and forms of dimorphic species that occur in the same locality. A comparison of the variation of different organs, for instance of wing-patterns and copulatory organs, reveals that at once." As for those who Eimer insisted agreed with his conclusions, Jordan thought this quite possible, "but I very much doubt," he wrote, "that a single one of them has examined the facts upon which the conclusions are based."[91]

In emphasizing the relation between the facts "*on* species" and conclusions "*of* species," Jordan's work served as a response to the insistence by naturalists such as Tutt and Günther on a dichotomy between systematics and biology. He admonished both sides of the supposed divide for implying that each somehow outclassed the other. For while he conceded that the mere description of new species does "not help the least to solve the all-governing questions of evolution, but add simply more 'species' to the hundreds of thousands of 'species' already made known," he defended the work when done well. "Although nowadays the recorder of facts, the diagnosticist, does not rank high in science," he noted, "every theory in Natural History depends especially on the correctness of the facts furnished by the diagnosticist."[92] A knowledge about species, built upon the careful study of variation, served as a foundation for good classification. And good classification served in turn as the basis for scientific (rather than arbitrary) theories regarding evolution, since they formed the basis on which generalizations could be tested.

This was the main point of putting all of those insects beneath the microscope, the tedium of book research required for naming, and the meticulous comparison of minute characters. In other words, the study of subspecies, in ensuring accurate classifications, provided the foundation for "philosophical" natural history. Careful systematists, Jordan argued, were "all-important in the science of life, as supplying sound criteria, where otherwise a lively imagination might run wild and substitute plausible assumptions for facts."[93] Thus, all the detailed, meticulous descriptive work on species appearing in both *Novitates Zoologicae* and other natural history journals was critical to the broader project of not just ordering animals, but explaining the mechanisms by which they evolved.

AMASSING THE *CONCRETA*

Jordan's excursion into debates over species concepts (the *abstracta*) was certainly in part a lecture to "biologists" not to ignore naturalists' data; namely, the *concreta* of individual specimens. But he later described his statements as "mostly written in defense against the disapproval by older colleagues who seemed to resent the evaluation of new discovered 'small' differences in the study of variation and greater precision in the nomenclature of varieties."[94] In other words, the paper was as much a methodological lecture to his fellow systematists as a lament regarding biologists' ignorance of naturalists' data. For even as he insisted that "diagnostic work is the true basis of evolutionistic theories and hence of the highest importance," he quickly added that "the record of facts must be exact." And he quoted the insistence by a hero of biology, Thomas Henry Huxley, that "the record of facts is not scientific if the facts do not permit of the drawing of general conclusions." Jordan firmly pointed out that the biologist August Weismann could not be faulted for a theory he had developed using an inadequate classification erected by lepidopterists. The blame, he wrote, "is much more on the side of the systematists" who had based their classifications on a single character.[95] To serve as reliable censors of conclusions, Jordan wrote, systematists' observations must be exact, and therefore minute, since exactness cannot be attained without minuteness of inquiry. This exactness might inspire the systematist to check conclusions under the microscope, or focus his attention on the slight differences between forms from different localities. In either case, Jordan believed that systematists could claim their proper place in biology only if they and their extensive network of collectors reformed their diverse ways.

Thus Jordan's excursion into philosophical natural history sent him straight back to Tring's crusade to amass more and better material. In order to determine whether variation was geographical, the systematist must examine many organs, species, and individuals. As a result, Jordan conceded, "finality, even if the classification is restricted to a small group of beings, entails such an enormous expenditure of energy that it can be approached only gradually in the course of time by continued co-operation between the various lines of research."[96] Jordan's answers to the questions "how to do entomology?" and "why make species?" entailed slow, meticulous, patient, permanently tentative work, constant access to material and advice from colleagues, a library, type specimens, and more and more specimens.

He and Rothschild demonstrated what such "scientific systematics" could look like when one had access to time and material when, eight years after publishing the Papilio revision, they completed "A Revision of the Lepidopterous Family Sphingidae" in 1903. The revision became the defining work of the Tring Museum among entomologists, providing, like the "Revision of the Papilios," the widely called for correlation of naturalists' dispersed facts. It drew on a variety of approaches in the naturalist tradition, combining the explanatory framework of evolution theory, the access to far-off landscapes, meticulous study of morphological detail, and the endeavor to find the order in nature into a breathtaking and synthetic analysis of a single family of insects. Fifty years later, one specialist cited it as "possibly the finest example of a taxonomic monograph that has ever been produced in Lepidoptera."[97] The "Revision of the Sphingidae" demonstrated what could be done through harnessing the new focus on geographical variation for the old purpose of naming and ordering the natural world. And this time they could direct readers to Jordan's 1896 paper, criticized by a disappointed Günther as an unfortunate excursion into "biology," as justification for their emphasis on naming geographical varieties. That Tring's methods had not caught on, despite Jordan's theoretical justifications, is illustrated by another long defense of naming geographical varieties in the opening pages of the 1903 "Revision."[98]

This time, in defending the study of geographical varieties, Jordan and Rothschild placed the old effort to describe every species—and Rothschild's long series of specimens—firmly within the context of improving the basis of facts upon which to build solid generalizations. Knowledge about relationships and the origin of species could not be established by a priori reasoning, they insisted. Because species arose via evolution, they resulted from accidental combinations of both the nature of the animal and its changing environment: "That means we do not a priori know that what holds good in all the cases examined is true also in every case not yet examined." Any premises formed in order to establish natural relationships were in turn better established "and hence the conclusion is the more likely to be correct, the more species have been examined." Furthermore, the stability of the genera, tribes, and subfamilies depended on the knowledge of the extent of variation of the species.[99]

Lest a reader find in such high-minded defenses of Tring's methods a self-serving excuse for amassing a huge collection and describing hundreds of new species, it is important to point out that Jordan emphasized the danger of having *too many* species. Since an understanding of the "geographical origin of the various members" depended on a knowledge of the relationship of the various species, one must compose the genera as accurately as possible. A genus with twenty species in it, half of which were really geographical varieties, told one nothing compared to a genus containing two species, each with a number of geographical varieties. Guided by this precept, and based on the careful analysis of both type specimens and long, well-labeled series, they reduced the total number of previously recognized species of Sphingidae by about 35 percent (describing 125 species as new).[100] This reduction in the number of described species eloquently reflected the insistence by reformers such as Jordan that the description of new species not form the major incentive of systematic work.

The amount of material and work required to complete both the Tring revisions and develop accurate generalizations from the facts amassed was immense. "In interpreting these facts of characters presented by the individuals," Rothschild and Jordan wrote, "one starts with the assumption that what has been found to be true in the necessarily limited number of specimens investigated, holds good also in the vast multitude of individuals not compared." The possibility of error with this method could be decreased by the comparison of many individuals. "How large it should be," they warned, "nobody can predict." But of one thing they were certain: "To ascertain the extent of variation of the chief classificatory unit, the species, the material can never be too extensive."[101] One might counter that surely alternative methods could be found to such resource-intensive research. But Jordan believed that his demonstration of the importance of geographical varieties demanded that museum research on specimens be continued, since geographical variation could not be studied in a laboratory. There, biologists drummed variation out of organisms in an effort to control the chaos of natural variation. If species indeed arose via geographical variation accompanied by some kind of mechanical isolation, then the study of geographical distribution with careful attention to natural relationships provided a critical path toward the solution of the origin of species. And only well planned museums, accumulating specimens based on certain principles, held such records.

The revision demonstrated how both building and using such a collection to make the species "good" absolutely depended on the efficient use of a diverse network of correspondents and suppliers. Though the Tring Museum contained nearly sixteen thousand specimens of Sphingidae, Jordan once more sent letters of inquiry to private collectors and museums, and visited others, in order to obtain more material. Tring's "appeal for help" on the group had been met "with the greatest liberality" from dozens of private collections and national museums throughout Europe. Only two potential correspondents failed to reply: "The names of the addressees may be passed over in silence," they noted, but the implicit reproach was severe.[102] The methods Jordan and Rothschild followed for these revisions, from the accumulation of long series to the accurate comparison with types, absolutely depended on the cooperation of their fellows, who sent specimens as loans, exchanges, purchases, and gifts.

Although many workers expressed delight at what Rothschild and Jordan could do using Tring's collection and network, some entomologists were dumbfounded to the point of dismay at the detailed work involved. One, taking note of Jordan's close attention to genital armature in order to distinguish geographical variation from other kinds of variation, confessed to Jordan that he had "no idea that the 'fundamental' part of classing depended so much on the 'tail end' of a moth."[103] Others found the amount of facts required utterly distressing. Lord Walsingham no doubt had work such as the Tring revisions in mind when he once confessed that his own views were too broad and old-fashioned; "I am fully aware that I must take a back seat in the presence of those who ... seem capable of interpreting the significance of the most minute differences and weaving them into a thread on which hangs a doubtless well revised classification."[104] Though used in all the Tring revisions, decades later the focus on the microscopic examination of the genitalia still caused consternation among those facing the study of large collections. W. J. Holland, an ardent admirer of Jordan's systematic work, wrote in 1924 that he suspected he had "lumped" some of Jordan's new species due to his lack of microscopic examination. He, too, confessed that despite his respect for Jordan's painstaking labors, "the worry of attacking the thousands of specimens which are assembled under this roof with the help of a binocular microscope and dissections in order to determine the specific identity, is, for a man who is living upon 'borrowed time,' not a most attractive prospect."[105]

Even if one had the time, training, and inclination for the meticulous preparation of slides necessary for the comparison of genital armature, the number of specimens required for "Jordanian systematics"[106] proved daunting. W. J. Kaye exclaimed to Jordan that a glance at Rothschild's collection of Sphingids "spoils the ordinary amateur and indeed makes the public collections look weak!"[107] Rothschild's power to amass a "scientific, world-wide collection" obviously improved his and Jordan's ability to make sound pronouncements in matters of genera. For example, the Tring naturalists could afford, literally, to insist that delimiting genera in cosmopolitan families demanded a worldwide collection.[108] But many simply did not possess the means to create such collections. Theodor Eimer betrayed a piqued resentment of this fact when he wrote, "I perfectly agree that even grave mistakes may innocently occur to somebody who is not in a position to have such collections at his disposal as Mr. Rothschild."[109] Even Jordan's good friend Charles Oberthür let a tone of frustration escape when responding to Tring's revisions of some of his conclusions regarding genera. In conceding that "the incomparable collections" of Tring naturally conferred superior authority on the authors of those revisions, and taking due note of the fact Rothschild and Jordan had three times the number of specimens of the particular group concerned, Oberthür noted sarcastically that no doubt the reader must conclude "Tring Museum über alles!"[110]

In response to their colleagues' grievances, Jordan and Rothschild conceded that the research and methodological ideal they created was perhaps too lofty for many. Still, they thought this no excuse why "an admirer of the frail and beautiful children of Nature should not try to advance from the position of a distant amateur to that of an intimate amant."[111] But limited finances proved only one reason why collections could not always meet Tring's standards. The existence of quite different goals in taxonomic work meant that even those with access to the largest museums in the world cheerfully (and sometimes not so cheerfully) ignored Tring's appeals for certain methods. In the Insect Room at the British Museum, for example, strict requirements for both efficiency and speed often dictated research methods and priorities, making it difficult to follow Jordan's proscriptions for completing scientific systematics. George Francis Hampson, who frankly explained that he believed two specimens of each species "quite sufficient" to carry out a classification of moths, admitted that he saw the value of Tring's practice of providing full descriptions of subspecies under "separate headings." But he excused himself from doing so in his own work on Indian moths on the grounds he had limited printing space and had to economize in every way.[112]

Hampson's statements proved an early example of how different priorities influenced the kind of work entomologists thought both feasible and useful. Over time, Hampson, working within a different kind of collection and with different aims, grew increasingly frustrated with Tring's high standards. "I do not at all agree with you about not attempting to define genera except in a monographic work," he wrote Rothschild. "We cannot all write nothing but monographs."[113] Rothschild and Jordan insisted, by contrast, that one should not expect too much from a catalog of the type produced by Hampson, since "even the best is full of errors, as a cataloguer of insects cannot possibly have intrinsically worked out all the groups catalogued."[114] Decades later, N. D. Riley would describe the difference between the two kinds of work when urging the trustees of the British Museum to accept a condition of Rothschild's bequest that they not see his long series of specimens as duplicates to be disposed of at whim. Riley explained how the British Museum's entomological collections had been formed to provide an accurately named reference collection. By contrast, Rothschild's collection had been formed in order to study the local, geographical, and individual variation of species, "not as an end in itself, but as a means of shedding light upon the problem of the evolution of species."[115]

Rothschild's and Jordan's papers often bemoaned the fact that many entomologists and collectors could achieve more than they did, especially with the most popular groups of insects. "Everybody has Amazonian Papilios," they lamented, "but nobody has long series from a sufficiently large number of localities."[116] They knew from experience how puzzling lists of names of Lepidoptera and localities could become if the compiler of the list had neglected the geographical distinctions and identified the insects carelessly. "Such lists obscure the composition of the fauna," they wrote, "and therefore, instead of being a contribution to our knowledge of the insects and their distribution, hamper the student in understanding the facts of distribution, variation and evolution, which stand all in very close connection."[117] Not surprisingly, they inevitably raised the hackles of less privileged entomologists by their idealistic demands of detailed locality labels and long series. Foetterle, working in Brazil without a library and access only to the "poor" museum in Rio, accepted their criticisms with grace. But he gently reminded Jordan:

> There you sit in a beautiful museum in the midst of a rich library, which contains everything written since the beginning of Entomology, with material at hand which few mortals have at their disposal. Treasures are stored close by that over the course of decades have been arranged by an expert hand. Besides this men of science stand ready to give any information desired. With such means it is easy to work. But put yourself in my position![118]

Jordan, stuck for much of his time at his curator's desk and the museum's corridors, had the opposite problem. Certainly, Rothschild could send out collectors to obtain more material from inadequately represented regions, exhorting them to explore mountain ranges or islands that might hold intermediates between "species." And eventually a loyal network of friends developed that loaned specimens to Jordan, convinced by the first revision that their material would be put to good use. W. J. Holland, for example, packed up Sphingids from the Carnegie Museum and sent them to Jordan even though he hated to send them across the seas. (Perhaps in recognition of this sacrifice, Jordan named a species after his friend.) But surrounded as he was by Rothschild's unsurpassed library and collection, including hundreds of type specimens, Jordan wanted specific things from the field. He needed specimens that he could associate closely with their environment, or the "altered conditions of life," with good labels and collected over a geographical range. In his letters to correspondents—from museum curators to those in the field—he constantly exhorted naturalists to improve the state of data arriving from the field. He gently tutored the Rev. Miles Moss in Peru, for example, not only on how to send the best Sphingidae and Saturnidae, but how to breed forms, figure the larvae and pupae in water color, and trace their identity to the imaginal condition.[119] He even pasted examples of butterflies that he wished Moss could breed at the top of letters.

The revisions demonstrated what taxonomists could do through such careful management of a natural history network. They cleared up much of the chaos amassed by the naturalist tradition's diverse ways and provided a foundation for future work. Yet doing systematics "scientifically rather than arbitrarily" ultimately depended on the material available, and even Tring had to wait years before the collection on some groups seemed adequate.

In the meantime, Jordan brought his conclusions regarding the role of geographical variation to an international audience on the occasion of a birthday celebration for Ernst Ehlers, a fitting tribute to a mentor who had included a role for natural history museums in zoology. Here, in a paper published in *Zeitschrift für Wissenschaftliche Zoologie* in 1905, Jordan again recounted his exchange with the mineralogy student, but insisted that, whether it was Darwin lamenting the time and trouble spent deciding whether a form was a species or the director of a certain large museum who refused to purchase a new butterfly on the grounds that the collection already had many red specimens, Jordan insisted that the poor opinions of systematists did not annoy him. He regretted only that the meaning of systematics for a correct understanding of the evolution of the living world was still so little recognized. He then outlined his argument, based on his detailed morphological work on long series of specimens, that geographical variation represented the key to the origin of species. Now, he added a new example of those insisting, by contrast, that species could arise *per saltum*: the Dutch botanist Hugo de Vries. No doubt with the recent interest in de Vries's book, *The Mutation Theory*, in mind, Jordan noted that he repeatedly found statements of fact in arguments for and against evolution that had long since been directly contradicted by the solid observations of systematists.

Yet again, this excursion into debates over the mechanisms of evolution returned Jordan to the campaign for better collections. He pointed out, for example, that the detailed study of natural variability could be accomplished only by using a large series of specimens. There were still far too many collections that provided an inadequate representation of the variability of the species concerned. No existing collection, for example, contained enough specimens of butterflies from the different areas of Europe to permit a truly in-depth study of their geographical variability, and he urged taxonomists to "specialize in this regard still more."[120]

It was another ten years before Rothschild and Jordan considered their own material adequate for commencing the second installment of their revision of swallowtail butterflies, the "Revision of the American Papilios." By then, one might assume Tring's working principles needed little introduction or defense. But again they launched into a lengthy justification of the study of subspecies and the use of trinomials. Just as, they explained, Linnaeus had rendered chaos into order with his binomial nomenclature, modern systematists must "follow him by bringing order into the chaos of varieties." They recognized it would take a long time for all systematists to learn to use the new nomenclature. "The more is it necessary," they concluded, "to bring the matter again and again before their mind."[121]

Here, as in his earlier writings, Jordan insisted that using trinomials was not simply a matter of nomenclature. The principles behind the new system of naming varieties related directly to the kinds of facts the naturalist must amass in order to do scientific work and, most importantly, understand evolution. Yet as the constant pleas for more and better material attested, the specimens coming in were still inadequate, the coverage over geographical areas too small, and the practice of those in the field and museum far too disorganized. To really remedy this situation, the naturalist tradition itself needed some ordering. Only then could entomological systematics continue to fulfill Jordan's insistence that it entailed much more than sticking insects between two bits of glass. And so, he decided to set the microscope aside and harness all those hours devoted to correspondence to a new ordering effort; namely, the lives and work of his fellow entomologists.

FOUR

Ordering Naturalists

Jordan's fellow systematists were astonished by the Sphingidae revision. "It quite takes my breath away," wrote an entomologist, confessing that his head got "fairly 'addled'" at the thought of how much work had been involved.[1] Ernst Ehlers acknowledged his copy with the exclamation, "What an abundance of work went into it!"[2] W. J. Holland believed it "the most scholarly revision of the whole subject which has yet appeared."[3] T. A. Chapman found the amount of material involved "unprecedented" and praised the accuracy and originality of the work.[4] W. J. Kaye found the synonymy so thorough that he was sure the species would be kept clear for the future.[5] The Belgian entomologist Guillaume Severin wrote that if only each family could be worked through in so complete and exact a manner, entomology would be much better off.[6] Holland's review in *Science* described the "Revision" as an *opus magnificum*, every page providing evidence of the "most painstaking and minute research." As a result, Rothschild and Jordan had brought into systematic review the work of one hundred and fifty years.[7]

Many lepidopterists set to work rearranging their collections based on the revision, their cabinet drawers becoming, as Jordan had hoped, more accurate reflections of the relations between forms. W. L. Distant reported that after rearranging his specimens according to the revision, he could see affinities where before he "could but recognise the utmost diversity in type and structure."[8] Jordan must have been pleased with this testimony. And if J. Butterfield, who thought the manner in which they had treated the subject would have "far-reaching effects upon future systematic work in entomology"[9] was right, then their reform by example would lead others to follow suit. "To students of the Lepidoptera," wrote Harry Eltringham, "the publications of Messrs. Rothschild and Jordan have furnished an example of perfection."[10]

The revision proved a wonderful testament to the kind of work Rothschild's collection and Jordan's methods could produce. As the Edwardian days of England passed, entomologists wondered what Rothschild and Jordan would work on next. Would their next installment on the Papilios be finished soon? If, as rumored, they had decided to tackle the Saturnidae, would they keep their correspondents informed? Would they, despite the infringement on their valuable time, identify the butterfly enclosed?[11] Those entomologists who were more philosophically inclined, listening to the increasingly strident debates over the mechanism of evolution, may have speculated whether Jordan would contribute further words on the subject along the lines of his 1896 paper.

But Jordan had, as he later put it, "thrown his stone amongst the giants," and from then on he generally kept his comments on evolution hidden within detailed accounts of species. What he ultimately decided to spend his time on reveals much about the state of entomology and systematics at the turn of the century. As others debated the implications of the rediscovery of Gregor Mendel's work for evolution theory, the rise of a new discipline called "genetics," and whether "Darwinism was dead," Jordan focused first on amassing the material required to complete good revisions and monographs. Second, he attempted to organize entomologists internationally in order to address the stark difference between the ideal role he saw for entomological systematics in biology and the reality posed by a diverse community and tradition. Both these ordering endeavors aimed at improving, as Jordan put it, the *concreta* upon which the *abstracta* depended.

MEN OF TWO CLASSES

Much of Jordan's "ordering" efforts entailed either organizing insect specimens or his fellow entomologists. But, before examining his campaign to order the latter, it is worth foreshadowing here that eventually broader twentieth-century ordering efforts would influence his seemingly more mundane campaigns. As a reformer rather than a revolutionary, Jordan's vision for improving the naturalist tradition entailed building on prior traditions even as one gently reordered priorities and methods. By contrast, revolutionary visions of new organizations of both men and knowledge loomed on the horizon, both of which can be introduced by an examination of, at first sight at least, an ally in Tring's endeavor to ensure a role for natural history museums in the new century.

In sending his thanks for his copy of Rothschild's and Jordan's 1906 "Revision of the American Papilios," Alfred Russel Wallace contemplated the Tring revisions within their context as material for the study of evolution when he praised the revision for giving "such valuable material for dealing with problems of Geographical Distribution as well as Evolution generally."[12] There was some irony in Wallace's appreciation for the masterful works of taxonomy coming out of Tring. As an ardent socialist, Wallace had long since been immersed in a campaign for radical changes of British society aimed explicitly at destroying the economic system upon which Rothschild's wealth, and therefore the Tring collection, depended. Indeed, the very year Jordan took up his position as curator to the Rothschild Museum, Wallace had attributed Britain's "social quagmire" to the influence of millionaires, speculators, and money lenders.[13] Though Wallace never mentioned the Rothschilds by name in his numerous indictments of exorbitant wealth, in 1885 he had castigated "immoral" foreign loans that financed war and conquest.[14] He specifically cited the Suez Canal as an example, an effort that had been financed by the London Rothschilds.[15] To socialists such as Wallace, the Rothschilds' millions inevitably symbolized the worst and most powerful excesses of capitalism.

Class tensions occasionally approached the walls of the museum, as when, sometime in the 1890s, workers laid off from a Tring boot factory dragged Walter Rothschild from his horse at a meet of the Rothschild Staghounds and "manhandled" him.[16] Believing firmly in the traditional, paternalistic duty of the upper classes to establish charitable organizations for the less fortunate, Lord Rothschild and Emma Rothschild subsequently bought the struggling factory and organized free meals for the children of those left unemployed. But many critics of the status quo found such palliative efforts woefully inadequate and argued that now was the time for revolution. Wallace, for example, insisted that the ability to amass great fortunes simply be destroyed. Furthermore, he held that the necessary reorganization of society and wealth depended at base upon the "adoption of a great ethical principle ... that the unborn have no exclusive rights to property, and its full development in the proposition that all inequality of inheritance is unjust." Wallace cheerfully described how abolishing the inheritance of wealth would lead to the disappearance of millionaires. Gone with them would be the ceaseless demand for "the enormous mass of toys, and jewels, and tasteless frippery now made chiefly to tempt the idle rich to expend their unearned money."[17]

Not surprisingly, Wallace himself did not add "butterfly specimens" to the list of luxuries that, with more equal distribution of wealth, would have given way to purchases of "the ordinary necessaries and simple comforts of life for the mass of the community."[18] But did Lord Rothschild, keeping a close watch on socialist arguments and despairing of his eldest son's inclinations, find any incongruity in the fact that Wallace did not include his son's birds, butterflies, and beetles within the list of idle "toys" of the rich? The irony continued in Wallace's strong critiques of imperialism—the "plundering and blundering over the whole globe"—even as he called for naturalists to amass more information on animal and plant distribution and collect long series of specimens.[19] When he lauded the work of ornithologist J. A. Allen, for example, in collecting long series of specimens,[20] Wallace praised work that had tracked the paths of the "great railway-making mania" of 1867–1875 that he elsewhere cited as a primary cause of the increase of wealth in the hands of a few and "a coincident increase of want for the many."[21] In thus paying tribute to the "immense accumulation of *facts*" that had convinced naturalists of the universality of individual variations demanded by Darwin's theory,[22] Wallace was paying tribute to naturalists following the paths of imperial and industrial expansion, both dependent on an enormous centralization of wealth.

Ultimately, the socialist and anti-imperialist sentiments expounded by Wallace at the turn of the century gave voice to seeds of discontent sown in the "belle époque." In their powerful impact on the world after the coming world war, this discontent and the responses to it would one day profoundly influence naturalists' ability to amass specimens for museums. But in the meantime, Wallace had another class warfare on his hands, one that would also have a great influence on the future of natural history. This particular fight began in earnest when, in 1900, William Bateson found in Mendel's systematic studies of discontinuous traits not only confirmation of his own views but the basis for a whole new way of ordering the study of biological life. By 1908, Wallace expressed dismay at "the complex diagrams and tabular statements which the Mendelians are for ever putting before us with great flourish of trumpets and reiterated assertions of their importance."[23]

The controversy that ensued may seem strange to twenty-first-century biologists accustomed to explaining evolution in terms of both natural selection and Mendelian genetics. But in contrast to the post-1930s consensus that both Mendelian genetics and natural selection are required to explain speciation, it was not clear in the first decades of the twentieth century precisely how Mendel's conclusions could be reconciled with a theory of evolution that emphasized small, continuous variation. Indeed, in their enthusiasm for discontinuous evolution, Bateson and others often portrayed Mendelian genetics as providing an alternative theory of speciation. The resulting so-called Eclipse of Darwinism pit selectionists such as Wallace and E. B. Poulton against Mutationists and Mendelians, who downplayed the role of both natural selection and continuous variation.

Bateson's campaign for discontinuous variation called into question more than just the sufficiency of natural selection. He repeated with renewed vigor his doubts of the continued relevance of the natural history collections in which many Darwinians worked. In his 1902 *Mendel's Principle of Heredity: A Defense*, Bateson complained that

> if a tenth part of the labour and cost now devoted by leisured persons, in this country alone, to the collection and maintenance of species of animals and plants which have been collected a hundred times before, were applied to statistical experiments in heredity, the result in a few years would make a revolution not only in the industrial art of the breeder but in our views of heredity, species and variation.[24]

Bateson coined a new term, *genetics*, to inspire this experimentally based revolution.

Clearly, the debate over the nature of variation and how evolution happened entailed debates over where and how nature should be studied, and by whom. Bateson described how the problems of variation "attract men of two classes, in tastes and temperament, each having little sympathy or even acquaintance with the work of the other." Laboratory men cared little for the superficial and vague study of living things out of doors, convinced that the "closer they look the more truly they will see." While "with the other class it is the living thing that attracts, not the problem." Bateson noted that "each class misses that which in the other is good" but, in his enthusiasm for experiment, he did not explain precisely what virtues he thought the laboratory "class" missed in the naturalists' approach.[25]

Wallace and Poulton, firmly committed to natural selection and the study of geographical variation, balked at Bateson's claim "that the survey of terrestrial types by existing methods is happily approaching completion." Poulton responded that "these words will sound somewhat ironical to any naturalist who has had to do with museums, and knows something of the difficulty in getting material worked out."[26] Wallace, having already, in 1895, criticized Bateson's strident claims regarding the importance of discontinuous variation, had even less patience: "The claims of the Mutationists and the Mendelians," he wrote, "as made by many of their ill-informed supporters, are ludicrous in their exaggeration and total misapprehension of the problem they profess to have solved."[27]

Immersed in a battle between scientists, here, too, Wallace fought against claims that a certain "class" be in charge of the future. Though in his position as curator to one of the wealthiest families in the world Jordan certainly sat on the wrong side of one of Wallace's fights, as a naturalist convinced of the importance of geographical variation, he was clearly an ally in Wallace's battle against men such as Bateson. "Systematists have proved by their minute research that geographical variation is the rule and not the exception," Jordan and Rothschild insisted in the 1903 "Revision of the Sphingidae," "and they may be justly proud of this result of their untiring labours. Curiously enough, non-systematists do not generally seem to be aware of this result, nor to fully comprehend its bearing on the theory of descent."[28]

In his campaign for a more biological entomology, Poulton certainly found in Jordan a valuable ally (enough to complain heartily when Jordan's 1905 paper appeared in German rather than English).[29] In his presidential address "What Is a Species" before the Entomological Society of London in 1903, Poulton called on Jordan's 1896 paper as evidence of the continued importance of systematics and museums. Jordan's careful attention to how to use a definition of species based on biology rather than morphology would help "the museum become a centre for the inspiration of researches of the highest interest to the investigator himself, of the greatest importance to the whole body of naturalists."[30] A few years later, composing an attack on mutation and the "extravagant claims of Mendelians" for the preface to a collection of essays on evolution, Poulton notified Jordan that he would be quoting his and Rothschild's statement that the student of geographical distribution of varieties was compelled to smile at the upholders of *per saltum* evolution "entire," since "it exactly expressed what I feel to be the truth."[31] (Poulton would use the quote again in his 1931 address "A Hundred Years of Evolution" to the British Association for the Advancement of Science's section on Zoology.)[32] Poulton also paid tribute to the Zoological Museum at Tring as "pre-eminent for exact and thorough researches" into "the geographical distribution of species and the changes on the borders of their range," research to which the "Batesonians" foolishly paid no attention.[33]

But while Wallace cited Poulton's *Essays on Evolution* as providing the best explanation of Mendelian germ theory and praised the Hope Museum as illustrating mimicry and continuous variation,[34] he never referred in print to Jordan's work on geographical variation; nor did he ever mention Rothschild's long series illustrating continuous variation. Wallace knew Jordan well enough, as his acknowledgment of the 1906 revision shows. But perhaps Rothschild's curator had not gone far enough in his defense of Darwin's legacy: Jordan tended to maintain an open stance on the question of natural selection. Or perhaps to highlight Rothschild's collection in the same way as Poulton's may have been more than the staunch, anti-capitalist Wallace could stomach. True, he left museum curators out of a long list written in 1913 of the "innumerable parasites of the ever-increasing wealthy classes" that included

> the builders of their mansions and their factories; the makers of their furniture and clothing, of their costly ornaments and their children's toys; the vast body of their immediate dependents, from their managers, their agents, commercial travellers and clerks, through various grades of domestic servants, grooms and gamekeepers, butlers and housekeepers, down to stable-boys and kitchen-maids, all deriving their means of existence from the wealth daily produced in mines, factories and workshops.[35]

But to any one lacking sympathy with natural history, by these criteria Jordan walked the halls of a museum similarly built upon the backs of the oppressed laboring classes. Wallace could apparently live with the contradiction. Whether they were reordering the relative status of different sciences, or society itself, others might not be able to suffer the irony so blithely.

ORGANIZING ENTOMOLOGISTS

In his 1896 paper "On Mechanical Selection," Jordan had laid the blame for the problems of taxonomic entomology squarely at the door of his fellow entomologists. Clearly, no matter what kind of revolution—socialist or experimentalist—was called for by critics of museums and "species-makers," systematists needed to reform their ways or be swept away by the changing tides of science and society. The manifestos added to the revisions, and Jordan's 1896 and 1905 papers, were one way of inspiring entomologists to adopt certain methods and principles in order to make systematics more scientific. But the matter was too urgent to rely on the written word alone. Jordan would organize the first International Congress of Entomology in 1910 in order to unite entomologists behind the campaign to reform natural history so that it could survive the imminent class struggles within science. In the end, the organizations would provide venues through which entomologists could attempt to adjust to the class struggles in society as well.

By the time he decided that entomologists required their own international organization, Jordan had inadvertently, using the infrastructure and stability of the age, established an ideal foundation for doing so. Through his constant letter writing and visits to other collections, he had made many entomological friends during his first decade at Tring. He had also created his own private set of associates keen to stay in his good favor both out of self-interest and—often—good will, fully aware that scientific work depended on common interests, mutual aid, and access to other naturalists' knowledge and specimens.

In drawing on his reputation and harnessing his cosmopolitan network for a new purpose, Jordan joined hundreds of other men in the effort to organize information and manpower as the new century dawned. For entomologists were not the only ones calling for reform in the interests of synthesizing the overwhelming amounts of data amassed by nineteenth-century producers of information. At almost the same time, Walter Rothschild's younger brother Charles turned his attention to the Rothschild bank's accounting practices, which had become chaotically inefficient. Charles carried out a full-scale review of these practices in order to avoid duplication and improve efficiency through better organization.[36] Jordan, dealing with units of exchange very far down the list of goods produced by international trade, set out to improve the efficiency of those studying insect specimens.

In turning his attention to international organization, Jordan answered contemporary calls for "something to be done" among entomologists. Entomologists such as F. Frost, for example, wondered whether they were not so prohibitively many in number, and chaotic in direction, that the huge amount of work done had resulted in very few concrete contributions to science.[37] William Sharp lamented entomologists' increasing specialization: "First we had Naturalists, Ray and Linnaeus; then came Entomologists, and we still call ourselves by that name, although there are really hardly any Entomologists now extant. We are Lepidopterists, Coleopterists, Hemipterists, and the like."[38] In the face of such specialization, Louis B. Prout urged that naturalists conscientiously maintain contact with their fellows. But such a scheme struck J. W. Tutt—crusading for a complete reform of how entomologists worked—as "too quixotic."[39] A decade later, in 1905, Tutt was still lamenting that too few entomologists worked at "classifying these facts before they become too overwhelming, and in rescuing the grains of wheat from the bushels of chaff in which most of the facts are buried."[40] Rothschild and Jordan had worked through the rather chaotic entomological literature at a great cost of time and labor, resignedly noting that "we have patiently to bear the fruits of the sins of our forefathers in science.... One may kick, but one has to suffer."[41]

As we have seen, others remained less convinced regarding the virtue of suffering and called for radical changes that struck at the heart of how entomologists had proceeded during the preceding century. The year Jordan arrived at Tring, the entomologist David Sharp issued a manifesto for public museums based on the fact that good scientific work depended on large collections, and it was therefore "not worthwhile for a private individual to make a collection of insects." So far, Sharp insisted, private collections had not proved their worth, for despite the "enormous amount of enthusiasm, labour, devotion, and study bestowed on Entomology," little had been effected toward what entomologists required; namely, the formation of a collection of insects useful to solving the biological questions of the day.[42] Rothschild and his curators would soon prove Sharp wrong, although no doubt the wealth backing the museum would have only emphasized Sharp's point. In any case, the state of things led to indignant criticisms of the average entomologist within the pages of the reform-centered *Entomologists' Record and Journal of Variation*—and these writers were sympathetic to the project of entomology. Jordan already knew well enough how an unsympathetic zoologist (or mineralogist) would respond to the state of affairs. How a skeptical politician glancing over the coffers of the British Museum might view matters was also, of course, cause for alarm. Clearly, entomology, too, required more organization to improve the utility of its results, the efficiency of its workers, and the stability of its conclusions in order to ensure taxonomy's continued relevance.

The entrance to the twentieth century found many fields similarly struggling to deal with the legacy of the nineteenth century's "age of specialism." The enormous growth in knowledge, institutions, and disciplines inspired hundreds of efforts to unify and standardize workers and methods. As a response, "progress through rationally guided organization" became a guiding motto for the era. As one German politician wrote in 1911, "Everywhere it is recognized that only firm, well-ordered collective action can produce influence and success."[43] Convinced by this paradigm of reform, Jordan had been struck with the idea of "the usefulness of an association of some kind aiming at international collaboration of Entomologists" in 1905.[44]

Of course, he could have chosen to organize British entomologists; they were chaotic enough. But motivations existed to look beyond British shores, not the least of which was the constant need to correspond with a small network of experts scattered over the globe in order to build Rothschild's "all world, scientific collection." This network emphasized the benefits of international cooperation with every letter or package of specimens (by contrast, foresters, focused on cultivating national resources, would not establish an international congress until 1926). Just as the success of N. M. Rothschild & Sons had depended on building an international financial network along which information and funds could be mobilized quickly, the Tring Museum had built unprecedented revisions through using a cosmopolitan network of fellow naturalists. Those entomologists who owned or curated large collections (e.g., W. J. Holland at the Carnegie Museum of Natural History, Charles Oberthür in Rennes, and Walther Horn in Berlin) formed only the most prestigious of a wide network to whom letters came and went in the process of amassing enough specimens to complete scientific systematics. Jordan exchanged specimens ("We have a long series—we'd gladly give up a few in exchange for X"), asked for clarifications regarding descriptions and names and compared notes with fellows specialists regarding confusing forms. Combined with endless hours studying specimens in the museum and poring over two hundred years' worth of literature to get the nomenclature sorted out, trying to figure out what species entomologists had meant and what names should be applied to what, this international correspondence formed the foundation of the robust systematics Jordan exhorted his colleagues to pursue.

Jordan's efforts to organize entomologists also took place within the broader context of an internationalist movement based on the explicit premise that people from different nations could—indeed, must—come together and cooperate in the interest of progress. This premise supported a range of idealistic endeavors, from the League of Nations to Wilhelm Ostwald's Esperanto movement. If Jordan succeeded, his international association of entomologists would join dozens of international congresses being convened in many realms of knowledge between the years 1900 to 1914, inspiring one historian of such efforts to call the period "the golden age of internationalism."[45] International congresses were convened on subjects ranging from criminal anthropology to experimental and therapeutic hypnotism. So many of these meetings were being held by the turn of the century that one announcement in the *Zoologist* began, "Another so-called International Congress! This time it is on sea-fishery and oyster-culture."[46] As the president of the 1895 International Congress of Zoology meeting in Leiden, F. A. Jentink, announced with macabre flair, "The Congress bacillus has infected Society. From high to low, no class has escaped."[47]

The efforts of Wallace's fellow socialists inspired the most famous of these international meetings. The International Workingmen's Association had given way to the Second International, both of which launched calls for reorganization and internationalism on a grand, stunning scale. Keen to inspire workers to feel loyalty to class rather than nation, the revolution required a whole series of international meetings and congresses for implementation.

Next to the socialists, those who studied the natural world took up the internationalist call most ardently. When the famous Second International held its congress in July 1889 in Paris, the Exposition Universelle was playing host to dozens of other congresses, including the first International Congress of Zoology. Scientists, having long since established relatively comfortable spaces within bourgeois society, obviously put forward more modest proposals for change. But when zoologists first met in Paris, they thus joined a broad and pervasive movement aimed at dealing with some of the legacies of the fast-paced, rapidly changing nineteenth century through international cooperation and exchange. The zoologists' portrait of the problems facing the coming century differed profoundly from those who thundered from the socialists' pulpits in Paris that year. But then, the list of members and patrons for the International Congresses of Zoology was quite unlike that of the socialists' meetings. The proceedings of naturalists' congresses reflected the strong ties of natural history with capital, land, and leisure. Heads of state arrived in full regalia, monarchs sent telegrams of support and served as patrons, and barons, lords, and princes graced the list of members. At the International Congress of Zoology in Moscow in 1892, Grand Duke Sergei Alexandrovich, the son of Emperor Alexander II, attended the meeting as patron, and the congress closed with the ponderous notes of the national Russian hymn, played by a military orchestra.[48] (The grand duke would be assassinated by revolutionaries in 1905.) In attending the 1901 International Congress of Zoology in Berlin, Jordan joined hundreds of other zoologists in being greeted by heads of state, city officials, and other "persons of distinction" and sat through ceremonies peppered by the music of military bands.[49]

With friends in high places and all the assurance of a stable world in which they would be allowed to pursue their interests, zoologists had organized their first congress largely in response to the practical problems facing the discipline, particularly the chaotic state of nomenclature.[50] Other congresses had set the pattern of procedure. A report on the 1896 International Bibliographical Congress described a typical meeting: "The Congress dined well, listened to fine speeches, and then, before separating, certain rules were made, regulating in some way the workers in other countries."[51] Debates over whether to use information-processing systems such as the Dewey Decimal System, how to create international bibliographies, and how to establish standardized rules for exchanging information took up much of congress members' time. The president of the 1895 zoologists' congress announced that members would examine a range of problems facing zoologists, from the tremendous increase in poor descriptions for the sake of vanity to the need for consistent nomenclature and better bibliographical organization.[52] By then, the international congress of zoology had become the accepted forum for proposing and announcing reforms to deal with such issues. For example, zoologists adopted a resolution at the 1895 congress protesting recent prohibitions by the Universal Postal Convention against sending animals and insects, living or dead, by mail.[53] One member brought forth a proposal for an International Colour-Code for Zoogeographical Purposes.[54] And O. C. Marsh spoke on the need for international criteria for designating type specimens.[55]

The congress's establishment of an International Commission of Zoological Nomenclature (ICZN) reflected participating zoologists' attention to standardization. As the strident complaints in the Tring revisions testified, naturalists were burdened by the constant need to navigate a wide variety of naming practices from both the past and the present. The American zoologist Austin Clark once captured the resulting frustration when he explained the difficulty with his particular group of choice, the crinoids, as due to two causes. First, "the natural cussedness of the beasts" ("this beast has no systematic ethics at all," he cried of one species, "but varies in the most ghastly manner; everybody has had trouble with it!"), and second, the fact so many names had been misapplied by his predecessors.[56]

Jordan had spent time enough on the practical ramifications of variation in both insects and entomologists, poring through Rothschild's library to sort out synonymy. "The Natural History of the animal being the subject of our science," he and Rothschild wrote in one of the revisions, "the accessory subject of nomenclature should never have assumed such magnitude. It is a waste of energy."[57] Keen to help redirect this energy, Jordan surely paid close attention at the Fourth International Congress of Zoology, in Cambridge in 1898, the first he attended, as G. F. Hampson reported on his survey of lepidopterists' opinions on nomenclature. One of Hampson's questions read: "In the event of there being any clear consensus of opinion on the above subjects by a majority of those to whom they are to be submitted, would you be willing to adopt their decision and abide by it?" J. H. Durrant, who processed the answers on Hampson's behalf, later confessed that he had "not yet quite recovered from the mental exhaustion consequent upon an attempt to ascertain the opinions and to tabulate the replies to this question. I retain only one impression, each member is willing to accept the decision of the majority *provided he is in it*."[58]

In witnessing such efforts, Jordan surely had evidence that instilling some degree of unity among entomologists would be a daunting task. But he had also seen how his colleague Ernst Hartert's position as secretary of the Fourth International Ornithological Congress that met in London in 1905 provided a venue through which Hartert could campaign for consensus—at least among ornithologists—on matters ranging from nomenclature to theory. Hartert had, for example, delivered a paper that attempted a bit of both. Entitled "The Principal Aims of Modern Ornithology," it placed Tring's emphasis on original, detailed labels and their importance to the study of geographical variation before the congress's international audience. "It is now widely understood that dreary species-mongering is not the gist of science," Hartert began. He then laid out Tring's case that the study of geographical forms had a direct bearing on ornithologists' understanding of the evolution of such forms and the relationships of forms to each other. Thus, geographical forms should bear names.[59] A field trip to Tring concluded the argument.

It was during the summer of 1905, as ornithologists convened at Tring, that Jordan was struck with the idea of forming an international congress for entomologists. In November 1906 he began canvassing the more prominent and well-connected entomologists throughout Europe, most of whom were already either friends or acquaintances. He wrote first to Christopher Aurivillius, of the Swedish Museum of Natural History in Stockholm, Karl Heller, of the State Museum of Zoology in Dresden, the German dipterist Theodor Becker, Charles Oberthür, of Rennes, and Guillaume Severin, of the Royal Museum of Natural History in Brussels; then to Eugène Louis Bouvier, of the National Museum of Natural History in Paris, Ignacio Bolivar, of the Museum of Natural Sciences in Madrid, Frederick Dixey, of Oxford, the Dominion Entomologist of Canada, James Fletcher, Raphael Gestro, of the Natural History Museum of Giacoma Doria, in Genoa, Walther Horn, of the German Entomological Institute in Berlin, Anton Handlirsch, of the Natural History Museum of Vienna, G. Horváth, of the Royal Hungarian Society of Natural Sciences in Budapest, Hermann Julius Kolbe, of the Berlin Zoological Museum, Edouard Everts, of the Dutch Entomological Society in Rotterdam, J. C. de Meijere, of the Artis Royal Zoo in Amsterdam, Friedrich Ris, of Rheinau, Switzerland, Yngve Sjöstedt, of the Swedish Museum of Natural History in Stockholm, A. von Schulthess, of Zurich, and Max Standfuss, of the Federal Polytechnic Institute in Zurich. It was a formidable list, representing the foremost specialists on a range of groups and institutional leaders from across Europe.

Jordan argued in his letters to these entomologists that, despite the best efforts of museum curators and entomological enthusiasts throughout the nineteenth century, insect collections and entomological knowledge seemed as disordered and unstandardized as ever. The discipline was plagued by a lack of trained entomologists, meager governmental support, exponential increases in the number of known species, and an overwhelming amount of unidentified specimens. He urged upon his colleagues the need for establishing a formal international entomological congress that could work to avoid repetition, disseminate research results more widely, and facilitate agreement on a common international code of nomenclature. "At the Congress," he explained, "where nobody's attention would be averted by other interests, general questions bearing on entomology might be discussed and thereby the efforts of collectors be guided into channels where their labours are most needed."[60] At the congress, in other words, the content of letters advising collectors in the field and the various manifestos urging entomologists to do scientific systematics could be delivered efficiently and in person. The campaign to reform entomology would be organized and united.

To convince his fellows to join him, Jordan cited more motives than simply the wish to improve collections. These additional concerns tie the congresses to a broader fight among professions for status and resources amid rising state support for science. In his letter to Raphael Gestro in Genoa, he introduced his campaign with the justification that because entomology had become a kind of stepdaughter to zoology, its position did not correspond with the importance of practical and philosophical results of entomological research:

> The public entomological institutes are generally provided with insufficient means and an inadequate staff. The governments do not take sufficient interest in the entomological departments of the museums. On the other hand the entomological public also is often indifferent or even hostile to the museums. This public (the "collectors") lacks often the scientific spirit necessary for good work.

As a result, a general organization of the "entomological World" was needed, he insisted. He asked Gestro for his opinion and also asked him to send the names of a few Italian entomologists to whom he could write to explain the endeavor and ensure widespread support.[61]

Jordan's justifications for an international organization of entomologists reflected his deep concern regarding two legacies of the naturalist tradition that plagued the endeavor to describe, order, and ultimately explain organic beings: the diversity and isolation of entomologists. He had specific problems in mind when describing the effects of isolation, all closely tied to the nature of his day-to-day work on Rothschild's collection: the tremendous amount of time required to clean up synonymy due to entomologists' varied ways and dispersed literature; the threats to museums and collections by the most vociferous of the young experimentalists; the disconnect between the specimens coming in from the field and the requirements of adjusting taxonomy to evolution; the tensions between "official and private" entomologists. All these formed the backdrop to Jordan's efforts to deliver on entomologists' long-standing calls for reform.

The specific form of his justification for the congress was further influenced by his opposition to some of the proposals for change being voiced by others. In contrast to David Sharp's insistence that private collections cease, for example, Jordan thought that given the massive amount of material needed to complete scientific systematics, private collectors must be encouraged. But, as he had learned during his early work on the Stromeyer collection, the usefulness of a collection of specimens for these purposes depended on whether they had been amassed for the sake of science or amusement. One could have access to all the specimens in the world and find them useless for the questions of biology if they had not been collected and labeled in a "scientific" manner. An international organization of entomologists could provide a forum for channeling the time-consuming work of advising (or admonishing) each collector through an authoritative body that could direct priorities and influence methods.

THE END OF TRING'S HEYDAY

By June 1907, Jordan had received enough replies to announce that the proposal of a congress "for the discussion of all matters relating to Entomology and Entomologists"[62] had found favor in all countries. This "all countries" referred to those "civilized nations" that counted in the scientific world (no one, it seems, concerned themselves with equal representation of all nations, much less countries in the tropics who harbored most of the animal and plant species of the globe). Encouraged, Jordan next sent an announcement requesting the support of national entomological societies in which he described the purpose of the congress more broadly than in his initial letters, explicitly placing the need for reform in the context of entomology's importance to biology. The congress, he announced, would "promote the interests of entomological research, and therefore of Biology in general, by furthering cordial co-operation between the Entomologists of different countries." Members could discuss "questions of general entomological interest, thereby stimulating research and directing it into channels where it may be most fruitful or where special research is most needed." He even added a nod to the systematists' economic brethren, announcing that "Questions of Applied Entomology will likewise be dealt with in the discussions and lectures, the great experience gained by the devotees to pure Entomology being applicable with profit in economic and hygienic

Entomology."⁶³ Having harnessed the support of his closest colleagues, he then cast a wider net in an effort to appeal to those outside his own collection-based network.

The cast was successful. More than fifty-three entomologists replied to Jordan's appeal on behalf of their respective associations.⁶⁴ Official statements of support soon arrived from the major entomological societies of Europe and from British entomologists. W. J. Holland, of the Carnegie Museum of Natural History in Pittsburgh, Lord Walsingham, Percy Lathy, and Edward Meyrick wrote encouraging letters, and Malcolm Burr offered his assistance, eagerly inquiring about progress over the following months.⁶⁵ With all the energy of one sure of a stable future, and having received assurance from Belgian entomologists that they would undertake the local arrangements, Jordan began to plan the first congress, to be held in Brussels in August 1908.

No doubt many entomologists thought that with Rothschild's backing, Jordan and the Belgian entomologists would succeed at such an enormous undertaking. Those following Jordan's efforts must have been bemused, then, when in early 1908 his letters regarding the congress abruptly ceased and the summer passed with no grand meeting of entomological workers in the Belgian capital. Meanwhile, inquiries regarding Walter Rothschild's whereabouts began dominating Jordan's correspondence. Both curators responded to the stream of queries with vague replies that their boss had traveled abroad for an indefinite amount of time. Rumors abounded. One friend of the museum wrote to insist that he, for one, simply refused to believe the tales being told in "the City."⁶⁶

In his obituary of Walter Rothschild decades later, Jordan briefly and discreetly summed up the sudden change in the Tring Museum's fortunes that occurred in early 1908. That year, Walter Rothschild had retired from N. M. Rothschild & Sons, "having neither inclination nor ability for finance." In addition, the museum had decided to build new buildings (the first in Tring to have electric lighting installed) and to finance various explorations in North Africa. Jordan explained that the associated increase in expenses "rendered it financially inconvenient to continue the issue of extensive *Revisions* and *Monographs*," and that Tring thus discontinued the planned researches on the Saturnian moths that he had commenced in 1906.⁶⁷

But in recounting the change of affairs that occurred in 1908, Jordan told only part of the story. He did not mention, for example, that by this time, Walter Rothschild had provided two mistresses, Marie Fredensen (who would bear him a daughter) and Lizzie Ritchie, with their own flats in London. Or that an unnamed peeress with whom Walter had had an adulterous affair had begun blackmailing him. By 1906, Miriam Rothschild recounts, her uncle was "a very worried man" who could no longer bear to read his personal correspondence and began placing his mail, unopened, in large wicker baskets that he padlocked shut when full.[68] But early in 1908 Lizzie telephoned Emma Rothschild to demand whether she knew that her son had "contracted a morganatic marriage with a whore and is the father of her child?"[69]

Soon the family also learned from a chief clerk at the bank that Walter had mortgaged a family estate in Buckinghamshire in an effort to raise money, raised the loan on the museum building to £25,000 at 5 percent interest, and "had been speculating wildly and disastrously on the stock exchange."[70] When confronted, Rothschild confessed the presence of the wicker baskets, and his brother Charles, with the help of four clerks, began opening and sorting through the letters. Charles settled things with Marie and Lizzie (a house each and £10,000 per annum) and paid the mortgage on the museum. Walter Rothschild divulged nothing concerning the existence of the "charming, witty, aristocratic, ruthless blackmailer" who, as Miriam lamented, remained free to ruin him financially and destroy his peace of mind over the next forty years.[71]

With the survival of the museum at stake, Charles recruited Jordan and Hartert to help put the finances in order. The two curators obediently began writing to each of the collectors and natural history agents, insisting that the network halt any consignments, submit any outstanding bills, and close every account (even as Rothschild secretly sent letters to collectors insisting these orders were not to apply to his beloved live cassowaries).[72] Jordan diligently composed letter after letter informing collectors and agents that Rothschild had decided to stop receiving collections. He offered a range of excuses for the sudden change. To Fred Birch, collecting in Brazil, he explained that owing to the enormous influx of fresh collections and their inability to name and arrange "the mess of Lepidoptera" already at Tring, they were stopping all new purchases "for some time."[73]

Whether they believed Jordan's and Hartert's excuses or gave credence to the rumors, the collectors must have been dismayed. Although they usually traveled at their own expense, receiving at most a small advance payment, collectors inevitably budgeted on captures estimated over the next year or more. Panicked letters arrived from those who had received encouragement from Rothschild but now received word that no collections would be bought or expeditions funded in the near future. They listed the rarities in their collections, the varieties from unknown localities, and the unique and scientific nature of their finds. But nothing could be done.

Although at least one time-consuming but unsuccessful lawsuit resulted from the cancellations, the collectors and agents had no choice but to acquiesce. Jordan and Hartert helped Charles cancel all orders for books, sold or gave away Rothschild's live specimens, and distributed surplus specimens to willing buyers. Jordan regretfully began returning specimens that had been dispatched by friends and agents. Tring had decided, he explained to Otto Standfuss, "for several reasons the previous winter to buy for a few years nothing for the museum."[74] They must have been hard letters to write, for he was having to return the very material upon which he had hoped to establish the biological importance of both collections and systematics.

One wonders whether, as he was forced to call a halt to the activities of the collecting network that had been so central to doing "scientific systematics," Jordan secretly sympathized with Wallace's campaign against hereditary wealth. Wallace once wrote that "observation and reason alike prove that all inheritance which enables a man to live idly without giving any adequate service in return for his wealth is an injury to him who receives it, that it renders him the centre of a vast circle of evil influence through his numerous parasites and dependents, and that to permit it is one of the greatest of crimes against humanity."[75] In any case, dealing with the repercussions of certain members of Rothschild's "vast circle" must have been painful. Jordan particularly regretted the need to return a series of specimens bred by Standfuss because, as he noted, the series demonstrated a crossing quite contrary to what Hugo de Vries's mutation theory would have predicted.[76] But it was the financial problems of the museum that now determined the state of the collection, rather than Jordan's attempts to both temper the enthusiastic speculations of theorists and raise the status of the systematics of insects within the hierarchical order of sciences.

A few useful results came of the museum's misfortune, however. Miriam Rothschild recounts how Walter's mother, Emma, and brother Charles expressed their "silent sympathy" for Walter by asking Hartert and Jordan what the museum needed most, and, finding that the museum urgently required a separate wing for the insect collections, arranged for it to be built.[77] In addition, the need to get Walter Rothschild quickly out of England led to some rather extensive field trips for both curators, as the family packed him off to the sanctuary of Europe's imperial network in Algeria, first with Hartert as a chaperone, then with Jordan. The unexpected adventure resulted in a collection, of course (Hartert would later contribute an account of their expedition to *Novitates Zoologicae*).[78] When Jordan returned from his time abroad after a four-month absence from Tring, the insect building was almost finished. He was soon back at work at his museum desk, packaging the Aphiden and Hemipteren he had been able to capture on the journey to send to his friend Guillaume Severin in Brussels. "Unfortunately they are quite few," he wrote, since he had to spend most of his time during the trip dealing with butterflies.[79] The specimens served, no doubt, as in some sense consolation for having temporarily abandoned all the grand plans for a congress in Severin's city.

When entomologists back home who had seen Rothschild as both competitor and patron realized what had happened to the formidably focused research factory at Tring, many wrote to convey their dismay. Some, depending on their specialty, had come to rely on the Tring specimens as much, if not more, than the national collection in London. Just as the crisis unfolded, Guy K. Marshall, attempting to work out the mimetic resemblances of *Charaxes* butterflies but befuddled by his own specimens, had just recently written to Roland Trimen that perhaps he could learn something from Jordan, "as I believe they have received considerable numbers of this species in recent years from their collectors."[80] Now, as word spread about the Tring Museum's change of fortunes, Poulton sent his condolences, well aware of the effects the belt-tightening would have on Jordan's work. "I am extremely sorry to hear (as I had in many directions) of your opportunity for obtaining material being restricted, even though only temporarily," he wrote. "There are so many in this country who feel grateful to you for your work that you have I am sure many ardent well-wishers who sincerely desire that you may have every possible means for carrying on your researches."[81] Perhaps to help temper the blow and remind Jordan there were other ways to get specimens than relying on Rothschild's now limited allowance, Poulton offered to bring any specimens they might need from the Hope Collection when he next visited Tring.[82]

Indeed, the influence of the restricted museum budget on Jordan's work was tempered to some extent by the fact that he had built up a strong network of friends who sent him specimens for determination, either as an exchange or for no charge. Certainly plenty of work remained. Jordan remained busy, continuing to process specimens that came from acquaintances and moving the Lepidoptera into the new building. The eight months in Algeria had resulted in a collection of nearly two thousand North African bird skins and nineteen thousand insects. Between 1908 and 1914 Rothschild continued to take trips to Algeria, sometimes with Hartert, other times with Jordan. By December 1910, the museum's finances had been placed on more secure, although relatively restrained, foundations. Eventually the museum could once again fund expeditions to amass collections for the museum. They had enough to engage the entomologist and collector G. F. Leigh to go to the Comoros, for example, Jordan giving him instructions that they would take ten good specimens of each species, except Micros, at eighteen pence each. "If the species is variable," Jordan wrote, repeating Tring's usual refrain, "all the individual varieties must be sent for selection."[83] They also found the requisite funds when Albert S. Meek, one of their best collectors, finally gave in to Rothschild's and Hartert's constant pleas to collect in the Snow Mountains of Dutch New Guinea. Still, although the resources available to Tring continued to astonish other naturalists, the access to collectors all over the globe upon which the breathtaking Tring revisions had depended never again reached the level of the 1890s.

This interlude in Jordan's working life illustrates how the idiosyncrasies of particular institutions and personalities could strongly influence a naturalist's ability to do certain kinds of work. The museum correspondence for the first years of the twentieth century also contains early hints of broader shifts that would eventually circumscribe many naturalists' ability to amass huge collections. For example, the long-standing emphasis on collecting specimens faced a growing fear on the part of some scientists and sectors of the public that such collecting might lead to the extinction of rare species. Hartert reported from the International Ornithological Congress in Paris in 1900, for example, that the congress had a "stormy session on bird protection."[84]

The storm had been brewing for a while. Rothschild's estranged mentor at Cambridge, Alfred Newton, had been campaigning for a "close time" during the breeding season in Britain since 1868, and his efforts paid off with the Wild Birds Protection Act of 1880.[85] Across the Channel, the Dutch zoologist A. Jentink, president of the 1895 International Congress of Zoology held in Leiden, argued against the creation of new museums on the grounds that it was purely for the sake of such museums that collectors sometimes sacrificed the last individuals of a species.[86] At the 1910 International Congress of Ornithology, James Buckland criticized the collection of long series on conservationist grounds when he argued: "No addition can be made to knowledge by accumulating the skins of a bird which has been already named and classified times without number."[87] Ernst Hartert heartedly disagreed, of course. He often noted that specimens of the commonest species told ornithologists much more about patterns of geographical distribution. Although both Hartert and Rothschild expressed sympathy with concerns regarding extinction and overcollecting (and Charles Rothschild would become a leader of Britain's nascent conservation movement), they insisted that any protection laws make clear exceptions in the name of science.⁸⁸ They tried hard to distinguish their collecting, which they insisted took a limited series from throughout birds' ranges, from others who killed thousands of birds from single localities for things such as the millinery trade.⁸⁹ Hartert lamented that "'Bird-laws' in the wilds of Africa are nothing but a trouble and a difficulty for the conscientious collector for scientific purposes," while laws in England were "useless as long as every hedgerow and common is pilfered by our hopeful boys."⁹⁰

But not everyone could see the difference between pilfering by schoolboys and the exploits of a Rothschild. Walter once began a paper describing his newly acquired specimens of California sea elephants with a description of how, when in 1907 Charles Harris cabled that the supposedly exterminated animals had been seen on Gaudalupe Island, off the Mexican Coast, "I at once told him to get them." "After several weeks' hard work in hurricane seas," Rothschild wrote triumphantly, "Mr. Harris and his party killed fourteen Sea-Elephants, of which four entire bulls, three cows, and two bulls' skeletons reached England safely." The brief paper then listed the specimens' length, circumference, and the dates and localities from which Harris obtained the animals.⁹¹

As sensibilities changed, some began demanding that the scientific study of animals be based on what such organisms did in life, rather than how large they measured in death. In 1907, for example, Hartert received word that the colonial secretary of Jamaica, Sir Sydney Olivier, a Fabian, had refused to give Rothschild's collectors a license "to take for scientific purposes six each of the different birds protected by the law." Olivier defended his decision by insisting that all the species and characteristics of Jamaican birds had been completely determined and recorded. Furthermore, he insisted that the accumulation of stuffed specimens would not advance natural history in the slightest; progress could be obtained only through the continued "observation of living creatures in their haunts and habits."⁹² (In private the man who conveyed the news to Tring—a friend of Hartert's—J. E. Sherlock, apologized to Hartert for the island's "socialist governor," concluding with the news that, "License or no," he had secured a few specimens to send on the next boat.)⁹³

That this dressing-down as to what constituted scientific research came from a Fabian hints that at least some of the critiques of specimen-based natural history as science reflected an increasing tendency to question the authority and pastimes of the wealthy. In 1893, a notice had appeared in the journal *Natural Science* that the rajah of Sarawak, Charles Johnson Brooke, was "closing his dominion to collectors, owing to the depredations committed by orchid hunters and the like." "It is scarcely to be wondered," the contributor wrote, "that a man objects to have the rare and beautiful objects of the fauna and flora of his country carted off wholesale to gratify the passing whim of a moneyed class in another hemisphere."[94] Criticisms of how to do natural history were sometimes merging with condemnations of the leisured class—a foreshadowing of profound tensions that would face the naturalist tradition in the coming century.

Nor were the habits of that moneyed class all that was up for scrutiny at the turn of the century. Some called the very existence of that class into question, as the Russian Revolution of 1905 demonstrated so profoundly. While Britain's empire continued to serve as an outlet for discontent, the Boer War (1899–1902) and the Boxer Uprising (1899–1900) demonstrated the fragile nature of imperialist nations' command over foreign lands. Political trends closer to home also foreshadowed what the coming century had in store for the world in which Rothschild and others had amassed their collections. Although the Liberals swept the election of 1906, the new Labour Party gained twenty-nine seats, and the following decade saw sweeping reforms, as the threat of social and political revolution demanded a response. In 1911, the National Insurance Act was financed by an increased tax on the wealthy via death duties. In a development lauded by Wallace, these tax reforms relied on a sharp distinction between earned and unearned income. Although the initial reforms were relatively moderate, they incorporated nascent features of what, following World War II, would begin a transformation of British society that would eventually have important consequences for scientists.

Meanwhile, scientists' increasingly successful campaign for state support meant they were joining an array of professionals attempting to persuade "the rest of society and ultimately the state that his service was vitally important and therefore worthy of guaranteed reward."[95] Although as a curator to a private collection Jordan need not have campaigned for state support of entomology, his personal experience with the vicissitudes of private patronage no doubt bolstered his belief, noted in his original letters regarding the idea of an international organization of entomologists, in the importance of securing government support. Surely, enlarged patronage by the British state would provide more of the stability required to build a secure research program. It was an assumption made by many sciences prior to 1914, with eventually weighty results for the kind of work that could be done and the justifications that could be made for doing that work.[96]

And finally, concerns that would have a profound impact on the form of future generations of taxonomists' priorities and justifications began appearing in the letters of some collectors after the turn of the century. Writing from the

Sao Jacinth Valley, Minas Geraes, in Brazil, Fred Birch informed Jordan that he had found areas in the immediate vicinity of the towns "useless to an Entomologist" since the virgin forest had been replaced by coffee plantations and pasture. Each year, the axes of the colonists made the forest that remained on the higher slopes smaller, Birch reported, "so eager are they to demolish the perfect work of the ages and substitute therefor some tame fodder crop." Birch thought such injury might be forgiven had the men again used the land they laid desolate, but they preferred to cut and burn a fresh piece of forest each year. "The whole region is gradually becoming a bare grassy waste," he reported, "which the rains work havoc with in some years."[97] A few years earlier, Hartert had noted in a paper on birds collected by Heinrich Kühn on islands near Celebes, "It is quite possible, and even probable, that on account of the thick population and the destruction of the forests some interesting local forms have disappeared."[98] This was a challenge to species-making, indeed.

"SCIENCE KNOWS NO COUNTRY"

Naturalists' ability to adapt to the above changes would depend in part on whether they could establish a consensus regarding how the methods and goals of the tradition must thereby change. Those who studied insects, for example, would be profoundly affected by the challenges of amassing large private collections and the increase of state patronage. But Jordan had been forced to quietly set his plans for an international congress, one of the primary means of establishing consensus in the face of change, aside in order to help Charles sort out Walter's affairs. As soon as possible, Jordan recommenced his efforts to organize a congress of entomologists, inquiring about the possibility of convening in Brussels in 1910, a date that would happily coincide with the Brussels World's Fair, the Exposition Universelle et internationale.[99]

He arranged a three-day meeting in London at the offices of the Linnean Society to start planning. Here, the dispersed members of a network originally built for exchanging specimens met together face to face as Walther Horn arrived from Berlin, Pierre Janet from Paris, and Guillaume Severin from Brussels. They were joined by Frederick Dixey, Henry Rowland-Brown, George Charles Champion, E. B. Poulton, and Guy Marshall from Britain. With the date set for the first congress, Jordan began gently cajoling his correspondents to attend. "An ardent entomologist like you," he wrote to one naturalist who had requested names for his Papilios, "will surely attend this first international conference."[100]

Not everyone believed that an international congress would provide the best solution to entomology's problems. George Blundell Longstaff complained to Dixey when Jordan insisted he run for chairman of the British Committee of the congress, "Little knows he how I hate *all* congresses!" But having succumbed to Jordan's lobbying, Longstaff set about organizing the committee, distributing ten thousand pamphlets and arranging preparation meetings.[101] Longstaff did not say why he hated all congresses. Perhaps he simply did not like going to meetings. But his lack of enthusiasm hints that not all shared Jordan's belief that organization would solve the perceived crisis facing entomologists or that an "international congress" provided the best venue to supply that organization. Certainly, those who organized international meetings, which aimed at unity of purpose and rules, had to make assumptions about their colleagues that often flew in the face of the evidence. They had to believe, for example, that it made sense to bring men from different nations together in one building in order to discuss their varied interests. In other words, such international organizations depended on a firm assumption that governments and peoples could act constructively together for the common good.

In a time of rampant nationalism that within a decade would escort Europe into world war, this could seem a particularly utopian scheme. There were other ways, of course, through which men could be ordered and identity established; namely, loyalty to king and country. A month before Jordan arrived in Britain, Edward Buckell had spoken at the Entomological Society of London of the need for British entomologists to "take our proper place in the commonwealth of science," by which he meant falling into line with the same names of their brethren in other countries.[102] But some British entomologists had argued against sacrificing their national individuality in the interest of universal rules "regulating all workers." J. C. Warburg, for example, had argued that British entomologists should adopt the Continental method of setting specimens on high pins on the grounds that "it is . . . very material that lepidopterists all the world over should adopt a uniform system."[103] In response, W. H. Harwood ridiculed the "enlightened cosmopolitan," his call for universalism, and any associated criticisms of the "uninstructed insular eye." Entomologists who disliked the British method of setting their insects should, Harwood argued, "confine their exchanging to those whose methods of setting suit them, and not endeavour, vainly, to suppress other people's individuality in order to absorb it into their own." The calls for high setting on the grounds it would standardize practice across national boundaries meant naught to Harwood. "No doubt a common system of setting would have its convenience," he conceded, as would also a common monetary or fiscal system, or a common language; but we "hardened Britishers" do not feel disposed to "fall into line" with the rest of the "civilised world," upon a point where we consider ourselves far in advance of other people, and where progress in their direction would mean a movement towards the rear.[104]

Such attitudes once inspired Tutt to describe his compatriots as "peculiar in their tastes and insular in their habits with regard to matters entomological."[105] And the British were not the only ones who could be insular. After more than thirty years of entomological work, Augustus Grote concluded that the enormous confusion in his particular group, the Noctuidae, arose not only from a lack of attention to detailed anatomical structure, but "the chauvinistic spirit" that "fostered the growth of peculiar names in the different European entomological circles."[106]

This "chauvinistic spirit" fostered more, of course, than different names for the same insect. Rising nationalism in Europe was infiltrating all aspects of life as the twentieth century began. Even scientists found it tempting to abandon the cosmopolitan ethos of science and campaign for state support on nationalist grounds.[107] The British Science Guild, founded in 1905, drew explicit connections between science, economic strength, and political power. In 1910, as German mobilization raised tension in Britain, the Guild even held a contest for an essay on "The best way of carrying on the struggle for existence and securing the survival of the fittest in national affairs."[108]

In often portraying scientists as exemplars of internationalist values who could show the rest of the world how to proceed beyond competition to peaceful and productive exchange, Jordan joined those who took a stance against rising nationalism over the ensuing decades. Though he did not go quite so far as Wallace, who used the occasion of an international congress of workers in 1896 to call for the proletariat to "combine in a solemn promise" never to use their arms against their fellow workers of their own country or any other nation,[109] Jordan maintained the time-honored tenet that, as one of his fellow entomologists put it in 1905, "Science knows no country, its devotees are cosmopolitan and universal."[110] Some emphasized the practical motivations for such claims. In Moscow in 1892, the French zoologist Raphael Blanchard cited the need to exchange ideas and communicate discoveries in natural history and the consequent efforts to establish standard names for animals and plants to explain the fact that science provided a place where men of any race, nation, and belief could meet and fraternize.[111] Others had their sights far beyond science, as when at the 1895 International Congress of Zoology in Leiden, the interior minister M. S. van Houten announced that as a counter to dangerous patriotism, the alliance of scientists would prove a precursor to no less than the alliance of humanity.[112] Kakichi Mitsukuri, speaking "on behalf of the Zoologists of Japan," echoed such idealism when he hoped the International Congress of Zoology in Cambridge in 1898 would "be one of the strongest forces for drawing nations together, and making a brotherhood of the whole world."[113]

Acknowledgement of other models of interaction sometimes counterpoised these optimistic flourishes. At the congress in Cambridge, for example, an ardent homage to international exchange by the Dutch zoologist A. A. W. Hubrecht acknowledged that "the sounds of the big trumpet" seemed to be loud at the moment. He, at least, felt it necessary to remind the congress that "true science is most averse to the use of that instrument."[114] At the 1901 International Congress of Zoology in Berlin, as Anglo-German relations deteriorated rapidly in the face of the Boer War, Britain's P. L. Sclater apologized on behalf of his countrymen for their scanty attendance. Politics should not be a factor in science, he insisted, yet he feared the present political affairs had reduced the size of the British contingent.[115]

International political tensions were not the only reasons for division among zoologists, of course. They often had their own skirmishes over national influence and prestige. Even the efforts to establish an international code of nomenclature, a primary centerpiece of the first international congresses of zoology, were often under threat of shipwreck on the shoals of national differences. In 1895, German zoologists openly ridiculed the nomenclatural rules established by previous congresses led by the citizens of other nations. Their insistence on sticking to their own system, established in 1893 at a meeting of the Deutsche Zoologische Gesellschaft, almost caused Blanchard (who would serve as president from 1895 to 1919) to abandon the International Commission of Zoological Nomenclature altogether in frustration, although he remained "to protect French interests" for twenty years.[116] The ease with which zoologists such as Blanchard could move from idealistic, internationalist statements to protecting national interests is a reminder that those attending such congresses took it for granted that members came as representatives of nation-states.

Tensions between internationalist and nationalist rhetoric could often be ignored, of course, since the primary motivations of many congresses lay elsewhere. Historians have shown how congresses held prior to World War I primarily reflected the necessities of scientific communities, rather than broader cultural or political agendas, even as they drew on the language of internationalism to justify their efforts. The "strange mixture" of entomologists proved no exception. Jordan's primary concern was diversity of practice, not nations (though certainly the two could be closely linked). Some had hoped that the international congresses of zoology would confer some degree of unity. The entomologist Charles Stiles enthusiastically claimed at the 1895 International Congress of Zoology in Leiden that such congresses solved the "old question of philosophy, i.e. [how] to find Unity in Diversity." Enwrapped in the heady optimism of the final congress banquet at the 1892 International Congress of Zoology in Moscow, zoologist Alphonse Milne-Edwards had announced that given the powerful bonds of brotherhood established by the congress, "We will always have the same flag to unite us, to fight against error and defend the truth."[117]

Whether zoologists could actually unite under one flag amid rapid specialization and competition proved another matter. The gap between Bateson's and Wallace's "two classes of workers" seemed only to be widening, for example. Some naturalists even worried that if zoologists united under an experimentalist flag, "old-fashioned" dissenters might be exiled in the process. Writing of the United Nations that dominated the newsreels four decades later, in 1946, Jordan insisted that "Political *unity* cannot be achieved as long as politics are essentially national, i.e. *separatist*."[118] But to unite entomologists at the beginning of the century, Jordan separated them en bloc from the international congresses of zoology. On one level, Jordan's actions reflected the demands of his own vision for the future of entomology, which required that entomologists combine to standardize methods and aims. But on another level, and like nationalists competing for control over scarce resources, Jordan clearly believed that separatism served the interests of entomological sovereignty better than unity with the zoologists.

Between 1906, when he had composed his initial letters regarding an international organization of entomologists, and 1910, when the congress actually convened, the conviction that entomologists must establish a united front surely increased in the face of developments at the most recent International Congress of Zoology, which met in Boston in 1907. There, Guillaume Severin had first announced entomologists' plans for a congress of entomology. He was flanked on all sides by younger workers urging that traditional fields and institutions be abandoned. Bateson's triumphant address to the section on heredity set the tone. "We are trying for fresh points of attack," Bateson announced, and he cited the creation of sections for experimental zoology and cytology as evidence of "new methods and new hopes."[119]

The geneticist Richard Goldschmidt would later recall that the experimental zoologists attending the 1901 congress in Berlin could, for their single section, all sit around one long table.[120] Now, writing from Boston to his wife, Bateson crowed, "From our point of view the meeting has been a stupendous success. Heredity, Cytology, and experimental Zoology have kept the whole Congress. Nothing else has had any hearing worth the name."[121] Bateson had stood before the 1907 congress and happily declared that all that could be found by the methods of systematics and comparative anatomy had been discovered. He insisted that experimentalists and the field of genetics (a word coined by Bateson two years earlier) would henceforth solve the mysteries of life, including the problem of species. In *Popular Science Monthly*, T. D. A. Cockerell described Bateson's address as "most typical, perhaps, of the whole trend of zoological thought.... It dealt with the subject of genetics; the genetics of things, cells, individuals, species."[122]

Fortunately for naturalists displeased by this experimentalist turn of events, when the zoology congress next met in Graz, Austria, in 1910, the dominance of genetics had disappeared, reflecting the well-known tendency for the host country's priorities to dictate the congress program. At the congress in Monaco in 1913, U.S. zoologists again took note of the small number of titles in experimental zoology and genetics.[123] And of course the triumphant claims of new fields often concealed real cooperation and interchange with more "traditional" workers. When pressed, Bateson himself admitted that established fields must not disappear entirely since "In the wider survey which we are attempting we shall need of all these things."[124] Bateson, at least, was trained as a naturalist and attended Entomological Society of London meetings when he could. But others, including Thomas Hunt Morgan, who served as program chairman for the experimental zoology section at the Boston congress, had fewer ties to the naturalist tradition, and certainly less sympathy.[125]

In any case, rhetoric could be more important than reality when funding and the ability to recruit young scientists were at stake. Fearful of the effect Bateson's strident tone could have on natural history, Poulton publicly protested against Bateson's "dogmatism" and "contemptuous depreciation of other lines of work." Such talk might influence the ill-informed to believe both the ideas of Darwin and his emphasis on continuous variation were played out. "No man is likely to continue the labours of investigation with enthusiasm and persistence," Poulton wrote, "when he is convinced, or even half convinced, by the overweening assurance of another that his subject is barren and useless."[126]

In the midst of his address to the 1907 congress, Bateson had insisted that the section on heredity reflected a sign of the maturity of zoology, compared to the old days when men specialized according to the group of animals to which fancy or opportunity had specially attracted them. The next phase loomed, he announced—one in which divisions would be based on biological problems, rather than according to groups of the animal kingdom. Others agreed that amid specialization, zoologists could be united around central problems such as evolution. But naturalists hearing calls for unity along the lines of specific biological problems simultaneously heard calls to deemphasize, if not abandon, hallmarks of the naturalist tradition, including museums and systematics. The manifestos for new divisions and methods hint at why, three years later, despite Bateson's happy declaration in 1907 that division by taxa no longer met the requirements of modern biology, the entomologists formed their own congress. To an entomologist like Jordan, who had very specific ideas about how reforms must proceed in order to produce good, scientific work, it was by no means clear that congresses controlled by men such as Bateson, Thomas Hunt Morgan, and other crusading experimentalists would serve the interests of entomology.

Establishing entomologists' independence would avoid submission to the ambitions of those who held systematics and natural history museums in low regard. In other words, scientists *did* know countries, though the boundaries of these countries were drawn in the nebulous realm of identity formation and discipline building, rather than lines on maps. As for why the nation must consist of entomologists rather than taxonomists, perhaps the best explanation is simply that loyalty to certain taxa had long since provided the means by which natural history communities had ordered themselves, in the interest of specimen exchange. After all, in suggesting the idea of the congress, Jordan had written first to those upon whom he had depended to build Rothschild's collection of insects.

A "NATION OF ENTOMOLOGISTS"

In his letters to a select group of colleagues in 1906, Jordan based his arguments for an international congress of entomology explicitly within his concerns for the status of entomology, systematics, and natural history museums given changes taking place in zoology. Intent on sovereignty, Jordan, Rothschild, and 65 entomological colleagues from Britain crossed the English Channel in August 1910 to join 225 entomologists from other nations who attended the first International Congress of Entomology in Brussels. When Belgian entomologist Auguste Lameere of the Université Libre de Bruxelles and the Société royale belge d'entomologie opened the congress, he issued a manifesto that perfectly echoed the sentiments Jordan had expressed in his letters regarding the idea of a congress.

Lameere took the inadequate status of entomology among zoologists and within society as his theme. After listing both the challenges and successes faced by entomologists, he raised their pride by declaring entomologists outnumbered all other zoologists, boosted their self-importance by pointing out that the number of insects surpassed all other animals, and then roused their indignation by noting that the zoological congresses permitted entomology just a single section.

Lameere described how entomology, "the Cinderella of the biological sciences," was absent from academic zoological courses since the latter emphasized animals that could be sliced beneath the microtome and harvested in great numbers from the ocean. The modern zoologist, he explained, was above all a man of the laboratory, leaving it only occasionally for the sea, and then only to look at plankton. The entomologist, by contrast, was a "naturalist of the open air," who after looking at his insects with a microscope or magnifying glass seeks also to understand the animal as it *lives* in the woods and field. Yet general zoologists ignored the huge sum of knowledge accumulated by entomologists, with the result that hard-won knowledge on insects was absent from theoretical debates or "natural philosophy."[127]

Following this lament, Lameere declared triumphantly that the congress established entomology as an "independent and victorious science." He did so even while insisting that congress organizers had not been motivated by a desire to either compete with or decrease the importance of the zoological congress. Rather, the congress founders hoped to increase zoologists' awareness of the importance of entomology and in doing so help bring the knowledge of entomologists to bear on general questions of zoology. In stark contrast to Bateson's insistence that the time in which zoologists organized according to taxa had passed, Lameere insisted that just as botanists and zoologists had once divided into their own societies and congresses, entomologists, too, must follow the law of the division of labor. But he insisted without irony that establishing entomology's independence would bring the two categories of workers closer together. Two years later, at the second International Congress of Entomology, in Oxford, L. O. Howard made this nascent nation-building language explicit when he insisted that at the congress "they were not English, they were not Germans, French, Belgians, Austrians—they were the nation of Entomologists, a world-nation."[128]

Meanwhile, although the nation-building metaphor need not be taken too literally, the fact that entomologists drew on the powerful language of nationalism as they tried to establish a place for themselves within the world of science is telling. Entomologists could have drawn on other metaphors. Edwards had compared the International Congress of Zoology to a "living organism," an image that might, then, have led to odes on how one part of the organism (say, the entomologists) could not live without the other, and that all must work together in the interest of the whole animal. The metaphor of nations, by contrast, emphasized group loyalty against a common enemy. Given some of the most strident claims of experimental zoologists, the temptation to draw on nationhood as a model must have been strong. In his writings against imperialism, Wallace defined independence—that is "as regards communities or nations"—as "self-government as opposed to government by an outside power which has annexed, purchased, or conquered some smaller and weaker people."[129] It called on the language of self-determination and independence, a useful image for a community that felt threatened by other kinds of zoology, no matter what truces might be agreed upon in future.

When Jordan stood to give his own brief welcome, he backed Lameere's manifesto by explaining his motivations for organizing the congress. In attending the zoologists' congress in Cambridge in 1898 and Berlin in 1901, he had realized that "often important entomological work, large collections, and the names of efficient specialists, remained completely unknown to non-entomologists." Furthermore, he had been astonished at how isolated workers at various collections—both public and private—kept from each other. When entomologists did interact, "in all too many cases a certain animosity prevailed." As a result, and despite the enormous number of entomologists and *Entomophilen*, the tremendous amount of work that had been carried out had resulted in little scientific result.[130] As he explained to N. D. Riley decades later, this looming crisis in entomology had made it impossible to "deal with the huge mass of insects as was necessary for Entomology as a *branch of science.*"[131]

Improvement in the scientific results could not be accomplished, according to Jordan, without first improving cooperation amongst entomologists. Yet, as Lameere announced, entomologists had a wide range of methods, backgrounds, and goals. The number of sections at the first congress clearly demonstrated the diverse array of entomologists' concerns. Systematics, nomenclature, anatomy, physiology, ontogeny, phylogeny, oecology, mimetism, bionomics, zoogeography, sociology, ethology, museology, and medical and economic entomology were all included. Inevitably, many different visions of the future of entomology existed among members, from the future role of private collections to the stage at which one should begin generalizing from available facts, and these visions came to the fore as colleagues pressed Jordan to organize the congress with particular goals in mind. The Dutch entomologist Erich Wasmann urged that the sections of the congress be on subjects of wide interest, such as morphological adaptations, evolution, and phylogeny, topics that he noticed British natural history journals generally ignored. H. J. Kolbe submitted a plea that any publications resulting from the congress omit the activities of amateur collectors in order to ensure that the organization be "truly scientific."[132] Both Wasmann and Kolbe hoped Jordan would, in other words, design the sessions and any resulting publications with the scientific status of the discipline in mind. His colleagues' advice provided early indications that the congresses would play host to tensions within entomology, as various entomological leaders tried to carry their own agendas forward, and order the discipline accordingly.

For his own part, in composing his contribution to the scientific proceedings of the congress, a paper entitled "The Systematics of Some Lepidoptera which Resemble Each Other, and Their Bearing on General Questions of Evolution," Jordan continued his campaign to bring systematics within the purview of "Entomology as a branch of science." Well aware his audience of entomologists might need the reminder as much as any experimental biologist, he composed yet another reply to those who belittled the study of insect systematics. This time, he used the systematics of mimetic butterflies to analyze the various theories regarding speciation proposed by biologists, from orthogenesis to natural selection. In other words, he compared the *concreta* represented by his study of the butterflies with the various mechanisms being proposed for speciation. Once again, Jordan pointed out midway that he was "not defending or attacking any one theory, but only endeavoring to demonstrate the great importance of correct systematics for all theoretical considerations."

Even as he ardently defended work on museum specimens from critics, Jordan acknowledged the current situation of entomological systematists with humor. He began his address by admitting, "The subject of this address being given as including the systematics of certain Lepidoptera, I am afraid that some of the members of this Congress may have been frightened away." But he begged to differ with those "biologists of fame who, in their misguided wisdom, scoff at systematics as more or less fruitless." He took the opportunity, he announced, that the first congress of entomology offered "of stating emphatically that sound systematics are the only safe basis upon which can be built sound theories as to the evolution of the diversified world of live beings."[133]

Both Lameere and Jordan clearly had distinct priorities as they tried to reform their nation of entomologists in the interests of improving the scientific results of entomology and thus raise the status of placing insects between two bits of glass. After extolling the virtues of the legions of naturalists who cultivate entomology, including thousands of amateurs, Lameere turned to how, precisely, unity could be achieved. "What we are most in need of," he wrote, "is method." Isolated descriptions without comparative work should be prohibited. New species must be described within revisions of surrounding forms. The practice of publishing descriptions of new forms in small local entomological newspapers or, worse, the catalogs of merchants, must be abandoned. Most importantly, entomologists must remember that they no longer worked in the time of Linnaeus. Distributing specimens in drawers and inventorying new species no longer counted as science. The theory of evolution, Lameere insisted, had transformed classification into a search for the bonds of descent linking living beings. Now, answering the needs of science called for explicitly building a foundation for examining the mechanism of evolution. This meant doing careful work on phylogeny, studying geographical distribution in close concert with phylogenetic classification, and, finally, examining the differences that exist between species and their variations.[134]

The reins guiding this first congress were firmly, albeit "behind the blinds," in the hands of Jordan and his network. Thus, the problems that had arisen during naturalists' efforts to name and order large collections in the interest of science were paramount during discussions. While entomologists such as Poulton lectured on mimicry, the challenges of organizing and processing huge amounts of information dominated the first meeting. Entomologists discussed how to best index their literature, compile bibliographies, and house type specimens. The paper submitted by E. E. Green entitled "A Plea for the Centralisation of Diagnostic Descriptions" was a typical contribution. Green noted that the 1910 volume of the *Zoological Index* had contained fifteen thousand new names of insects. The descriptions, Green wrote, with which the systematic entomologist must make himself familiar, were dispersed throughout hundreds of publications, rendering his work so arduous as to be hopeless. He lamented:

> How much weary searching through out-of-the-way journals, how many tiresome journeys to public libraries, what annoying delays in the endeavour to obtain copies of obscure papers, would be avoided if every country had some recognized medium for the publication of all new diagnoses![135]

Since the chaotic state of the practice of naming insects had served as a primary incentive for the creation of better organization, the sessions devoted to nomenclature were particularly focused. No doubt the contentious nature of the debates at earlier congresses of zoology inspired Jordan and his fellow organizers to keep a tight rein on the deliberations. They asked members to focus on twenty-four detailed points circulated prior to the congress by Jordan and the Berlin entomologist Walther Horn. Each point arose directly from the kinds of decisions Rothschild and Jordan faced as they considered the validity of hundreds of names for the revisions. Congress attendees debated, for example, whether descriptions without figures should be considered valid, and whether to correct misprinted names. Not surprisingly, the liveliest debate occurred over the issue of whether entomologists should follow a strict rule of priority. (Even the Rothschild brothers came down on different sides of that issue.) In other words, Jordan and his friends hoped that entomologists at the congress would develop a system for countering some of the nomenclatural windmills to which they were forced to devote so much time—time that could otherwise go toward actually studying insects, whether in museum drawers or in the field.

Other discussions clearly reflected the influence of Tring's particular methods. The need to establish a consistent formula for naming geographical, seasonal, and individual varieties figured prominently, as did the need for detailed locality labels. As the hours passed, W. J. Holland suggested settling any remaining points raised in the Horn/Jordan proposals with a general resolution that the congress recommend the adoption by all entomologists of the rules of nomenclature adopted by the various international congresses of zoology. Horn proposed a telling (and successful) amendment that added the words "in so far as they are in accordance with the requirements of entomology."[136] The congress assigned Jordan the task of conveying entomologists' resolutions to the zoologists' meeting a few weeks later in Graz. Backed by this newly organized army, Jordan then prepared to represent his "nation of Entomologists" before the world's zoologists.

As a naturalist endeavoring to unravel creatures' "synonymy and life-history, and trying to bring the immense multitude of diverse forms into natural order,"[137] Jordan had found that naturalists themselves needed some ordering so that they could both do good work and garner some of the resources available to science. As Jordan had hoped, entomologists' discussions during the meeting focused on how to ensure that entomologists could efficiently and productively contribute their part to the grand project of describing, ordering, and explaining the natural world. But at least one entomologist expressed concern at entomologists' ardent declarations of unity and independence. Adalbert Seitz urged, for example, that any nomenclature committee established by the congress must work with the zoologists as well, since existing separation between the two disciplines had already created great difficulties.[138]

But standardizing the process of naming, which may indeed have depended on unity with general zoologists, formed only one part of entomologists' concerns. On a more practical level, the final points on the 1910 Horn/Jordan list had diverged from strict issues of nomenclatural procedure to include the infrastructure of specimen exchange itself, from how to determine the best way of designating and storing types to the need for a simple manual on scientific collecting for use in the field. These specific reforms reflected some of the initial inspirations behind Jordan's organizational efforts; namely, the need to ensure that the specimens arriving at Tring would help make systematics more scientific. Ultimately, the priorities of the first congress reflected Jordan's hopes that entomologists could eventually deliver on his insistence that work on specimens in natural history museums provided crucial data for understanding the origin of species. These reforms did not require close cooperation with zoologists.

All of Jordan's correspondence organizing the first congress, entomologists themselves, and the tiny specimens that formed the basis of their research, traveled via stable routes over land and sea. But other windmills, all too real, spun on the horizon. No amount of international organization among entomologists, whether for the purpose of standardizing names, making species, or theorizing about the origin of diversity could prevent Jordan's network and congresses from succumbing to the destructive forces of world war. Ultimately, naturalists' ability to adapt—both to the influence of Bateson's experimentalist "class of workers" and Wallace's "workers of the world"—would hinge on their response to the profound shifts set in motion by that war and the new world order it created.

Karl Jordan chose to sit before a microscope when Douglas Glass photographed him in 1957 for the Portrait Gallery of the *Sunday Times*. © J. C. C. Glass

Walter Rothschild at his desk in what he called "My Museum." © Natural History Museum, London

South door of The Tring Museum as it looks today. Jordan lived just a few blocks away on Park Street. © Natural History Museum, London

Entomology library and office. PH/34/7, Natural History Museum Archives. © Natural History Museum, London

A view of the collections in one of the entomological halls built in the wake of the scandal of 1908. PH/34/2, Natural History Museum Archives. © Natural History Museum, London

A drawer housing specimens of the garden tiger moth (*Arctia caja*) in Rothschild's collection, illustrating the variation displayed by a long series. © Natural History Museum, London

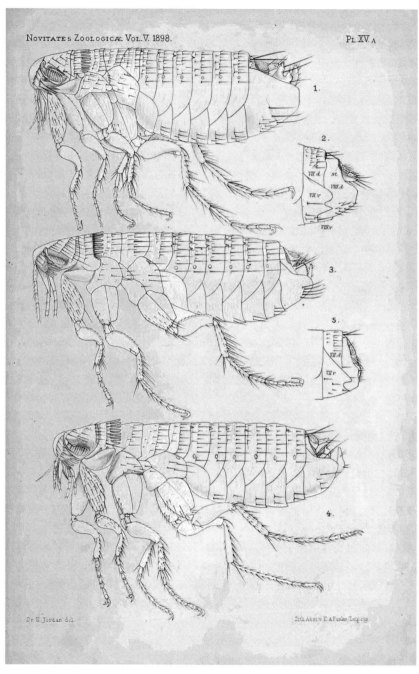

Drawings of Typhlopsylla fleas. Plate 15A from the 1898 volume of *Novitates Zoologicae*.
© Natural History Museum, London

Illustration from the 1896 volume of the Tring journal *Novitates Zoologicae* showing a Nymphalidae butterfly (*Anaeomorpha splendida*) surrounded by various species of Sphingidae. © Natural History Museum, London

Walter Rothschild (*left*) and Karl Jordan at the British Association for the Advancement of Science meeting in 1932 in York, northern England, where Rothschild, as president of Section D: Zoology, delivered his address "The Pioneer Work of the Systematist."
© Natural History Museum, London

The cover of the program of the First International Congress of Entomology in Brussels, 1910. The illustration of the woman with the flag was used by the Exposition de Bruxelles in general, but in retrospect the flag's winged knight fighting a black dragon seems apt, given the references to Don Quixote in Jordan's writings.

Detail of a photograph taken in 1932 when the International Congresses of Entomology met in Paris. (*Front row, from left*) Walther Horn, Karl Jordan, E. L. Bouvier, Ernst Hartert, L. O. Howard, Walter Rothschild, P. Marchal, and G. Horváth. (*Second row*) Hilda Jordan stands behind her father. Photograph from L. Berland and R. Jeannel, eds., *ICE*, Paris, 18–24 July 1932 (Paris: Secrétariat du Congrèss, 1933), 56–59

FIVE

A Descent into Disorder

By the time the second International Congress of Entomology convened in Oxford in 1912, Jordan had lived and worked in England for two decades. He had carved out a respected role as an "eminent scientific entomologist"[1] while helping to open up an insular community to a more international scene through both the example of his own work and the congresses. While diligently processing specimens, keeping up with museum correspondence, and carrying out the careful requirements of "sound systematics," Jordan helped an old tradition compete for the limited resources available for science. He thus joined reforming colleagues intent on bringing entomology and systematics safely into the twentieth century.

In mobilizing the best of the naturalist tradition, its journals, workers, societies, museums, collectors, and patrons, Jordan built on the legacies of nineteenth-century society, a time that one entomologist described as having been "emphatically the 'systematist's century.'"[2] Inevitably, a huge part of how systematists worked was part and parcel to that century. Cosmopolitan entomologists such as Jordan and Rothschild depended on the communication and trade networks of both international capitalism and Empire. In other words, Jordan's vision of how to order both entomologists and insects relied on the infrastructure of an age. From the liberal-parliamentary style of sorting out differences in congress meetings to the stable transportation network along which communication, specimens, and entomologists could move efficiently and securely, both endeavors depended on the permanence of that infrastructure.

This dependence was profoundly illustrated when, within four years of the triumphant first congress, total war descended upon Europe. The war destroyed the access to colleagues, collections, and specimens upon which Jordan's efforts to order both insects and entomologists relied. Meanwhile, as scientists campaigned for state support within increasingly enfranchised populaces, entomologists, like other scientists, found a potential ally in appeals to applied sciences such as agriculture and medicine. Yet this and other changes within biology as a whole called into question the place and methods of traditional disciplines, even while they offered new opportunities. The international congresses were the primary means Jordan and his colleagues had established to deal with these changes, and their disruption by the war and its aftermath did not help entomologists' efforts to navigate their specialty into a new age.

TELLING "WHICH WAY THE WIND BLOWS"

In the interests of ending the "crisis in entomology," Auguste Lameere had urged at the first international congress that entomologists "forget that we belong to different nationalities." In bringing the congress to a close, surrounded by the "many leaders of the Science, wearing the numerous brilliant orders conferred by Sovereigns and other heads of States,"[3] Lameere noted that although traditionally such meetings toasted the king of the country and although the members were representatives of various states, the entomologists had come to Brussels as faithful subjects of a queen, and he raised his glass to her—"Entomology!"[4]

In the months following the first congress, Jordan's initial calls for such unity on the grounds it could help secure more government support for entomology seemed particularly justified. As he had traveled to Oxford throughout 1911 in order to help with the preparations for the second congress, for example, the council of the Entomological Society of London was up in arms over the British government's decision to give land that had been set aside for the Natural History Museum to the Science Museum instead.[5] In May, the Entomological Society of London passed a resolution in protest.[6] In the end, the plan was scrapped, but the London entomologists still smarted from the fight. In his presidential address for the year, F. D. Morice noted that "even a straw may show which way the wind blows." He thought the government's consent to the "outrageous proposal to deprive the Natural History Museum of land . . . a black cloud . . . on the horizon" and a matter of "grave concern."[7]

In view of this and other clouds on the horizon, naturalists increasingly debated what should happen within the square footage devoted to the study of insects. As the first congress had drawn near, Ray Lankester, the academically trained director of the British Museum of Natural History, publicly decried the wealthy Lord Walsingham's insistence that the museum trustees provide a curator, an attendant, and special rooms for the collection he had given to the nation. Lankester argued that the cost of such provisions would exceed the market value of the collection. The entomological watchdog J. W. Tutt countered Lankester's opposition by pointing out that Lord Walsingham's conditions would ensure his specimens would not disappear into the oblivion of Kensington's basements. "The understaffing of this department is a public scandal," Tutt argued, and "work as hard as the present staff may, it is quite impossible to keep pace with everyday requirements."

Cognizant that such skirmishes indicated bigger fights, Tutt warned that the lack of support given to the national collection of insects would surely influence the public's view of systematics:

> With so many enemies attempting so sedulously to filch, for other purposes, the present government contribution to systematics, it behoves us to see that we are getting the maximum scientific value out of our marvellous entomological heritage.... Our enemies must have no ground on which they can make any stand against "systematics," which they pretend to despise, but without which they cannot themselves stir hand or foot.[8]

But men such as Lankester, entrenched in the defense of a gigantic institution subject to parliamentary oversight, had to consider how the provision of extensive resources to a single lord's collection of microlepidoptera might appear.

Lankester had quite a different response to those attempting to "filch" systematics; namely, an emphasis on the study of insects' utility for agricultural and medical entomology. But many British systematists had an ambivalent, if not hostile, attitude toward applied entomology. Traditionally, the leaders of British entomology, from H. J. Elwes to E. B. Poulton, had made little pretense to the "utility" of studying insects beyond vague references to increasing knowledge for its own sake. Many within Jordan's network made a clear distinction between "scientific" entomology and applied entomology. In blaming the government's lack of support for entomology on politicians' emphasis on applied sciences, Tutt wrote in 1901 that he hoped that one day

> the conditions of modern life, which are all in favour just now of the sciences which are purely utilitarian, shall not act against the true scientific enquirer, but put him in the same satisfactory position for real scientific work, as that in which they at present place his more fortunate brethren, the students of chemistry and physics.[9]

Tutt had a starry-eyed view of the position of chemists and physicists, who had their own troubles with tensions between so-called applied and pure science, but his comments reflected many entomologists' view that a focus on utility produced a different kind of knowledge than "true science."

In organizing the first congress, Jordan had taken a more ecumenical stance toward the various classes of entomologists that had developed over the course of the previous century. Between his first round of letters sounding out his closest friends regarding the idea of an international congress and the announcement sent to entomological societies and institutes, he had explicitly added applied entomology to the reasons given for an entomological congress. The stance was no doubt influenced by both advice from his more pragmatic friends and the fact that he had just witnessed for himself the enormous potential for cooperation between systematists and applied entomologists through his work with Charles Rothschild, studying fleas.

Charles had recently provided a stunning demonstration of the potential use of sticking insects between two bits of glass through his systematic work on various species of flea, especially the culprit in plague transmission, *Xenopsylla cheopis*. Following a plague outbreak at the turn of the century, a British inquiry, the Commission for the Investigation of Plague, which had been sent to India from 1904 to 1905, had enlisted Charles to sort out the distinctions between the various kinds of fleas studied by those seeking the key to plague transmission.[10] Together, Charles Rothschild and Jordan made sure plague workers got their identifications right. In doing so they helped prove that one of the major criticisms of the flea-transmission theory, that plague did not occur throughout regions in which fleas were found, was based on the assumption that all fleas in India were of one species. Rothschild and Jordan showed that in fact those places without plague were home to a different species of flea. Their work thus demonstrated an outcome of "sticking insects beneath two bits of glass" that Jordan had not anticipated when he supported Charles's interest in fleas, and one completely absent from his 1896 and 1905 manifestos for the importance of systematics to biology. In a 1908 revision of the group to which *Xenopsylla cheopis* belongs, Jordan and Charles explicitly pointed out that the study of fleas had "ceased to be the mere hobby of a small group of entomological specialists." The knowledge of fleas was now of the utmost concern to those working on the plague problem, and it had become imperative, to avoid confusion, to demonstrate precisely the characters by which the various allied species could be distinguished.[11] Accurate "species-making" had suddenly become a matter of life and death.

Jordan and Charles Rothschild had thus inadvertently strengthened one of the winds entomologists would have to navigate over the following decades, for it was largely based on the discovery of the insect transmission of disease, especially the role of mosquitoes in transmitting malaria, that governmental and public support for entomology grew rapidly during the first decades of the new century. The British government appointed "An African Entomological Research Committee," for example, in 1908 (with Charles Rothschild as a member) and the University of Cambridge hired a demonstrator in medical entomology in 1911. In the face of such developments, and although he had not actually mentioned applied entomology in his initial letters regarding his idea of a congress, Jordan announced in his 1910 address to the first congress that part of his inspiration had come from the fact "too few scientific entomologists (in Europe) are occupied with those insects which have direct meaning for economics and medicine."[12] And he made sure applied workers would be welcome to the congress, noting in the first circular that "anybody who takes an interest in any branch of Entomology, scientific or applied, may become a member."[13]

Given Jordan's and other entomologists' experience with professors of zoology who ridiculed the systematics of insects, the temptation to emphasize practical applications loomed large in entomologists' campaign for both prestige and financial support. After an address on the insect pests of the West Indies at the 1910 congress, at least one listener, F. Merrifield, noted with relief that no longer would young men attracted to the study of entomology be prevented from continuing because they thought it would lead to nothing.[14] Walter Rothschild even proposed at the first congress that the meetings be held every two years on account of the great number of questions, especially those connected with medical and economic entomology that were accumulating rapidly in number and importance.[15]

But Jordan, like other systematic entomologists who highlighted the importance of their work in identifying disease vectors, tended to emphasize how knowledge pursued for its own sake could have surprising results, rather than argue that entomology should be supported because it could help humanity. In announcing the foundation of the African Entomological Research Committee in the pages of Nature in 1909, F. A. Dixey insisted that "natural knowledge pursued for its own sake, without any direct view to future utility, will often lead to results of the most unexpected kind, and of the very highest practical importance." Dixey believed the demands "that both Governments, and such private individuals as have the means," support science well justified on these grounds.[16] Continental entomological leaders, too, shied away from emphasizing purely utilitarian justifications of their field. In his address to the 1910 assembly in Brussels, for example, Lameere acknowledged the recent impetus that the discovery of insects as vectors of disease had given to entomologists in the colonies of Britain, Germany, France, and Belgium. But he urged that naturalists never lose sight of the fact that there is no nobler goal than to contribute a stone to the edifice of general knowledge.[17]

At the second International Congress of Entomology, convened in 1912 in Oxford, the presence of different visions of entomology's future within what L. O. Howard had described as a "nation of Entomologists" were on full display. E. B. Poulton took the opportunity of his speech at the banquet to insist that "the primary, in fact the only real motive" of studying natural history, entomology, or any other science, "was that of finding out." Entomologists "worked because they were interested, and any further object, however laudable in itself, only tended to bias and mar the inquiry." He quoted Charles Darwin's insistence that the instinct for truth, knowledge, or discovery, provided reason enough for scientific researches. Poulton was sure that the quality of research declined as soon as physicists or chemists delved into practical work, and even went so far as to ponder whether "the scientific spirit was incompatible with the qualities required in an expert witness." Still, he consoled any applied entomologists present by noting that he thought entomology stood out as the one science in which practical application did not injure investigation, and he cited the work of L. O. Howard, W. M. Wheeler, R. C. L. Perkins, and Guy A. K. Marshall as excellent examples of science being carried out amid applied concerns. In contrast, Malcolm Burr pointed out, with "due deference to Prof. Poulton's remarks," that the importance of the Congress would be to show the public the practical importance of entomology. Only the latter, Burr insisted, could garner the requisite support for the discipline from those with the means to do so.[18]

To entomologists safely shielded by the nineteenth-century system of patronage, Malcolm Burr's statement may have seemed premature. Certainly Jordan had not based his defense of the systematics of Lepidoptera, given at the first congress, on the practical importance of entomology. And now, at the 1912 meeting, he and Poulton put forth a stunning vindication of Jordan's insistence that systematics was crucial to establishing a true portrait of the blood relationship between species used to study evolution. In his contribution to the 1910 congress, Jordan had argued, based on detailed morphological work and a study of variation, that *Pseudacraea* butterflies from West Africa, formerly described as seven distinct species, were in fact one species with various polymorphic, mimetic forms; with a long series of specimens one found intermediates that broke down the presumed boundaries between the forms.[19] Then, to "great astonishment and even some incredulity," as Poulton recalled it, Jordan argued that, based on structural features, four mimetic forms of *Pseudacraea* from Uganda, formerly described as different species, were also in fact one species.[20] He had, in other words, proposed to *unmake* a whole list of species.

Jordan had thus placed specific, predictive statements before the congress that could be tested by breeding experiments. At the 1912 meeting, Poulton informed the gathered entomologists that he awaited a cable announcing the result of such experiments by his correspondent, George Herbert Carpenter, in Uganda. These tests would either confirm or deny Jordan's controversial prediction two years earlier, for example, that the supposed species *Pseudacraea hobleyi*, *terra*, and *obscura* would in fact prove to be polymorphic forms of a single species. The cable arrived nine days after the congress, proving Jordan right.[21]

In lauding the vindication of Jordan's work on mimetic butterflies, Poulton focused on their relevance as further support for natural selection. Jordan, by contrast, focused on the implications for the role of systematics in testing alternative theories in general. For conscientious "species-makers," the episode certainly served as a beautiful example of how the scientific systematist *unmade* species when necessary. The predictions and the proofs highlighted the aim of the systematist; namely, as Jordan had stated in 1910, "the establishment, on reasoned evidence, of the degree of relationship between the forms with which he is concerned."[22] (Burr might have pointed out that the study also served as evidence that the expansion of government support for applied entomology could indeed serve the interests of systematics and so-called pure entomology since, at the time, Carpenter was in Uganda as a member of Royal Society's Sleeping Sickness Commission.)

In any case, Jordan's defense of systematics seemed completely lost to those immersed in building new fields. William Bateson insisted in a textbook on genetics published a year later, that "it is impossible for the systematist with the means at his disposal to form a judgment of value [of a species' status] in any given case. Their business is purely that of the cataloguer, and beyond that they can not go. They will serve science best by giving names freely."[23]

To set up his sweeping directive, Bateson prefaced the book by a lengthy description of how systematists worked. Most, he argued, still relied solely on observational methods laid down when naturalists believed species were fixed and varieties impermanent. "Sufficiently careful observation" provided the key to accurate separation. This assumption, he insisted, remained the guiding principle of systematists despite "the growth of evolutionary ideas." "There are 'good species' and 'bad species' and the systematists of Europe and America spend most of their time in making and debating them." As evidence that systematists misguidedly focused on species rather than varieties as the primary unit of importance, Bateson claimed that "the whole business of collection and distribution of specimens is arranged with regard to it." Collections themselves were almost always arranged in a way that masked variation, as when forms intermediate between species were placed in a separate drawer and given a third specific name.[24] In private correspondence Bateson was even more strident, exclaiming in 1919, "I often wonder what systematists will do with themselves when they discover Mendelism!"[25] (One wonders whether Bateson was the scientist who Jordan recalled exclaiming at an official dinner that Mendelism solved the problems of the origin of species and thus eclipsed the study of species produced by nature, concluding with a flourish, "We can now make species ourselves!"[26])

Had Jordan read Bateson's jibe about systematists discovering Mendelism, he would no doubt have replied, as would Poulton, that systematists knew quite enough about Mendelism. It was often discussed, for example, at the meetings of the Entomological Society of London. Unsatisfied with just discussions, at the 1912 congress R. C. Punnett had urged lepidopterists to "familiarise themselves with the Mendelian principles of heredity, and direct their experiments in accordance with them."[27] But thus far naturalists saw no reason to think that the enthusiastic statements of the Mendelians solved their problems. As Jordan repeatedly pointed out, naturalists' hard-won data often contradicted geneticists' enthusiastic claims regarding the origin of species *per saltum*.

Bateson, it turns out, was not entirely to blame for his misapprehension of what systematists were up to. As the continued appearance in *Novitates Zoologicae* of long exhortations for naturalists to change their ways demonstrated, not everyone adopted the methods that Hartert, Jordan, and Rothschild insisted were necessary to make systematics scientific. Not all of their friends, for example, abandoned the term *variety* in favor of a more careful differentiation between the different kinds of varieties. As Walter Rothschild wrote in 1911 (peppering his critique, as he often did, with boldface type), "unfortunately *zoologists in general* have continued to misuse the term **'Varietas'** so much that it is almost impossible to clear up the confusion."[28]

Furthermore, some of Tring's own correspondents still thought single specimens of each species constituted a "scientific" collection. Almost three decades after the editors of *Novitates* began their campaign for long series of specimens, Benjamin Preston-Clark, of Boston, who kept only a few specimens of each species, still believed his plan of completing "a perfect and complete collection of every known Sphingidae" could result in "a document of great scientific value, from the standpoint of distribution as well as determination."[29] Knowing that one of his favorite collectors, Miles Moss, had promised Preston-Clark some specimens, Jordan urged Moss to let Tring have them first so that the species could be "worked out."[30] "He spends a lot of money on his hobby, apparently," Jordan wrote a few months later, in a weighty choice of words.[31] The presence of men such as Preston-Clark made it difficult to convince nonsystematists that in fact some workers were paying careful attention to the implications of evolution. Their activity as customers of the same natural history agents and professional collectors used by Tring (Preston-Clark once boasted that he had five hundred collectors at work)[32] also made it difficult to convince the collecting network as a whole that distinct "scientific" methods of amassing long series of specimens must be followed.

Jordan's friend at the Berlin Zoological Museum, Hermann Julius Kolbe, had an answer to dealing with wealthy "hobbyists" such as Preston-Clark. He recommended excluding such "amateurs" entirely from congress publications. But Jordan had long since been convinced that, given the amount of work to be done and material needed, entomology could not afford to discourage its most ardent and resource-rich workers. In the interest of taking advantage of as much of the existing material in collections as possible, Jordan once began a series of papers on classification by first examining the antenna, in the hopes that he could thus interest the average lepidopterist "who naturally abhors all methods of research by which his specimens get damaged."[33] Decades later, he noted how, when ascertaining the limits of variability of any specific population, "here the enthusiastic amateur specialists come to the rescue."[34] Given the amount of *concreta* needed to do good work, the tradition could not afford to get rid of these ardent workers. Rather, Jordan insisted, like Lameere, that "more method" must be instilled in the work of both private and official entomologists.

Jordan knew that the first methodological "windmill" taxonomists had to conquer was nomenclature. All entomologists conceded that their nomenclature was, as Walsingham put it, under imminent threat of "being drowned in a butt of ridicule."[35] Standing before the 1910 congress Jordan had announced that the "philosophic aspect of systematics" he urged as the avenue to ensuring its place in science was indeed obscured by the "unduly great importance" attributed to the "mere giving of names." "Nomenclature," he maintained, "is the servant of science, but has in many houses the position of master."³⁶ In order to place the servant in its rightful place so that the real work could get done, Jordan and others campaigned for the international congresses of entomology to take nomenclature in hand and clear up the chaos produced thus far.

With these goals in mind, Jordan began writing to his international friends and colleagues immediately following the 1912 congress to encourage support for the formation of national committees for entomological nomenclature. He sent summaries of the resolutions taken by the 1912 congress's meetings of a newly formed committee, the Committee on Nomenclature, to national entomological societies for comment and recruited secretaries for each national committee. The work soon paid dividends. In view of entomologists' increasing organization, the secretary of the International Commission of Zoological Nomenclature (ICZN), Charles Wardell Stiles, urged in 1913 that additional entomologists be placed on the ICZN, precisely the sort of response Jordan had hoped for.³⁷ In other words, having their own congress and commissions had made entomologists a force to be reckoned with, at least in the realm of zoological nomenclature, since united they could threaten to rebel en bloc.

In 1912, Jordan became a permanent member of the International Commission of Zoological Nomenclature. He would remain a member for four decades, almost two of those as president. Jordan justified his commitment to the notoriously time-consuming and bureaucratic nature of the commission's work on various grounds. He once wrote, for example: "In the present spiritual turmoil in which humanity finds itself, one point of general agreement, like the basic principle of nomenclature, renders general agreement in other matters a possibility and gives humanity some hope."³⁸ This was an idealistic departure from the oft-cited, highly practical goal of clarifying the chaos of synonymy and establishing consistent naming practices.

In any case, the sources of disunity—the rising power of applied biologists and geneticists, the widely divergent methods of both collection and study, the different views regarding what constituted scientific systematics, the debates over the mechanism of evolution—all monopolized the limited time and resources available for naming, ordering, and explaining the natural world. Jordan, for example, devoted time and energy to fighting zoologists' nomenclatural windmills so that future generations could focus on the real work at hand, despite the fact this meant, as a colleague put it, "his microscope and camera lucida [were] buried beneath a never-ending cascade of correspondence, books, and incoming specimens."³⁹ Ordering life, in other words, required a tremendous amount of work ordering the names given to life by one's predecessors.

THE BALANCE OF EUROPE IS UPSET

A year after the Oxford congress, with resolutions on entomological nomenclature in hand, Jordan traveled in the company of Walter Rothschild and Hartert to the International Congress of Zoology in Monaco. At this gathering, zoologists witnessed how the parliamentary models of government followed by their congresses and commissions came with no guarantee that peaceful consensus would result from the deliberations. For these models could also be accompanied by the language of secession, revolution, and civil war.

Not surprisingly, it was the rules for naming that provided the excuse for adopting the more divisive tendencies of the nation-building metaphor. Zoologists at the International Congress of Zoology that met in Berlin in 1901 had finally adopted an International Code of Zoological Nomenclature after twelve years of debate. But when the text of the new code finally appeared in print in 1906, many who had not been at the Berlin meeting criticized the document's strict "law of priority" as stretching beyond the bounds of common sense. Smithsonian zoologist Leonhard Stejneger's account of the frustrating events in the Commission on Nomenclature's meetings described one zoologist's "declaration of war" and concluded "A row is imminent."[40] Discussions became so heated at the congress that ICZN secretary Charles Stiles lamented the disquieting rise in intemperate language during discussions of nomenclature, the use of which, he insisted, constituted a "departure from those methods adopted by professional gentlemen."[41]

As Stiles cast about for "professional gentlemen" to enlist as allies on the commission, Jordan arrived ready to present entomologists' resolutions on nomenclature. At the time, the commission needed replacements for two absent commission members and Stiles called on both Jordan and Walter Rothschild. A wealthy collection owner and a professional curator, the duo brought an alliance of two types of zoologist to the commission table. As Francis Hemming recalled the appointments:

> Rothschild, with his massive personality, his great wealth and his position in the world of affairs, Jordan with his distinguished record of performance in entomology, with his acute and logical mind and his conciliatory and friendly spirit, were highly contrasted figures but together constituted a formidable combination when they took their places as members of the largely discredited Commission.[42]

The Tring naturalists hardly arrived as strangers to the problems of the commission. Jordan had arrived ready to represent his "nation of Entomologists," some of whom would quite readily secede from the international rules altogether if the commission ignored entomologists' wishes. He and Rothschild set to work helping the commission establish a balance of power that would be acceptable to the majority of zoologists. "Strongly backed by his powerful chief Lord Rothschild," Jordan successfully proposed that the congress should give the commission plenary powers to suspend the normal operations of the rules in the interests of nomenclatorial stability when, in its opinion, strict enforcement of the rules would result in greater confusion than uniformity.[43] Notably, this compromise measure went against Rothschild's and Jordan's previous emphasis on strict priority, but they hoped it would be acceptable since ultimate decisions would be placed in an authoritative body, rather than in the judgment of individual zoologists.

Through their "act of zoological statesmanship," the new commission members from Tring tempered the row anticipated by Stejneger and brought the commission out of a crisis that had nearly made its international rules obsolete. They proved a formidable pair, uniting both routes to status—professional and moneyed—and inspired allegiance, cooperation, and deference from a range of naturalists. Jordan and Rothschild also secured an agreement by which all questions on entomological nomenclature would go to the International Commission on Entomological Nomenclature for comment.[44]

Diplomacy elsewhere was proceeding with less success, as international tensions arising from a fragile balance of power between Britain, France, and Germany reached dangerous proportions. In toasting the close of the International Congress of Entomology in Oxford in 1912, Wilhelm Junk, of Berlin, had attempted a bit of internationalist humor amid such tensions when he announced: "Perhaps it was a little dangerous that Englishmen should show Germans that England was such a beautiful country (laughter). But he was sure the danger was not so great, and that there never would be anything like war or such things, about which the German journals and the English journals so often wrote (Applause). He thought it was quite impossible." The comment belied the growing tension caused by the arms race between Germany, France, Britain, and others. Having witnessed the increasing nationalism among the public, F. Wichgraf, of Berlin, announced to the Oxford gathering of entomologists that "as far as he could see, there was nothing which ought to separate the two nations, except perhaps the English Channel (laughter)." The meeting must surely help to "establish a permanent understanding of friendship between the two great and closely related nations."[45] His words were greeted by enthusiastic applause.

Siding with the optimism of the Wichgrafs of the world, Jordan and his colleagues set plans in motion for the third International Congress of Entomology to meet in Vienna in 1915. This would give entomologists plenty of time to prepare a united front before the next International Congress of Zoology, scheduled for Budapest the following year. Throughout 1914, Jordan wrote to the heads of the newly formed national committees—among them, John Hartley Durrant, for Britain, Henry Schouteden, for Belgium, Antoon Oudemans, for the Netherlands, and Yngve Sjöstedt, for Sweden—regarding certain nomenclatural questions.[46] He sent Anton Handlirsch, who was in charge of arranging the Vienna congress, advice and made plans to visit entomologists in Vienna in person to help with the local arrangements. In working with Handlirsch to obtain support for the congress from the state, Jordan bowed slightly to the winds of change by acknowledging in these letters that they would probably have to constantly emphasize the "immense practical value of research on insects to governments."[47]

No matter what vision of the future of entomology was contained therein, the letters organizing the congress and its committee were all composed under the assumption that travel and communication would remain open and that men could continue building ties that united rather than divided. This assumption was roundly shattered with the beginning of the First World War. Desperate to maintain the network upon which taxonomy depended, Jordan proceeded to write all his letters in English and leave them unsealed in the hope that they would pass the censors. Still confident in the ability of postal networks to withstand warfare, five days after Britain declared war on Germany in August 1914, Jordan agreed to determine Curt Schrottky's South American Sphingidae, offering to send him the 1903 "Revision" in exchange for specimens from Paraguay.[48] But as the summer passed and borders closed, keeping friends and their collections safe took precedence over exchanging specimens and consolidating nomenclature.

Anxiety over the fate of friends quickly appeared in correspondence arriving via initially neutral nations such as Italy. Jordan asked entomologists in neutral countries like Sweden to send word to friends in Germany that he and his family were well. "Although most of our friends and acquaintances here treat my family as if nothing has happened," Jordan wrote to Sjöstedt, "the suspense is a great strain on the nerves of my wife, quite a number of our nephews being at the front and no news from our relatives obtainable."[49]

The Jordans and Harterts were not treated as though "nothing has happened" for very long, however. Miriam Rothschild later described how, "even in a small backwater like Tring, insults were shouted at foreigners in the streets and daily life was made difficult and at times painfully unpleasant."[50] (The Jordan family proved a target despite the fact he had taken British citizenship in 1911.) Workers at the Rothschild bank had to abandon German in conversation when someone erected a poster in the office from the *Daily Mail* with the words, "Intern Them All," and Walter Rothschild eventually resigned from the Tring town council when the village passed a resolution to similar effect.[51]

As groups from all walks of life, from suffragettes to socialists, declared their support for the war, naturalists had to decide how they would help. Within a month, Jordan was spending weekdays in London volunteering at the Red Cross, fulfilling his work as curator only on weekends and at night.[52] Jordan's frustration with the halt called to his international organizational efforts occasionally escaped in letters to overseas correspondents. "The equilibrium of Europe was so perfectly balanced that it had to get upset sooner or later," he wrote to Miles Moss in September 1914, "It is a disgusting state of affairs for an international like myself." Well aware that some intellectuals thought the rigor and competition of war would prevent Europe's degeneration, he confided to Moss that he found it difficult to adopt the "higher standpoint of the philosopher and to console oneself with the thought that hard blows and a thinning out will do good to Europe and stop or delay its decay."[53]

The thinning out soon became personal. Upon pages once devoted to documenting variation, the *Entomologist's Record and Journal of Variation* listed the names of those who had "given up the net to take their share in the terrible European struggle," including more than a dozen men from the British Museum's Entomological Department. Inevitably, the effort to improve the quality and quantity of entomologists' *concreta* soon suffered as a result. During his presidential address for 1914, G. T. Bethune-Baker noted he had expected to have more specimens to back up his conclusions, "but alas, the war entirely upset these arrangements, and now some of my friends are at the front."[54]

Lists of those killed and wounded, including Poulton's son, appeared all too soon in the entomological journals. Occasional contributions came in from fellows of the Entomological Society of London who were at the front. K. G. Blair wrote a resigned but observant discussion of the entomology of the trenches, carefully describing the various lice and flies that he and millions of other men came to know all too well.[55] Notice next arrived that he was convalescing in a hospital at Boulogne. Noel Sennett sent observations on ants, and D. A. J. Buxton submitted notes on Lepidoptera while stationed at Gallipoli.[56] Later, in 1917, Poulton shared observations on butterfly migration sent by W. Pendlebury, serving in the British trenches around Gaza.[57] As the war continued, the society received notice from E. M. Dadd, interned in Germany, that he and others interested in natural science had formed a natural history society with a regular course of lectures and a biological laboratory with microscopes. Would the London entomologists, he asked, send literature?[58] Colonel N. Manders sent observations from his station in Egypt, although the censors prevented him from stating the exact position from which he wrote. "I daresay," he wrote, "entomologists can make a fair guess from what I have written." Following this gesture toward entomologists' expertise on the ties between species and geography, he described the beautiful summer weather and the flora and fauna and then shared that he sat near a few dead men and a mule covered in mud, with Testaments and Bibles scattered among the debris.[59] Soon, Manders's obituary appeared after he was killed in the Dardanelles.

The members of Jordan's "nation of Entomologists" dispersed to the front. Walther Horn served at the German front as a regimental doctor, Paul Speiser as a doctor of a regiment of field artillery. Russian entomologist Andrey Avinov served his country with the Red Cross in Poland and narrowly escaped being killed by a bomb as Russians and Germans fought for Lodz. Amid this movement of manpower out of the halls of museums, the impact of the war on natural history pursuits must be acknowledged as more than just practical or economic. Confronted by the devastation of material, lives, and values, the French entomologist Charles Oberthür gave in to despondency for some months, delaying his famous series of *Lépidoptérologie Comparée*. When the books finally appeared, entomologists read a plea for forgiveness on the grounds that with the sad news of the "glorious, no doubt, but so grievous" death of a relative or friend, "one feels that all scientific labour becomes impossible, and that the publication of any work begun in more propitious circumstances must be postponed *sine die*."[60] Justifying the time and energy spent placing insects between two bits of glass and writing up one's conclusions was sometimes difficult even for the most committed entomologists, if circumstances destroyed the peace of mind required for careful and motivated attention.

Meanwhile, the breakdown in the infrastructure required for careful exchange of results further dampened any motivation to produce those results. The librarian of the Entomological Society of London noted that, even if one found the energy to publish, "very many of the Foreign Magazines are not coming to hand."[61] Trying to continue what work they could, yet unable to communicate with entomologists or publishers in Germany, Jordan recruited Swedish entomologist Yngve Sjöstedt to mediate the delivery of printed locality labels and Hartert's book on Palaearctic birds. "Kindly reply in English," Jordan cabled, in another attempt to avoid the censors.[62] Eventually they maintained contact with their German colleagues only through the mediation of friends in the United States.[63]

By 1916, the Entomological Society of London had lost almost all its younger men to war duties, although members prided themselves on their respectable attendance list "in spite of darkened streets and irregular train service."[64] As mobilization reached into every aspect of society, calling on each institution to stand and deliver, someone needed to remain vigilant for their field. When a government official suggested that the British Museum of Natural History be closed during the war, G. T. Bethune-Baker urged that a resolution be sent to the prime minister.[65] The council held a special meeting and drew up a protest, citing the important part the museum played in answering urgent questions "affecting the health of the Nation and its forces at home and abroad, its food supplies, its timber, and the raw material of its manufactures," all of course essential to the conduct of the war.[66]

C. J. Gahan ended his presidential address to the society for the year with a lament over the extremely low status of entomological research in government circles, made all too obvious by the (ultimately unsuccessful) proposal to move the collections.[67] In 1916 one observer, frightened by plans to disperse museum staff to war jobs, wrote to the *Times* listing every benefit provided by the museum that he could think of, from the study of anthrax bearers in camels to slugs devastating rubber plantations in Jamaica. The writer lamented that

> it would take too long . . . to explain to politicians what men of science understand well enough, that the systematic study of "cataloguing" of the varied forms of life, even the minutest, constitutes the indispensable foundation for all this economic knowledge. [Surely] these few instances are enough to show not merely the Imperial but the direct military value of the work being carried on by gentlemen whom our economists would take away to copy documents and add up figures.[68]

As entomologists mobilized to convince officials of the importance of their field to national interests, internationalist efforts to raise the status of entomology floundered. During the first months of the war, Austrian entomologist Anton Handlirsch had continued making arrangements for the International Congress of Entomology planned for Vienna in 1915. Spurred by Handlirsch's stubborn optimism, Jordan wrote Sjöstedt in October of 1914 "to ascertain the opinion of our colleagues on this matter." "Some people think that it would not be possible or advisable to hold the Congress at Vienna," he wrote, "on account of the hatred which this war arouses, and that a neutral country should be selected instead, Holland for instance. What do you think?"⁶⁹ To the Italian Raphael Gestro, he confessed he feared the congress planned for 1915 would have to be postponed indefinitely if the war lasted through the winter. "Although Entomologists, as a rule, are peaceful and sensible," he wrote, "the war is certain to rouse animosities detrimental to the success of the Congress."⁷⁰

Even if the war had ended quickly, relations within the "nation of Entomology" had already disintegrated, as identification with nation trumped individuals' solidarity as entomologists. In December 1915, Jordan lamented to Sjöstedt, "It is very sad that hatred and abuse is not kept out of science."⁷¹ Some of his closest entomological friends succumbed to such "hatred and abuse." Oberthür used the latest installment of his *Lépidoptérologie Comparée* to denounce the Germans. Their constantly manufactured lies, he wrote, had also led to the war of the 1870s and interfered with entomological monographs in France. "This is not strictly 'Entomologie,'" T. A. Chapman noted in reviewing the book, "but it illustrates how it is impossible in Entomology, or in anything else, to avoid the effects of the war, and to reflect on the indescribably abominable conduct of the Prussians."⁷² In defiance against rising nationalist sentiment, Jordan's good friend Horn, who served throughout the war, began printing "All men of science are brothers" in English at the top of his notepaper, despite the antagonism it caused from some of his more patriotic fellows.⁷³

Already-heated debates over nomenclature rules became early avenues for nationalist outbursts in Horn's fallible brotherhood of scientists. The curators of Walter Rothschild's museum, with their German accents, proved early targets. Sir George Hampson, a British Museum entomologist who often exchanged specimens with Jordan, took the opportunity of the war to aim some long-simmering barbs at the principled Tring naturalists. He raged against "INSIDIOUS GERMAN SPECIFIC POLYNOMIAL NOMENCLATURE" in the pages of the *Entomological News* of the Academy of Natural Sciences, Philadelphia, arguing against naming local, seasonal, sexual, polymorphic, and other forms on the grounds that subspecies did not exist in nature. He hoped that "one of the minor benefits of the present war will be that we in Britain will return to a simple binomial nomenclature and purge ourselves from this form of 'Kultur.'"⁷⁴

This time, Walter Rothschild replied, not Jordan. He tried to take refuge in the International Rules, reminding Hampson that not just the Germans but the international commission insisted that naturalists distinguish between the different kinds of varieties and that they name and describe subspecies or geographical races.⁷⁵ But an appeal to an international commission on which trinomialists, including Rothschild and Jordan, had a strong influence, was surely unconvincing to Hampson. He began prefacing his papers with the warning: "No quotations from German authors published since August 1st, 1914 are inserted. *Hostes humani generis.*" He was cheered on in his wartime patriotism by Lord Walsingham, whose insistence that "for the next twenty years, at least, all Germans will be relegated to the category of persons with whom honest men will decline to have any dealings" appeared in the pages of *Nature*. "At a bare minimum," Walsingham concluded, scientific men should ignore all papers published in German.[76] Hemming later wrote in grand understatement that "the outbreak of war in Europe in 1914, coming so soon after the close of the Monaco congress, was a severe set-back to the international regulation of zoological nomenclature."[77]

Certainly, resolutions by entomologists, united or no, went nowhere during the war. Rothschild's appeal to a virtually defunct International Commission must have seemed almost seditious to patriotic Englishmen such as Hampson. Meanwhile, as more philosophically inclined naturalists dove headlong into biological justifications of both peace and of war, according to their predilections, the relatively mundane spadework that Jordan insisted was absolutely required to do such "philosophical natural history" ground to a halt.

THE STANDSTILL

By September 1915, Jordan found it "more than doubtful" that the International Congress of Entomology would be able to meet in the near future. "This unfortunate war," he wrote to Sjöstedt from the Red Cross headquarters in Piccadilly, "has put everything into the background. The preparations for next year's Entomological Congress are at a standstill." He put on a brave face, certain that "passions will soon quiet down when the war is over, I think. Let us hope that civilisation will not seriously suffer by the war."[78] The wish encapsulated much. Jordan once noted that a culture's support for natural history seemed directly related to its degree of civilization. Others, too, had noticed the close relationship between "civilized" commerce and science; it was a theme occasionally touted by political guests at zoologists' congresses. During the International Congress of Zoology in Germany in 1901, for example, the mayor of Hamburg pointed out how "Science desires that the People forget boundary lines; so, too, does trade and shipping. They are to a certain extent the brothers and sisters of science."[79]

The influence of the war on natural history proved that the ties between natural history and trade were familial indeed. As we have seen, the natural history network that made the Rothschild and Jordan revisions possible followed the paths of empire and commerce. In Jordan's case, the ties to the prewar imperial and commercial organization were especially obvious. The Rothschilds had built their fortune on the ability of trade and commerce to dissolve national boundaries. The effect of the war on their financial network was immediate. In the first year N. M. Rothschild & Sons lost close to £1,500,000, commencing a decline in the family's wealth and influence that would last half a century. As historian Niall Ferguson explains, "The world in which the Rothschilds had thrived, came to an end in 1914." Cooperation with its Continental houses had ended with the outbreak of war. Ties with German banks were severed, overseas trade suddenly disrupted by blockades and submarines, and the monetary system based on the gold standard ceased. The railways that had been used to transport goods now moved troops to battle, not goods for trade.[80] If one substitutes the exchange of specimens and expertise for gold, the foregoing also serves as a portrait of the influence of the war on the world of specimen-based natural history.

All the organization in the world could not help systematics become scientific if men could not actually get to the field to obtain the long series of well-labeled specimens. The traditional collecting network in Europe ground to a halt during the war, its close ties to the stability of a peaceful Europe immediately apparent. Within a few weeks of Britain's entrance to the war, Jordan wrote Miles Moss that the insect trade "is practically at a standstill." Collectors in the tropics would soon be in financial difficulties, he warned.[81] Specimens already collected could not be shipped, including around six thousand specimens that had been en route to Tring from collectors in Algeria.[82] Jordan stopped sending specimens and manuscripts to correspondents on the Continent. Descriptions of new species especially risked destruction by the censor, he explained, since he wrote them in Latin.[83] Although a year and a half after the war began he wrote to Sjöstedt that "our work at the museum is going on as usual," he added that "we receive very little from the tropics. The Brit.Mus. has no funds for purchases and publications."[84] Six months later, Jordan wrote again, asking Sjöstedt if he could secure black pins since "the manufacturers of insect-pins are now employed in making ammunitions."[85] Walsingham, too, informed the American entomologist A. Busck that "everything has given way to the war and scientific work is almost entirely at a standstill." His "secretary," J. H. Durrant, gave more time to the Red Cross than to entomology, and another of his assistants had been at the front since the start of the war. In any event, Walsingham confessed his own "zeal for new discoveries" had been "damped out by age and circumstances."[86]

Work at the museum continued, but on a constrained level hardly comparable to the production of the 1890s. Cut off from Continental collections and colleagues, Jordan increased the usual caveats regarding the tentative nature of his taxonomic conclusions. "As the present state of Europe makes traveling outside England impossible for me," he wrote in a paper on *Somabrachys* moths, "I have not been able to consult any Continental collections."[87] Instead, he relied on the nine hundred specimens they had obtained from various parts of North Africa to analyze the group's individual variability (based on this study, he concluded that one group of specimens represented only three species rather than the many more described by Charles Oberthür, yet another example of "species-unmaking"). They could still attempt exchanges with those in allied or neutral countries despite the constant threat that parcels of specimens would be opened by the censors and the specimens suffer damage.[88] But Jordan stopped work on some manuscripts, including his contribution to Adalbert Seitz's monumental *The Macrolepidoptera of the World*, when war broke out. Without access to the network required to ensure scientific systematics, the time would have been wasted, the results serving only to muddy a field already overburdened by work on an inadequate foundation of facts.

As the war immobilized the European natural history network, American naturalists looked on from their summit of wartime isolationism with occasional impatience. C. S. Baker complained that the war interfered with his efforts to work up and publish his results on Malayan entomology since he could not communicate with many Continental entomologists. Still, activity did not shut down completely. Baker expressed surprise that of some fifty entomologists cooperating with him before the war, about twenty-five had been able to maintain some degree of activity despite wartime chaos and restrictions on travel and trade.[89] N. D. Riley would later describe the war as having produced a "crop of collections" for the British Museum of Natural History since some men, such as Carpenter and others, had the leisure of far-off colonial posts to collect specimens.[90] (Whether experts would be able to process them, given the existing backlog, was another matter.) Some colonials even found that their scientific work and field explorations progressed as rapidly as usual—"perhaps more so," compelled as they were to remain in distant corners of the Earth "and use our utmost endeavors in connection with agricultural education and development."[91]

The reference to agriculture and development is significant; it provides a hint as to the future grounds on which work could get done in an environment of restricted resources and highly managed priorities. Entomologists could not agree what these hints would mean for the future of entomology. Some scientists, such as Arthur Schuster, warned their fellows "not to beat the utilitarian drum too loudly" despite the incentives of wartime.[92] But others enthusiastically trumpeted the importance of their field to the war effort. A triumphant notice appeared in the *Entomologists' Record* of 1915, for example, on the role of the British Museum in aiding the War Office to protect biscuits supplied to the army from attacks by insects.[93] By the end of the year, the war mongering prompted Jordan to confess to Miles Moss that if the war "goes on much longer there will be no money left for a pure science which does not refer to the killing of man."[94]

Meanwhile, and just as "J. P. Morgan succeeded N. M. Rothschild as the linchpin of war finance,"[95] the power to drive entomological priorities and goals was moving to the other side of the Atlantic. This shift would ultimately seal the increasing importance of applied entomology within the discipline as a whole, the Americans having long been at the forefront of this field. Even an internationalist such as Jordan lamented the evident movement of scientific authority and resources from Britain to the United States. He urged Moss to make sure he exhausted all other sources of support for publication of his work on Sphingidae before going to the United States, since "it looks like a *testimonium paupertatis* if [to get funds] Englishmen have to apply to U.S.A."[96]

The Americans, by contrast, were quite amused by the changing circumstances, particularly when, after the war, Europeans had to visit America to consult important collections, rather than the other way around. "It will be rather sporty," William Schaus, of the National Museum of Natural History in Washington DC, confessed to entomologist C. H. Lankester in 1923, "to have the English entomologists come over here to examine Old World types."[97] Schaus, at least, found the change encouraging evidence of the rise of New World entomologists. U.S. entomologists urged Schaus to help put the United States "on the entomological map" by obtaining collections before Europe could "cinch the job."[98] A number of "patriotic Americans" helped him secure the Paul Dognin collection in 1924, with its thirty-five hundred types, despite European entomologists' last-ditch efforts to keep the collection from crossing the Atlantic.[99] Schaus had been sure that Rothschild would buy it and had enlisted the help of friends throughout the country to find American patrons to scoop the wealthy naturalist from Tring.[100]

Like the fate of the Rothschild bank, the factors involved in the rise and fall of Walter Rothschild's museum proved both systemic and idiosyncratic. Ferguson attributes the decline of the Rothschild house in part to the fact the generation that followed N. M. Rothschild felt little inclination toward finance. On the death of his father in 1915, Walter had succeeded to the title, but he had stopped even pretending to work at the bank in 1908. His father had long-since pinned his hopes on his younger son, Charles, who diligently attended to his duties in London despite the fact that he, too, loved natural history.

No doubt Jordan, too, placed his hopes for Tring and much of its research program on the younger, more responsible Rothschild. But in yet another example that science, like family banks, depends much on the vagaries of human fortune, events would soon destroy Jordan's plans for the future of the museum. Shortly after Lord Rothschild's death, Charles Rothschild had developed postencephalitic parkinsonism. His doctors insisted he be taken to the mountains of Switzerland to recover, in spite of all the inconveniences and dangers of wartime travel. From Miriam Rothschild's account, we have the eloquent picture of a desperate family convening at the home of Charles and his wife, Roszika, at Ashton Wold to discuss what should be done. In the end, Roszika decided that Jordan must accompany Charles on his journey. When she anxiously asked Walter if Jordan could leave, "Walter never discussed the problem or any of the implications, because he never discussed anything. He merely said: 'Go.'"[101]

Jordan and Charles left for Switzerland on January 2, 1917, on a journey they thought would last only a few months. Jordan locked manuscripts up in drawers, not expecting Walter might be pressed by the authors to publish them before he returned home. In Switzerland, he and Charles collected fleas, beetles, and Lepidoptera, visited Swiss entomologists, and hiked for miles under the assumption that exercise and mountain air would do Charles good. By July, the patient had indeed made progress. "Here we are collecting ardently," Jordan observed to Miles Moss, "which is a very healthy occupation, for us, not for the bugs." But he warned that "under the circumstances, tropical entomology must do without me at present."[102] Walter, busy with war work yet trying to arrange various groups in the museum, admitted finding Jordan's long absence annoying. He confessed to Hartert that while he did not "grudge my brother anything that may conduce to his complete recovery," it was "most inconvenient and embarrassing" that Jordan was away.[103] He set aside specimens that required careful examination of the genitalia as he tried to write up material they had collected in Algeria, treating some forms as species "till such times as Dr. Jordan and Professor Dixey can find time to work out the races."[104]

A year later, Jordan and Charles had not returned home, and Rothschild, writing to Hartert from Piccadilly, betrayed his frustration; "I hope to get to Tring on Friday. I do not know what we shall do only having Fred Goodson for the amount of stuff to be set is enormous, at least 200,000. Doncaster's people are all gone and Miss Sharpe's setting is vile."[105] Rothschild's lament regarding Doncaster, one of the biggest natural history agents in Britain, shows that more than just a museum curator had been temporarily lost. Even had Jordan been present, it was not clear how much a curator could actually have done without the support of natural history agents, collectors, setters, and access to foreign correspondents and journals. Meanwhile, Jordan tried to maintain his correspondence with friends in the field such as Miles Moss, but letters went astray, perhaps finding "a resting place at the bottom of the sea or in the waste-paper basket of some censor." "To work up a year's arrears," Jordan wrote Moss in 1917, "will occupy me for some time!"[106] (He and Charles would not in fact return home for two years.)

Writing from Switzerland, Jordan did his best to help Hartert with the administration of the museum, reviewing consignments and sending advice regarding potential purchases. He warned that G. Melou, for example, seemed to think that collecting masses of common things would bring him a good return, and suggested they ask him to limit his consignments to thirty specimens of each species. (Indeed, Melou would soon file suit against Rothschild on the grounds he had been misled to think extensive collections in Madagascar would be purchased at a good price.[107]) Jordan sent packages of pins, warning Hartert that "everybody is short of them. Please ask our people not to waste any." And he listed the Swiss entomologists and ornithologists he called on during their travels, particularly those who could still provide bird specimens.

Though they were in a neutral land, wartime regulations of human interactions occasionally governed who they could visit, even for this famously internationalist pair of entomologists. Speaking of an acquaintance, Jordan wrote how he and Charles would have to avoid meeting him "as it is considered for an Englishman a very grave breach of etiquette to associate with a German, during the war I suppose."[108] Yet despite this evidence to the contrary, the war could seem far from the mountains. Hints of Alfred Russel Wallace's socialist visions of a pacifist future even appeared, a position virtually extinct from the rest of Europe as workers volunteered to fight their proletarian brethren in droves. Jordan recounted how in August 1917 they encountered

> a large gathering of "Frei Jugend," a socialist section of young people of both sexes; their main doctrine is opposition to military service. Most of the young fellows exhibited cards with "Nieder mit dem Militarismus" on them, and many of the Ladies and their gallants wore phrygian caps. There was also speechifying,

among other things free love being advocated, which rather spoilt (for us) the enjoyment of the pretty picture presented on the green meadows by the red banners, caps, and sashes of the damsels and young men.[109]

The young socialists had reason to be both so festive and bold, even in the face of the war's destructive effects on idealistic calls for the workers of the world to unite. For they picnicked just six months after Russia's February Revolution had resulted in the abdication of Tsar Nicholas II. Though the resulting provisional government was dominated by the bourgeoisie, Vladimir Lenin's radical Bolshevik Party was gaining in popularity. Increasing evidence existed, at least in Russia, that what had seemed the utopian visions of the socialists, communists, and anarchists might actually become reality. Eventually, Russia plunged into civil war, a war that would end with the foundation of the Soviet Union in 1922.

With the West thus polarized into distinct ideological, political, and economic spheres, one embarking on a completely new economic and political reorganization of society and the other fiercely defending the old, testing the meaning of Wallace's ideal world for the future of natural history became a possibility. Many feared the worst. The German dipterist Paul Speiser confessed in awkward English to his colleagues in Sweden in 1920 that he feared for his collections should the "bolschweists" advance further. "Who will know," he moaned, "what will be coming?!"[110] Certainly, the immediate results of the Russian Revolution did not bode well for a tradition so firmly tied to the economic structure of the immediate past. Vladimir Nabokov, born into a wealthy St. Petersburg family, suspected he might have ended up a full-time lepidopterist rather than the famous author of *Lolita* had it not been for the Russian Revolution.[111] But his family had been forced to flee Russia in 1919, their estates confiscated by the state. Andrey Avinoff, who would become director of the Carnegie Museum of Natural History in Pittsburgh in 1926, also hailed from a wealthy family closely tied to the ruling Russian nobility and had to escape to the United States. As Jamaica's Fabian governor had hinted when he tried to prevent Rothschild's collectors from securing specimens, and in contrast to Wallace's tolerance of these particular beneficiaries of a highly stratified society, naturalists would not necessarily be given special privileges as the workers of the world reordered that world. The age in which the naturalist tradition had flourished was indeed being transformed.

RECOVERING FRIENDS, COMMITTEES, AND CONGRESSES I

From the little information she could obtain from Jordan regarding his time abroad with her father, Miriam Rothschild learned that her father's illness made him so deranged at times that Jordan had occasionally even feared for his life, suspecting that in such a state the naturally mild Charles could become murderous.[112] But Jordan constantly and patiently calmed his friend over a period of two years, and when the borders slowly opened up at the war's end in 1918, they both came home. They returned to a grief-stricken town that had lost more than a hundred of its young men to the war. Hartert and his wife, Claudia, had lost their only son, killed in action fighting for the British in October 1916.

Everything, it seemed, had changed after the war. The boundaries on world maps had once again shifted, this time in transfers of entire territories, both in Europe and abroad, and especially in Africa. Germany lost the empire its zoologists had explored so enthusiastically. Its economy at home lay in ruins, as exchange rates fluctuated wildly. Russia was in the midst of civil war. And the British government faced four million demobilized men who returned home weary and watchful as to whether the new age marked by the horrendous war had been worth the terrible price. What they, and their girlfriends and wives, thought of this new age now mattered even more, for in 1918 Britain had increased the franchise by two million men and six million women.

Once he and Charles arrived back at the museum, Jordan focused on assessing the damage to the natural history network and helping friends in need. When the British took an entomologist named Peters prisoner and confiscated his seventy-nine chests of specimens and belongings, Jordan and Charles fired off letters to get the collection released. Charles also fronted the £250 required to return the collection to its rightful owner.[113] The letters that eventually arrived from Germany contained grim news. Jordan's entomological friends there suffered with the rest of the populace from the inflation gripping the country, and a lack of newspapers heightened the anxiety. The mathematician and sometime-entomologist Eduard Study wrote long letters from Bonn describing the state of chaos in Jordan's homeland. Smuggling was rife, gangs of armed robbers attacked in broad daylight, and the railroads had been halted for six weeks.[114]

Eventually a letter finally arrived from Walther Horn, who lamented that surely his wife could have expected a better fate in her old age than to return to the work of her early life, but Frau Horn now bent over the wash barrel, often alone.¹¹⁵ In thanking the Jordan family for care packages, Horn sadly confessed that he feared his own institution, the Deutsches Entomologisches Museum, would not survive the postwar period simply because there was no one to replace him.¹¹⁶ Meanwhile, "invested funds melted away, the rich became paupers in a night and thousands of marks would not buy a single meal." Horn's museum could not even afford adequate heating, despite its ample endowment. He had to dismiss the staff, and ultimately saved the enterprise from ruin only by begging financial assistance from his friends from other countries. (He also dropped the name Museum in favor of the more scientific-sounding Institute.) "For scientific work I have hardly any time," he lamented as he tried to keep the museum afloat.¹¹⁷

Like so much of the infrastructure of the previous age, the systems naturalists had developed to exchange information and specimens came out of the war severely damaged. At the most basic level, the easy communication and travel naturalists had enjoyed during the previous era had ended. Jordan once reflected that conditions for traveling and collecting would never be as pleasant or as easy as they had been prior to 1914. Consider, he proposed, travel in Europe: "In 1907 I was suddenly asked whether I could go to France 'the day after tomorrow.' I simply packed a suitcase and went. No passport, no identification paper, no inquiry about how much money I had with me, and food and accommodation in hotels everywhere."¹¹⁸ Now, travel for both naturalists and packages proved inconvenient and even precarious. As late as 1920, the American James Rehn informed Yngve Sjöstedt in Sweden that he could not send some Australian material requested, as considerable opposition existed in America to shipping unworked material overseas, "on account of losses during the war of specimens and publications, particularly the latter." Even though, as he conceded, these conditions no longer existed, "most European conditions are still so uncertain and unsettled, and in many sections hazardous, that many institutions will not loan material overseas until more normal times."¹¹⁹

Those willing to send specimens faced a tremendous loss of the manpower required to process letters and pack specimens. Eduard Study reported that the Berlin museum had had no assistants to process their dozens of unorganized collections for some time. Although he admitted his private complaints seemed a small matter compared with the general misfortune, he confessed that "the fact that the English collections remain inaccessible is a heavy obstacle."¹²⁰ Jordan felt the same about the collections on the Continent, as he and Rothschild tried to start their long-planned work on the silkmoths, or Saturnians. "My work proceeds very very slowly," he wrote in 1921 to W. J. Holland in

Pittsburgh, "If travelling to Europe were as easy as before the war, I should have no difficulty in going about inspecting the types scattered over many collections, in France and Germany especially. But under the present circumstances the crossing of frontiers is an unpleasant task, particularly if one has to take some specimens for comparison."[121]

Ease of travel and physical access to collections were not the only things curtailed. Efforts to standardize nomenclature and avoid synonymy were engulfed by the void of four years during which entomologists could not obtain literature from many colleagues. "Every work must remain very fragmentary..." one entomologist warned, "because literature is wanting nearly wholly, especially of the years 1914 to 18."[122] In particular, taxonomists faced the arduous task of figuring out what people had described during and just after the war. The *Zoological Record* had not appeared since 1915, and obtaining copies of literature from many countries proved impossible. As soon as he could get a letter to Germany, Jordan began tracking down rumored publications, asking Ernst Ehlers, for example, whether he knew what had come of the results of a German expedition to New Guinea after the members had been captured during the war.[123]

To make matters worse, some naturalists refused to acknowledge or cite work done in enemy countries both during the war and after. As late as December 1920, the Entomological Society of London and the Zoological Society of London would still not exchange journals with former enemies (although Jordan had hopes that when Rothschild took over the presidency of the Entomological Society in 1921, he would be able to change the policy).[124] Sorting out and avoiding the creation of synonymous names had been difficult enough, without a war. Now, taxonomists intent on revising a group of insects had to face more than four years during which the new literature might as well have been spread to the four corners of the Earth.

As internationalist politicians such as President Wilson fought for a League of Nations to provide the basis of a new, postwar international order, internationalism proved as hard to recover in science as in any other realm of life. "We are letting Wilson air his idealism as much as he pleases," George Hampson wrote to the American William Schaus, "but shall keep Germany down now we have got them under and I hope make her pay a very large indemnity."[125] Hampson's inability to let bygones be bygones became official policy among some scientists. The International Research Council (IRC), for example, founded in 1919 by the scientific academies of the major Allied nations, followed the example of the Treaty of Versailles in excluding the former Central Powers from membership.

Following the lead of the International Research Council, many international organizations that reconvened after the war excluded the Central Powers on the grounds that German scientists had subordinated their research to the interests of warfare. Of course, by those criteria scientists—including entomologists—from every nation were guilty, although working on agricultural pests or disease-transmitting insects no doubt seemed innocuous compared with what some chemists had been up to. Intent on recovering scientific internationalism, William Bateson and others protested against the IRC's exclusion statute. Bateson was dismayed, upon arriving in Brussels as a British delegate at the council's inaugural conference, to find that "every scrap of common sense is gone." He concluded that the IRC had become yet another occasion for exploiting science for "chauvinistic purposes" and joined a Royal Society resolution boycotting the newly formed International Union of Biological Sciences unless the IRC removed the exclusionary statute.[126]

Jordan, intent on reestablishing the International Congress of Entomology, kept a close eye on these developments. Like Bateson, he refused to follow the postwar trend of excluding the former Central Powers. As early as 1917 he had been corresponding with those friends on the executive committee of the International Congress of Entomology who lived in neutral countries regarding the prospects for reconvening the congresses. Each expressed fear that holding the congress too early, before the passions ignited by the war had cooled, would run the risk of a permanent division on national grounds. Jordan agreed a congress that was not truly international could only be damaging to entomology. He called for patience until the time when scientists once again acknowledged that science cannot be limited to individual countries or peoples.[127] He ultimately delayed the third congress until entomologists of all nations could officially attend. In the mean time, he had to find a new venue since to ask entomologists to convene in Vienna, as originally planned, hardly seemed politic.

Jordan also set to work recovering the various immobilized conversations regarding zoological nomenclature by distributing a circular asking the members of the various national committees of entomological nomenclature for comments on the old bugbear of priority. As secretary of the ICZN, Francis Hemming later described how, although the issue had supposedly, with Jordan's and Rothschild's help, been resolved at the Monaco congress, the break in communication caused by the war "robbed the decisions taken at Monaco of much of their practical value."[128] Indeed the status of the International Commission on Zoological Nomenclature's judgments in general had become precarious. John Merton Aldrich, assistant curator of the Division of Insects at the Smithsonian, dismissed any authority the commission might once have had, for example, since it was "virtually a committee" of the now defunct zoological congresses.[129] But Jordan persisted in his efforts to reopen the old avenues of communication among entomologists so that they would be prepared to bring their wishes to the table when the world became sane once more.

In the midst of trying to reestablish commissions and connections with long-lost friends, Jordan lost two of his most ardent allies. His wife, Minna, who had once been "content to sit by his side and help him with the labeling of specimens" for half the night, passed away after a debilitating illness in 1925.[130] In 1923, he had also lost Charles Rothschild, who, despite the efforts of both Jordan and physicians, had never fully recovered from his illness. On October 12 he locked the door to his bedroom and shot himself. Miriam Rothschild, a teenager at the time, recalled how Jordan was never the same after the loss of his friend.[131] She vividly remembered Jordan, having rushed immediately to Charles's home at Ashton Wold, sitting at the dinner table, "in another world, staring off into space."[132] He returned to his desk after a few days; to the routine correspondence of the museum, international congresses, the journal, collectors, drawings, and to Walter Rothschild. "But," Miriam wrote, "a permanent shadow had been cast over the scene." Although Jordan published half of his 460 papers, singly- or coauthored, after Charles Rothschild's death, she insisted that the "truly creative period" of his life had abruptly ended in 1916 when Charles had first fallen ill.[133]

If Jordan's creativity is judged solely by his production of either revisions or generalizations based on those revisions, then the work he completed after Charles's death may indeed seem less impressive. Judged by twentieth-century standards of science that emphasize theoretical innovation, his 1896 and 1905 papers certainly stand out. But Jordan insisted that the quality of such work, for example, his conclusions regarding geographical variation, was inextricably bound to the meticulous work of revisions and monographs. And given this close dependence of robust generalization on a foundation of facts, Jordan had seen the need to order entomologists themselves grow even more pressing, as entomologists faced new obstacles to doing good work. Thus, within a few months of his friend's death, Jordan renewed his effort to organize the first postwar entomological congress in earnest. By then, ten years had passed since the start of the war. Jordan had heard that a successful international meeting on phytopathology had taken place in neutral Holland in 1923 without excluding former enemies. With this evidence that perhaps entomologists, too, could meet each other once again, he began canvassing the executive committee for potential dates and venues.

The entomologists Malcolm Burr, of England, Walther Horn, of Germany, Pierre Lesne, of France, and Henry Skinner, of the United States, joined him without reservation, but Guillaume Severin, of Belgium, refused. According to Jordan, he did so on grounds of "political hatred."[134] In an effort to dampen such resistance, several entomologists in England and the Continent argued that the next congress be held "in a country which remained neutral in the war."[135] Jordan eventually decided on Switzerland, though he had to write long conciliatory letters to Anton Handlirsch, in Vienna, explaining this move. He hoped, Jordan wrote, that the congress would be able to convene in that city after passions had calmed down (he also offered to pay all expenses for Handlirsch and his wife, if they would attend the Swiss congress).

During Charles Rothschild's convalescence, Jordan and Charles had periodically met with Swiss entomologists and talked about what should happen after the war. Now he wrote to Anton von Schulthess, inquiring whether he would agree to a meeting of the congress in Zürich in 1925 and act as its president. He then began writing to entomologists throughout Europe, most of whom responded in favor of an early meeting.[136] By April 1925, Jordan had sent four hundred invitations to entomologists and institutions all over the world.[137] "The response to our invitation has been most encouraging," Jordan reported to W. J. Holland. He admitted that, unfortunately, the French and Belgian entomologists continued to take the view that Germans must not be invited to international congresses (eventually, Lesne, too, declined to attend). Officially they "hold aloof," Jordan explained. But he made a point of insisting this had been their own decision, rather than due to an official rule of the congress that a particular nation be excluded.[138]

In the end, the French and Belgian entomologists protested by not attending the congress and withdrawing from the executive and permanent committees. Jordan and his fellow organizers passively rebelled against the sentiments represented by both the boycott and entomologists such as George Hampson and Lord Walsingham by printing a larger number of papers than usual in German, including Jordan's. Friends found it inspiring, if not almost amusing, how Jordan ignored opposition to the congresses, once evidence that all countries could be invited had been obtained. "To him national rivalries in this field," Riley wrote, "like personal rivalries, were abhorrent, and he deliberately ignored them."[139] Years later, Jordan confessed he thought the entire episode "best forgotten."[140]

"THE REQUIREMENTS FOR A THOROUGH INVESTIGATION"

More than national boundaries were at stake in entomologists' attempts to successfully navigate the turbulent twentieth century. Jordan had founded the international congresses of entomology in part to help publicize the importance of their work and conclusions among general zoologists. He and his fellows also hoped to assert entomologists' independence from trends in zoology that threatened to engulf traditional institutions and endeavors. At the first congress, Auguste Lameere had launched a justification of their rebellion, largely on the grounds that modern zoologists knew little of insects in the field. When the third congress finally convened in Zurich, Karl Hescheler, of the Zoological Museum at the University of Zurich, also played on anxieties over entomologists' place within zoology as a whole. But Hescheler used a different tactic to emphasize entomologists' importance. Of the four hundred thousand known animal species, he noted, three-quarters were insects. If the current fears over the fate of the "so-called higher animals" proved to be well-founded and they all became extinct, then naturally all those zoologists concerned with vertebrate animals would become extinct! "Zoology," he concluded triumphantly, "would return again to its favorite objects of the 17th and 18th century."[141]

Yet many entomologists knew that, fifteen years after the first congress, Lameere's high-handed dismissal of the zoology of the lab had become untenable if the meetings were to remain relevant to a modern age. And clearly, even if zoology returned to insects as its favorite objects, it would most definitely not do so via the favorite methods of a previous era. Indeed, the entomology that would be conveyed by the proceedings of the 1925 congress in Zurich already looked quite different to that gracing the congresses of 1910 and 1912. As a member of the Laboratory of Zoology and Comparative Anatomy at the University of Geneva, the chemist Amé Pictet would use his time at the congress podium to deliver an ode to the experimental method, for example. He conceded the importance of "observational" entomology, defined as the determination, classification, and arrangement in collections of insect forms, and the study of geographical distribution. But insects, he noted triumphantly, had ceased to provide material exclusively for museums; they had become, like rabbits, guinea pigs, and rats, animals of the laboratory. Thus, Pictet announced, entomological research could now share fully in research on the problems of evolution. As evidence, he cited a case from the field of genetics. The most beautiful discoveries of the new century in that field, he noted—the discovery of the role of chromosomes in heredity and sex-linked traits—had been based on the study of insects. Pictet's paper on sex ratios and parthenogenesis contained numerous references to the work of T. H. Morgan, Richard Goldschmidt, and E. B. Wilson, as well as to the contents of the *Journal of Genetics* and the *Journal of Experimental Zoology*.[142]

Specialists on various insect groups had been keeping a close eye on developments in these laboratories of genetics for some time now. Following a paper on the devolution of wing structures in cockroaches, one member of the Oxford congress of 1912 had wondered whether "Mendelian research would not lead to clearer results, as it was obvious that the form and size of the wings depended on certain units present from remote times." Yet the answer that had met that prewar suggestion had been brief and clear: Cockroaches took a long time to reach maturity and thus "did not lend themselves well to Mendelian research."[143] Even as members mused on this response, Thomas Hunt Morgan's fly lab in New York was two years into developing a pioneering way to study variation and heredity based on data that could be amassed even in relatively straitened times.

Morgan's work on sex-limited inheritance was summarized for readers of the *Entomologist's Record* by H. J. Turner in 1917. Seven years earlier, Turner wrote, Morgan had begun work on *Drosophila*, which, due to its frequent generation time, willingness to breed in captivity, and inexpensiveness, allowed his team to surmount some of the difficulties that had plagued work on other organisms. Over a period of seven years, Morgan's group had bred half a million flies, producing an amount of data on genetics that surpassed that obtained from any other animal or plant. Turner quoted William Bateson's impression of the groundbreaking work coming out of Morgan's lab:

> Let it be explicitly said that not even the most skeptical of readers can go through the *Drosophila* work unmoved by a sense of admiration for the zeal and penetration with which it has been conducted, and for the extension of genetic knowledge to which it has led—greater far than has been made in any one line of work since Mendel's own experiments.[144]

Naturalists such as Jordan already had ample experience with Bateson's enthusiasm for the methods and conclusions of geneticists. And it had left them wary of the claims coming out of laboratories such as Morgan's. But, contrary to the tales later told by biologists that naturalists remained ignorant of advances in genetics, entomologists reported on the more complicated view of heredity established by Morgan's lab early on. After all, as Poulton noted, the fact Morgan worked on an insect placed that work "directly within the survey of the

Entomological Society." With this as his excuse, Poulton used the pages of the *Proceedings of the Entomological Society of London* to suggest to entomologists that they might not have to make a choice between Mendel and Darwin. He summarized Morgan's results on the seven gradations of color between white and red eyes, "each heritable in the Mendelian manner," and noted that "by means of these graded changes one could obtain, by the mutationist's own statement, the continuously graded results which selection actually gives. What more can the selectionist ask?"[145] In 1918, Poulton quoted a letter he had received from Herbert Spencer Jennings that demonstrated the ardent Darwinian was not misinterpreting geneticists' conclusions. "We feel that we have here in America," Jennings wrote, "in Morgan's *Drosophila*, a sort of machine for grinding out answers to all sorts of questions in genetics, and now that that question of inheritance of small variations has been put to it, it yields an emphatic affirmative answer."[146]

Jordan witnessed this machine first hand when the recently formed Genetical Society held its eleventh meeting at Tring in 1922. As one attendee, F. A. Crew, reminisced years later, Morgan and A. H. Sturtevant attended the meeting "loaded with hat-boxes filled with Drosophila cultures and with lots of microscopic slides displaying the chromosomes." Hearing the two Americans share their work on the mutants of *Drosophila melanogaster*, "the enthralled audience" was immediately convinced of the extraordinary impact the Morgan group was having on genetics, all in turn dependent on the mutability, reproductive habits, and simple chromosome constitution of their organisms of choice. As Crew recalled, "No one in the United Kingdom had been using an animal or plant that could possibly have yielded in so short a time the genetical and cytological information from which emerged the Theory of the Gene." Crew also recounted how, at some point during the meeting, attendees went on a tour of the Rothschild Museum, where Rothschild and Jordan gave a "truly remarkable demonstration of polymorphism and geographical variation in Lepidoptera"[147]—a tantalizingly vague portrait of Jordan lecturing the American geneticists on the museum's work.

Jordan no doubt repeated his insistence that entomologists in museums could contribute to the grand aim of explaining the diversity in nature by demonstrating, as chronicled in his 1896 and 1905 papers, that through detailed morphological work and the collection of long series, useful generalizations could be made regarding speciation. Five years prior to the arrival of the famous geneticist and naturalist Theodosius Dobzhansky at Morgan's lab, perhaps Jordan shared his conclusions that geographical variation provided the key to the origin of species. Poulton would later bring to the fore cases in which geneticists' results confirmed conclusions already obtained by naturalists, especially his colleague at Tring, Karl Jordan.[148] But even if Poulton or Jordan could convince a staunch experimentalist such as T. H. Morgan that work done in museums had indeed proved useful, inspiring him to be a loyal supporter of Tring's natural history–based research program required more than a conversion to concepts such as the importance of geographical variation. Tring's methods of amassing long series and doing detailed comparative work, a practice intimately tied to both Jordan's vision of why and how to do "species-making," relied on access to a huge number of specimens from all over the globe. Rothschild's collection, for example, relied on enormous resources that must have staggered the Americans. (Certainly it impressed another attendee, the geneticist Richard Goldschmidt, who recounted how, when he declined a glass of whisky, "a very noble looking butler" replied "in a grand style, 'Sir, think it over, you have never tasted such a whiskey, and you will never taste such again.'"[149])

Obtaining the kind of material required to accurately assess variation—in turn the basis for both good species descriptions and the accurate classifications required for good theoretical work—had always been difficult. Now, for many entomologists in Europe, it seemed an impossible directive, like so many things in 1918, of a lost world. Thus, the fate of the program Jordan outlined for doing good, scientific systematics and contributing to biology in general, depended not only on convincing other entomologists to follow suit but on the persistence of the context in which the natural history network could amass large collections. This network, as a system developed over the course of the nineteenth century, proved only as permanent as the society from which it came.

British prosperity had depended in turn on "a complaisant Empire and on a peaceful world in which trade and capital flowed easily and whose center remained in the delicate complex of brokers, banks, merchant houses, and insurance and shipping companies in the City of London."[150] The scientific productivity of the Tring museum, from the boxes of specimens arriving and being sent out for exchange to the voluminous correspondence between collectors and entomologists, relied on the same peaceful trade in specimens, correspondence, and publications. Indeed, Jordan's methods bound the work based on collections even more intimately to the existence of this complex infrastructure.

During the postwar period, entomologists' limited access to material and the continued chaotic state of nomenclature, both exacerbated by the isolation of the war, made it extremely difficult for those who agreed with Tring's methods to follow them in practice. The American entomologist T. D. A. Cockerell described in 1922 how even those far from the economic distress of postwar Europe found it difficult to carry out Jordan's standard of research for the study of geographical variation. Within Colombian butterflies, for example, one found not only an enormous diversity of species but considerable local diversity. Some races even kept to particular mountain ranges or valleys. Cockerell agreed with Rothschild and Jordan that to develop a thorough understanding of such phenomenon, one needed "collections from all over the country, with long series of specimens—enough from each locality to show the true character of the insects, and avoid taking aberrations for subspecies." But this ideal proved extraordinarily difficult to fulfill, Cockerell noted, as he contemplated "the requirements for a thorough investigation of the Lepidoptera."[151]

Meanwhile, even among Tring's collecting network, some entomologists rebelled against the idea that a thorough investigation could be based on dead specimens. Jordan's good friend Miles Moss permitted himself a long lament in a paper on Sphingidae printed in *Novitates Zoologicae* that he had no information on "what the caterpillar looked like, what it fed upon, whether there was anything beyond the ordinary in its method of pupation or the egg-laying of its mother, or indeed anything at all about its habits as a living organism." Despite having been a collector for more than three decades, Moss proceeded to strongly criticize the driving assumptions behind the creation of large collections. "I have come more and more to regard a big collection," he wrote, "with feelings akin to dismay. Though it be the outward and visible result of years of patient toil ... it can impart such limited information about that great world of life which lies behind it."[152]

Jordan, who encouraged Moss and found these comments acceptable for publication in the journal of, by then, the largest private collection of natural history in the world, firmly believed that in contemplating a dead specimen the taxonomist actually studied insect *life*. He insisted that the work done in museums could be central to an accurate understanding of both the order in nature and the development of theories of how that order came about. The belief that such collections could be much more than "a cemetery of corpses" had been one of the driving forces behind Tring's call for better locality labels and long series. Jordan held that specimens in a walled roomed could reflect the field if amassed in certain ways, and thus the relationship between insects and their environment could be clarified. Indeed, Jordan had staked his own effort to raise the status of systematics on the careful study of geographical variation, associated nomenclatural reforms such as trinomial nomenclature, and the requisite study of long series of specimens from all over the world. He insisted that through adopting such methods systematists could then play a central role in both checking and developing increasingly accurate generalizations about the natural world.

Allies such as the American entomologist T. D. A. Cockerell joined Jordan's campaign for the continued importance of collections of butterflies, moths, and beetles to which so many naturalists had devoted time and resources during the preceding century. Cockerell certainly thought the difficulties of a "thorough investigation" based on insects, including the difficulties of "space and cabinets," could be overcome. "Even the largest Lepidoptera take little space compared with vertebrates," he wrote, "and are relatively easy to obtain, and it ought to be unthinkable that the great and wealthy United States cannot afford what is, for such a country, the merest trifle."[153] Cockerell had "the great and wealthy United States" to complain to, with at least the potential of convincing those with the means to support his work. Entomologists living in postwar Europe could not say that supporting systematics, particularly along the resource-intensive lines described by Jordan, would be for their countries "the merest trifle."

SIX

Taxonomy in a Changed World

Jordan and his fellows worked hard to recover their research programs, institutions, and networks after World War I. In doing so, they joined other scientists casting about for support and legitimacy amid limited resources. One of the most prominent shifts that took place within British entomology in the postwar era involved a new willingness to appeal to the utility of studying insects in the defense of man's welfare. "Applied entomology" reflected a utilitarian wing of the naturalist tradition that had thus far played a relatively small role in both British and European entomology. But during the first years of the twentieth century, and especially as recovery proceeded after the war, a whole series of new institutions and government positions reflected the increasing dominance of applied concerns.

As new opportunities and challenges arose for taxonomists as a result, they greeted the increased support available for the study of insects with both hope and concern. Suddenly, governments and the public—two targets of the publicity Jordan hoped would result from the congresses—were interested. In his excitement that young entomologists would now have a reason to stick to entomology, Frederick Merrifield had noted at the Brussels congress of 1910 that "we now know upon reasonable authority that the demand for qualified entomologists exceeds the supply." But doing taxonomy in the service of applied entomology influenced both the kind of collections amassed and the knowledge produced. As a result, the rise of applied entomology raised basic questions of identity and purpose for the tradition. What would be the role of systematists, for example, given new priorities and patrons? Were taxonomists searching for the fundamental order in nature or were they clerical staff working to provide identifications for other, supposedly more profound, work? Would accepting the task of the latter prevent them from doing the former, and ultimately turn them into species-makers indeed? And who, amid newly powerful institutions and disciplinary leaders, would get to say?

Competing visions within the naturalist tradition reached far beyond the occasional power struggles of so-called pure taxonomists and applied entomologists. Everything, it seemed, was up for reevaluation, and every answer chosen determined how naturalists would be spending their time as years passed. After the war, entomologists joined dozens of scientific disciplines in the contest for the limited resources available for research. All had to adjust to new values by which sciences and pastimes would be judged and who they would be judged by. These changes placed increasing pressure on systematists to—as one applied entomologist explained—"put their house in order."[1] And again, Jordan and his taxonomic friends cast about for organizational solutions to the problems they saw arising for the increasingly harassed "species-makers" in this new world.

THE RISE OF APPLIED ENTOMOLOGY

The rise of one field does not necessarily engender a challenge to another. But when manpower and money are limited, decisions to support one kind of worker or institution inevitably reduce resources elsewhere. After the war, the manpower and money that had traditionally supported scientific research became not only limited but in some places completely disappeared. The old patterns of international trade—the lifeblood of the British economy and fortunes such as the Rothschilds'—had been fractured, and despite a brief boom immediately after the war, by 1921 a marked economic depression set in. Inevitably the prosperity of the upper classes, the traditional patrons of natural history, suffered in the chaotic postwar world.

Decreased government resources influenced sciences such as entomology in Britain because by the end of the war the primary channels for the patronage of science had shifted from private sources to the state. The effort to gain government patronage had been active for decades. In 1906, for example, E. Ray Lankester insisted that scientists stop relying on the liberality of wealthy members of the scientific community "for the development of the army of science, which has to do battle for mankind.... The organization and finance of this army should be the care of the state."[2] But it took a world war to deliver this state-supported, scientific army. By 1919, T. D. A. Cockerell reported for his fellow Americans on how, although Britain had long been the "land of amateur naturalists" with little government support for natural history, the "organisation of British science for public ends" was "going forward with extraordinary vigor." Cockerell attributed this change directly to the war, as did many British entomologists.[3] Entomology had received "a great stimulus as a result of the

Great War," noted Imperial Bureau of Entomology chief S. A. Neave in his 1936 presidential address to the Royal Entomological Society of London (in 1933, the year of the society's centenary, George V added *Royal* to the society's name).[4]

The war had destroyed, along with so many other beliefs, at least one assumption on which the call for government patronage had been based; namely that "the means and lives of nations are alone commensurate" with the long-term research needs of modern science.[5] But four years of unprecedented economic and social mobilization had also provided a source of funding for those sciences that could demonstrate their relevance to the war effort. Many scientists and government officials worked to maintain these ties after the war ended. Meanwhile, an enormously expanded British electorate had resulted in demands that, as H. G. Wells insisted, the scientific community thenceforth recognize that it depended on a democratic political system for support.[6] Clearly a reorientation, if not a recreation, of priorities was in the works.

Often the call for reevaluation entailed much more than a simple redistribution of meager funds. Values were changing, as men and women looked back with shock at the past four years and tried to find in the previous century the sources of the horrors of the war. These reevaluations reached into the depth of politics and social life, including science. In their overtures to the state after the war, for example, scientists had to contend with a rising Labour Party driven by explicitly socialist ideas and spurred by discontent among workingmen. The Liberal Party had been controlled by wealthy elites, providing a context in which gentlemanly naturalists had thrived. Science had been hallowed in their world, so long as gentlemen scientists accepted the secularization and professionalism instilled by T. H. Huxley and company. But their kind of professionalism had often emphasized science's purity, even as it pried open government coffers. By contrast, Labour Party rhetoric indicted the ruling classes (and their pastimes) and glorified working-class heroes. It extolled activities that would bring immediate benefits to "the people," and ridiculed the indulgent luxuries of the upper class.

Natural history disciplines and institutions found themselves in a particularly sensitive place amid these shifts. Their museums had been formed within a society dominated by the rule of the landed and capitalistic classes. Their research programs, based on the accumulation of thousands of specimens, had been tightly tied to that society's economic and social structure, a point not lost on contemporaries. The American entomologist Austin Clark, for example, attributed his British colleagues' focus on "pure entomology" to a society in which the laws of inheritance had created a tremendous concentration of power and wealth. "Those who inherited the power," he explained, could occupy themselves with unpractical affairs, since their power conveyed a legitimacy "on anything they might choose to do." By contrast, American inheritance laws meant his compatriots "were judged according to the ability of their pasttimes to contribute to personal or national wealth."[7] Clark thus tied the character of British natural history directly to the laws of inheritance that Alfred Russel Wallace had placed at the root of British society's "social quagmire." And while Wallace had tolerated the contradiction of refusing to indict naturalists for the injustices of Victorian and Edwardian society, it now became evident that some saw natural history collections as the luxury of a bygone, self-indulgent age. Amid the postwar calls for thrift and protective tariffs, and as lowered wages resulted in crippling strikes, attempts to garner support for natural history museums for their own sake could seem shockingly self-indulgent.

Pressed by the need to find new sources of support according to postwar standards and values (and for some entomologists, personal commitment to making knowledge useful), the wind that would now dominate entomology had blown in with a vengeance during and after the war. By 1933, during the Entomological Society of London's centenary celebrations, most of the tributes sent to the society made reference to this new climate. "Since the beginning of the present century a great change has come over the status of entomological science in the public eye," announced one congratulatory letter. "Previously regarded as a harmless but useless pastime, it is now realized to be of vital importance to the community because of the damage that the unchecked ravages of insect pests may do to health and to commodities."[8] In focusing on the danger of insects to both human health and agriculture, entomologists, in other words, could now urge their importance to both socialists' concern for the community and capitalists' interest in commodities. This was a powerful justification indeed.

As an entomologist who had almost ended up a "professor at some High School of Forestry," Jordan knew the applied element of the naturalist tradition better than many of his British colleagues. Germany had a long-standing practice of including entomological knowledge in foresters' scientific toolbox, and forestry academies owned collections of beneficial and harmful insects. Still, the study of forest insects entailed a highly specialized form of applied entomology. It was the United States who entomologists widely acknowledged as the leader of applied work on insects in general, especially the study of agricultural pests. There, the rise of agricultural colleges over a wide territory had meant entomology courses had needed teachers since the 1870s. The Hatch Act of 1887 had provided $15,000 to each state to set up agricultural experiment stations, establishing an institutional framework in which entomology flourished. As the result of a few highly publicized successes in controlling insect pests, L. O. Howard could note triumphantly in 1894 that "economic" entomologists had "justified our existence as a class."[9]

The "class" of economic entomologists arose at different times in different countries. During his address at the first International Zoological Congress in Boston in 1907, Howard had noted the scarcity of such workers in most other civilized countries, including Britain and Germany. Although in 1894 naturalist F. Enock urged Britain to follow "the example set by the United States government in having an entomological section attached to the Agricultural Department," he saw no chance of the "culpably ignorant" government taking "up the matter."[10] Most entomologists conceded that any advance in Britain in economic entomology had been due to Eleanor Ormerod—and she, as a member of the upper class, had always refused a salary as degrading. In England, the total amount of funds spent on agricultural entomology at the turn of the century remained the same as in 1840.[11] Meanwhile, those who wished for more state support lamented the great difference in entomological resources available in the United States when an insect seemed to be at the root of some garden or hygiene problem. "In America," wrote one frustrated entomologist, "they would dispatch a man to the spot with orders to stay there until he had found it out."[12]

Not everyone agreed that the way toward more support for entomology should be through its application to agriculture. As we have seen, when Jordan arrived in Britain in the 1890s, those in charge of natural history journals did not even consider economic entomologists to be scientists. William Sharpe thought economic entomology "no part of the scientific extension of Entomology," but rather a department of Agriculture. "Attainment of ultimate truth only," he insisted, "can worthily be called Science."[13] When E. B. Poulton proudly announced at the 1912 congress that a friend at the British Colonial Office had once stated he knew "well that in an enthusiastic naturalist he would also secure a better public servant,"[14] he was referring to naturalists' moral worth and diligence rather than any particular data they could bring to bear on colonial problems of agriculture or health.

At the same congress, Malcolm Burr had adamantly countered Poulton's insistence that any motive other than truth for its own sake sullied scientific work. Now, the reentrenchment of resources during and after the war forcefully justified Burr's claim that practical need would inspire the most support for the discipline. Soon European entomologists began catching up with their heavily applied cousins across the Atlantic, establishing state-supported institutes for agricultural and medical entomology. When Howard had summarized the state of economic entomology in Britain in 1907, he could cite only dispersed and unorganized efforts, while noting that "official recognition of this science in Great Britain is slight." But within a decade, applied entomology in Britain had made enormous strides, following on the heels of the 1909 Development Act, which increased financial support for agricultural science.

The act was one of a series of unprecedented social reforms instituted by David Lloyd George and Winston Churchill (accompanied by both the Labour Exchanges Act and the National Insurance Act).[15] Cited as the first attempt at a national redistribution of wealth (long called for by Wallace and company), these acts were part of the reigning Liberal government's "People's Budget," a set of reforms explicitly aimed at stemming the tide of revolution that would soon engulf Russia. As Wallace had wished, increased taxation on the wealthy formed the financial backbone of the reforms.

It is important to note that the ability of the naturalist tradition to harness the new emphasis on applied concerns for both prestige and financial support arose in part because the new differed less from the old than it may at first appear. The British Museum had long been enlisted in the service of government enterprise through the identification of imperial natural resources. And as a state-supported institution, receiving funds even earlier than the physical sciences, the museum at South Kensington had always called on its usefulness to the state to justify its existence.[16] Thus, the museum simply fulfilled one of its primary justifications when capitalizing on the increasing concern with insects as vectors of tropical disease and agricultural ruin. But never before had the museum entomologists been at the center of such tangible links to the welfare of empire. Pronouncements such as Winston Churchill's 1907 claim that "Uganda is defended by its insects" hinted at the enormous potential for entomologists to take advantage of the colonial endeavor as a source of both resources and social justification for the discipline.[17] The potential became real when Secretary of State for the Colonies Joseph Chamberlain's movement to establish colonial administration on a scientific basis resulted in the Colonial Office's appointment in 1909 of a Committee for Entomological Research, to be headed by Guy A. K. Marshall (in 1913 this would become the Imperial Bureau of Entomology, in 1930, the Imperial Institute of Entomology, and in 1948, the Commonwealth Institute of Entomology).

Committees were not the only things being created. By 1911 more than thirty posts for economic entomology existed throughout the Empire. In London, the elevated status that applicability conferred on entomology as a central defense of Empire was reflected at the British Museum of Natural History by the promotion in 1913 of the Insect Room to the level of Department of Entomology.[18] When the traditionally "purist" Entomological Society of London took up new quarters a few blocks from the museum in the early 1920s, its fellows noted with transparent relief that the presence in the same building of the Imperial Bureau of Entomology "would ensure and enhance the importance of utility of 41 Queen's Gate as the head-quarters of our science in the British Empire."[19]

Increased support from the state inevitably meant fiscal oversight by politicians and taxpayers. U.S. entomologists had long been accustomed to the challenges of explaining their chosen field to such attentive judges. As August Busck explained to the no-doubt-bemused Lord Walsingham in 1916, "Everything here must have a more or less economic aspect in order to get the congressional appropriations."[20] Closely dependent on such appropriations, American entomologists kept a close eye on the public's perception of entomology. Smithsonian entomologists, for example, balked when a newspaperman asked why an entomologist of the Smithsonian Institution had traveled more than nine thousand miles to gather a few thousand specimens of diptera. "Why Go So Far to Look for Lot More Trouble?" the article headline asked: "The Smithsonian professor gives out the discouraging news that there are about 150,000 different kinds of diptera still unknown to science. But why worry about them if they are unknown? The two that are best known can give trouble enough."[21]

J. W. Tutt had acknowledged in 1895 that the public generally viewed entomologists as "widely different from other men, totally immersed in his subject, and stupid to the highest degree in all matters else."[22] But it had not mattered so much in the 1890s. Now even British entomologists had to worry about the public's perception of their field. Driven by necessity to abandon idealistic statements about pure knowledge, entomologists—including Jordan—capitalized on the postwar potential of applied results to further justify the "pure" wing of the tradition of naming, describing, and ordering species. Surely, they felt, a central complaint in Jordan's original justification for the congress—the lack of government support for entomology—could be addressed through convincing the public and politicians of the important services that entomologists could provide for the state.

Yet taxonomists in the United States had long struggled with the role they had adopted, both officially and unofficially, as providers of identifications for applied workers and other biologists. Indeed, as British entomologists had looked with envy at the amount of government support granted to their colleagues across the Atlantic, American entomologists, pressed by correspondents for identifications, sometimes expressed astonishment at what their British friends could spend time on. William Schaus, of the U.S. Department of Agriculture's Bureau of Entomology, after receiving a request from a British entomologist in 1923 for information on the migration of butterflies, wrote sarcastically: "The English are fortunate to have so much spare time on their hands that they can devote their study to the subject."[23] A year later, when asked for information on aberrational varieties on which so much had been written in Germany and England, he replied shortly, "No one in the Museum has had time to devote to these experiments."[24] In 1928, he admonished a colleague pressing him to complete a volume; "You surely understand that I have an enormous amount of work to do for the Bureau of Entomology and requests for identifications from all over the world."[25]

S. A. Neave captured the resulting tensions between "applied" and "pure" workers when he later explained how "the recent meteoric advance in the attention given to insects, more particularly on the economic side" had "somewhat thrown out of gear the relations between it and the other branches of the science, more particularly the taxonomic one."[26] Taxonomists had entered the century overwhelmed by the work required to clear up the errors of prior workers and process new material. In 1923, A. J. T. Janse still bemoaned the amount of work at the British Museum required to correct the poor systematics of earlier workers. "They require at least a dozen people to deal with it," he wrote, "or they should limit themselves to one twelfth of the work, that is asked now."[27] Now, increasing support for agricultural and medical entomology inevitably led to more and more requests for staff entomologists to identify the little beasts concerned. Complaints grew louder as the years passed. By 1930, a British carcinologist, Keeper of Zoology W. T. Calman, described how "It is a common experience with us at the Natural History Museum to have some mangled fragments of an animal brought in by a practical man who expects to be supplied with the name of it while he waits. I am afraid that he often goes away with a low opinion of our competence."[28]

In 1921 Fred Muir, assistant entomologist for the Hawaiian Sugar Planters' Experiment Station, complained about receiving material "from all over the place" for determination. He insisted he had been trying to get up to date, "but the more I send back the more comes along!" He would have to put a stop to it soon: "Some kind of international understanding will have to be found! One cannot refuse to do it yet one cannot spend his whole time on it."[29] Jordan's initial justifications for international organization had been based on raising the status and power of entomologists within zoology as a whole. Now, as taxonomists became swamped by entomology's new-found importance, the incentives to organize internationally were strengthened even as the environment in which they must find a place for work on species was being transformed.

"SOMETHING AMISS"

This new environment was brought home to Jordan when, in the summer of 1925, in the company of his daughters, Ada and Hilda, and Walter Rothschild, he crossed the English Channel and finally made the journey to Zurich for the Third International Congress of Entomology. The account of the first evening's meeting described the feelings of the organizing committee as they waited for the congress members to arrive. Jordan, Walther Horn, Anton von Schulthess, and the other Swiss entomologists who had organized the congress stood anxiously in the hall reserved for the opening reception. Would all who had promised to attend actually come? With relief, they watched as the stairs to the hall filled with entomologists from near and far. Old friends reunited, and correspondents met face to face for the first time. The president and his helpers relaxed: "the success of the congress was secure."

The memory of the war defined the first meetings, as statements of internationalist sentiment took on a whole new meaning. The president of the Swiss Entomological Society, Arnold Pictet, announced with old-fashioned internationalist fervor that the gathering expressed the scientific solidarity of all nations, one of the highest representations of the progress of man's intelligence and conscience. Firmly aware of recent evidence to the contrary, Schulthess pointedly concluded his welcome with the hope that no allusions within or outside of any of the sessions would disturb friendly agreement and impair the congress's proceedings.[30] Jordan took the opportunity in his paper on fleas to emphasize the "drawbacks of nationalism in science" by noting that the demonstration that fleas transmitted bubonic plague from rats to humans had appeared as early as 1902 in Russian, but the lack of a summary in French or other Western language had prevented this conclusion's acceptance abroad.[31] Jordan and Horn even turned the appeals to applied entomology into a new argument for internationalism. "Entomology is a national economic factor of great importance," they wrote in their introduction to the *Proceedings*. While a small number of skillful and awkward leaders had led their peoples to fight against their own species, an infinitely larger army of insects destroyed lives, property, food, and forests with no attention to either national feelings nor a sense of justice.³² Entomologists, they insisted, must follow nature's example, and abandon national feeling and petty political disputes.

Couched in internationalist rhetoric or no, the new-found interest in entomology as relevant to human health was on full display at the Zurich congress. In his welcome address, Karl Hescheler noted how the world war had shown that many problems of human health were in fact problems of entomology, and as a result, the international congress had become a matter not only of promoting the science of entomology but the well-being of mankind.³³ H. S. Fremlin, of Britain's Ministry of Health, gave a long list of threats posed by insects to human health and agriculture, citing in particular the huge tracts of fertile land in Africa rendered uninhabitable due to the presence of the tsetse fly.³⁴ Addressing the general session, the American "King" of economic entomology, L. O. Howard, in stark contrast to Poulton's insistence on tight distinction between pure and applied, tried to maintain the portrait of unity in a new way. He pointedly insisted that

> all workers in Entomology are really economic Entomologists, whether they are willing to admit it or not. Insects are the worst rivals and enemies of the human species, and we must learn absolutely everything about all of them. Therefore the work of the taxonomists, biologists, geneticists and all the rest of them is helping to solve the multifarious problems that confront the economic Entomologist.³⁵

Jordan could agree with much of that statement, particularly given the role his and Charles Rothschild's taxonomic work on fleas had played in demonstrating the role of *Xenopsylla cheopis* in plague transmission. In working on flea taxonomy, clarifying the systematics, and making sure plague workers got their identifications right, Rothschild's work had demonstrated that "the study of systematics is of fundamental importance for applied Entomology, as it is, in fact, for all biological research." Yet he made a point of noting that, for Charles Rothschild, the investigation of "the number of species, their distinctions, variation, distribution and relationship," had been "a matter of love."³⁶

The place of the taxonomist may have thus seemed secure given the clear importance of work on certain insects to the century's new priorities and values. Certainly, the work on fleas had provided a good example that Jordan's rule—that "science must not prejudge anything and must not be negligent anywhere"—applied to "species-making" as well. Someone had to figure out "what was what" in the fight against insect pests and insect-transmitted disease, by naming, describing, and ordering the insects in question. But limited resources demanded prioritization. The discovery of the role of a particular flea species in the transmission of plague had certainly justified entomologists' attention to this group. Channeling the efforts of collectors to "where their labours are most needed" had become important indeed. But defining "most needed" would now be done according to quite different criteria than that used to revise or monograph a colorful group of butterflies or fascinating group of beetles.

First and foremost, applied concerns influenced what kind of insects could be studied in depth. As a result, museum taxonomists soon keenly felt the prospects and the pressures of their new-found importance. Entomologists who had taken economic jobs, such as Alfons Dampf, the chief entomologist of the Federal Office for the Protection of Agriculture in Mexico, became so occupied with dangerous insects (in Dampf's case, "the fruit fly problem") that they had to warn correspondents requesting more traditional specimens that they could unfortunately "pay no attention to butterflies."[37] Even when one focused all one's taxonomic time and effort on insects of economic interest, the work required to identify and order the beasts concerned proved daunting. If one added Jordanian standards to how exactly one completed the taxonomic study of a group, it became overwhelming.

The resulting tensions between systematic workers and those who wished to have their organisms neatly and quickly named in the course of applied work in medicine or agriculture was on full display at the congress. R. T. Leiper, of the London School of Hygiene and Tropical Medicine, used his time at the Zurich congress's podium to chastise entomologists for being indifferent to the needs of medical entomology and pleaded for "a closer working connection between pure and applied Entomology." Rothamsted Experimental Station entomologist C. B. Williams, who had worked on both sides of the "class divide," countered that any failure on the part of entomologists to supply information to the medical men arose not from lack of interest but rather from "lack of time and money to do more than they already had to do."[38]

Discussing his work on fleas at the congress, Jordan insisted that good systematics was not only critical to the study of the origin of species but that it could only be done through meticulous, detailed comparative research.[39] But the time and resources required by Jordan's ideal taxonomist did not operate according to the time requirements of applied questions. Indeed, to those overwhelmed by both the obligations of government service and insect diversity, Jordan's program for a reformed, scientific systematics could seem the luxury of another world. In 1927, American economic entomologist Elmer Darwin Ball protested against Tring's calls to name subspecies with trinomials, for example, as follows:

> If we knew our bugs one half as well as the bird men do their birds, we might talk intelligently on sub-species, geographical races, varieties, etc., but in the present condition of entomology, with the overwhelming amount of unworked material, the large areas from which no material has ever been taken, and especially in groups where food plants may be the major factor in changing the character of the insect, we are not in position to make these discriminations.[40]

Pressed for time and overwhelmed by the number of unnamed forms, California Academy of Science entomologist E. P. Van Duzee expressed similar exasperation with arguments that detailed morphological work, such as the study of the genitalia, be required to determine species. "If I can not determine a species without dissection," he wrote in 1935, "it can stay put just where it was."[41]

Jordan had tied the status of systematics as a science tightly to certain methods of doing research. Taxonomic entomologists' struggle to adopt such methods amid an increasing work load thus became an issue of whether taxonomy would be a science at all. When Jordan's good friend Walther Horn addressed the Zurich congress, he described the resulting situation with a large dose of gloom in a lecture entitled "On the Plight of Systematic Entomology, with Special Regard to the Conditions in Germany, and Suggestions for Reform."[42] What, he asked, would be the future of systematic entomology if past trends continued? Faced by a chaotic literature and a mass of undetermined material in collections, Horn attempted to explain the source of systematists' troubles. The war, he announced, had put a halt to utopian reforms. The decline in personal fortunes, the constant pressure to make money that had caused men to sacrifice their love of science, and "last but not least," the stamp-collecting approach popular among many who studied insects were all to blame.

To confront this somber situation, Jordan and Horn set in motion a plan to help achieve the international understanding for which Fred Muir had pleaded. Prior to the Zurich congress, they circulated, with Horn as author, a prospectus outlining major points for taxonomists to consider in the section on systematics and geographical distribution. In particular, Horn wondered whether a unified statement might tend to a better understanding and international cooperation between taxonomists and those for whom they make determinations.[43] As a direct result of Horn's suggestions for reform and as befit a deliberative body styled on parliamentary lines, the section drew up the "Horn-Escherich-Nuttall Resolution," (German entomologist Karl Escherich and Cambridge University entomologist G. H. F. Nuttall had joined the call for change) for the congress's approval. The section belonging to the "species-makers" thus came before entomologists as a whole to register their complaints and urge reform.

The opening statement of the taxonomists' resolution placed the problems created by the rise of applied entomology front and center:

> An enormous expansion in Applied Entomology has recently taken place throughout the world. This has necessitated, as a first step, the exact determination of an immense number of insects. The result has been to show the utter inadequacy of the present means for undertaking such work.... Everywhere the systematic specialist is over-burdened, and his load has now become an intolerable one.

This new challenge allowed entomologists to return to an old complaint; namely that in academic circles, both systematic and applied entomology had not received due recognition. "Entomology has long been the Cinderella of the Sciences," the resolution insisted, echoing a point made by Auguste Lameere at the first congress. Systematic entomology in particular continued to be regarded as "definitely inferior" to other branches of zoology. As a result, professors advised postgraduate students not to undertake research in systematic entomology on the ground that it led nowhere, creating, the resolution authors insisted, a grave danger to science, both pure and applied.

A series of recommendations followed: that universities establish chairs of "Systematic Entomology"; that good systematic work qualify candidates for degrees; that the number of entomological assistants in museums be increased in direct proportion to the amount of species work entailed; and that candidates with experience in systematics be appointed to administrative posts in museums. Meanwhile, in an inspiring display of solidarity, the section for applied entomology passed its own resolution firmly in support of the concerns expressed in the "Horn-Escherich-Nuttall Resolution," on the grounds that sound insect control depended on good systematics.[44] Jordan must have been pleased by this show of unity—a specific and clear effort to move the discipline forward. It served as a fine example of how changing priorities might be harnessed to serve old traditions as well as new ones. The congress unanimously agreed to the resolutions at its general meeting, providing an excellent example of how efforts to remedy the "crisis" could be channeled through these international meetings.

Yet in working to extend the spirit of the resolutions into action, Jordan faced competing interests between the various "classes" of entomological workers. A few months after the congress, he joined a discussion on "The Place of the Systematist in Applied Biological Work" held by the Association of Economic Biologists, an organization founded in 1904 as part of economic entomologists' campaign for status, legitimacy, and state support.[45] Jordan attended the meeting with the Zurich resolutions in hand. But he arrived as an outsider, standing before an audience who perhaps associated him with a dying patronage system and tradition, rather than an army of "independent" entomologists united by the International Congress of Entomology. The members of the association had their own concerns, their own identities and patrons, and their own visions of the future.

They even had their own ideas about species. As he listened to the speeches and prepared to make his pitch for more international organization, Jordan's old fears that other workers paid little attention to the work of systematists must have returned in force. For while Jordan had emphasized in all his work that an objective definition of species could in fact guide species work, he now heard speakers wondering how one could classify species "when one neither knows what constitutes a species nor can define its limits in any particular case?" Various contributors, from a range of institutions, made it clear that they thought systematists described arbitrary units. After one commenter spoke, H. A. Baylis, of the British Museum, touted an old line: "Probably many of us have long wanted to know what a species is. Now we are assured by an eminent authority that it is merely what a competent systematist says it is." But that paled in comparison to one speaker's insistence "that the numerous subdivisions that have taken place in biology have made systematic enquiry obsolete."[46]

In rising to put the Zurich congress resolutions before the meeting, Jordan ignored the archaic definitions of species. He simply announced that, in consideration of what he had heard, "there is something amiss in the relation between systematics and applied biology." He proceeded immediately to the organizational, rather than intellectual, requirements of a remedy, briefly repeating his old defense of systematics on the grounds that it provided the basic foundation for doing good work in other fields. He firmly insisted that given the resources available to systematists, they simply could not provide the information applied biologists desired. "The student of applied biology requires information from the systematist which the systematist cannot fully give," he explained, for systematists could "not possibly cope with the enormous demands made on them." He then presented the "Horn-Escherich-Nuttall Resolution" as aimed at ameliorating the "utter inadequacy of the present means" for undertaking the work required of systematists by the enormous expansion of applied entomology. "Now the question arises," he said, "as to whether we are satisfied with having expressed these opinions or whether we wish to see the recommendations carried into effect." He called for the association to join the resolution and appoint a member to the committee tasked by the entomological congress to take "the matter in hand, keep the agitation alive and enlighten the public."[47]

As the comments of association members demonstrate, it seems that some biologists needed as much enlightenment as the public. After Jordan finished, William B. Brierley, of the Rothamsted Experimental Station, requested that the discussion return to the present-day state of systematics and not some ideal state of things. "If we stray from what is and discuss what might be," he urged, "we shall lose touch with reality and take refuge in a variety of Utopias." But Brierley had a different vision of entomological reality than Jordan. Brierley defined a systematist as "one who classifies and gives names to specimens," and viewed systematists' primary goal to be the provision of identifications for economic workers. Apparently trying to console systematists' feelings of neglect, Brierley claimed that

> the more successful the systematist is, the less necessary does he, personally, become, for once his diagnoses and keys are printed other people can use them as well as he can and themselves classify and name their specimens. The systematist, in fact, works towards his own elimination and should rather hold it a source of pride than regret that he no longer occupies the pre-eminent position as in the younger days of biology.

Perhaps in deference to Jordan's presence, Brierley admitted that "of course the pioneer and the rare man gifted with the *systematische Blick* will always be eminent." But he admonished those who gave systematics more importance than it deserved. "Like all tools ... it must be used only for its special and legitimate purposes and kept in its proper place"; namely, to provide names for use by applied and biological workers, "the defined end objective" of systematics.[48]

Jordan believed, by contrast, that, done well, systematics could be as biological as the next science, and that its end objective was not simply to provide names for applied and biological workers. He and Rothschild had written during their defense of naming varieties in the 1906 *Revision of the American Papilios*, that "the describing and cataloguing of 'species' are certainly the basis of systematics, but also the lowest degree in this science. After that comes classification, or in other words, research in relationship."[49] Indeed, Jordan often blamed both the low status of systematics and its poor results on a naturalists' misguided emphasis on completing the catalog of names. Furthermore, he believed that if systematists' only aim was to confer names, then the species would indeed most likely be arbitrary constructs "made" by the taxonomist—just as the old jibe of "species-makers" implied.

VARIOUS UTOPIAS I: THE ITHACA CONGRESS

The Fourth International Congress of Entomology, held in Ithaca, New York, in 1928, exemplified the opportunities and challenges facing taxonomists amid such different ideas about the purpose of taxonomy. In recounting the meeting's results in the pages of *Science*, G. W. Herrick concluded that whenever "the jingoes of this country" talked of war, entomologists would remember the men they had met at the congress, and thus be loath to join in international quarrels. "Unquestionably every such international meeting of men from different countries," Herrick concluded, "whereby they come to know each other as human beings, tends away from war and toward peace."[50] Like Herrick, in planning the congress Jordan focused his attention on establishing unity across the national divisions highlighted by the war. But American entomologists, more isolated from European politics, had their eye on a different kind of balance of power as plans for the congress proceeded. As a result, the congress demonstrated how utopian resolutions sometimes belied very real quarrels within the "nation of Entomologists."

The quarrel began when Jordan's good friend W. J. Holland, director of the Carnegie Museum of Natural History, roundly criticized both the choice of Ithaca as the site of the meeting and the increasing dominance of his "economic" colleague L. O. Howard during the congress preparations. Amid competing claims for what entomology should be, Holland saw much at stake in both the choice of the congress's venue and its organization. As early as 1925 he had written to Jordan that he hoped the congress organizers would settle on Pittsburgh—a city, he argued, that was "more important entomologically than New York" since its collection of exotic Lepidoptera was vastly superior to anything in the National Museum in Washington.[51]

But of course entomologists held widely divergent views regarding the criteria to be used for defining "most important entomologically." A few years later, Holland received word that Jordan, Howard, and Walther Horn had chosen Ithaca, New York during planning meetings held at the International Congress of Zoology in Budapest in 1927. He dashed off a long letter of protest to Jordan, in which he complained that "the International Congress of Entomology consists of yourself, Walther Horn, and L. O. Howard, and all the rest of us are expected to be your benevolently inclined assistants." In effect, he noted, this means that "Our friend, Howard, can now apparently say with propriety, as Louis XIV said of himself: 'l'Entomologie, c'est moi!'" Holland drew a clear analogy to the changing values of the age regarding the basis of just governance when he concluded,

> Really all joking aside, don't you think that it would be of some advantage to the cause of our favorite science for the imperial leaders therein to endeavor to awaken and preserve an interest in a larger number of persons.... I never have known an institution to succeed, the officers of which have undertaken to "go it alone," without enlisting the sympathy and help of the good people about them.[52]

Jordan, noting that "he would be happy if in 13 years time he could say *Donnerwetter* in such a hearty way to anybody who had done something with which he did not agree," tried to explain the reasoning behind the intense concentration of authority in the preparations for the congress. "The Entomological Congresses are still in their infancy and require looking after by willing friends," he wrote. "That nurses are a bit domineering in the nursery is quite true." But Jordan defended his approach in no uncertain terms. The domineering nurses, he explained, were needed "for the sake of pure entomology, which stands in danger of being swamped by applied entomology."[53] Reading between the lines to discern the real source of Holland's anxiety, Jordan assured him that while the Ithaca congress would be in Howard's hands so far as applied entomology was concerned, pure entomology would be in Holland's hands. He knew as well as Holland that placing L. O. Howard "as King" would imply that economic entomology *was* entomology.

True to the tradition of congresses to reflect the priorities of the host country, applied entomology indeed did rule the 1928 congress. In traveling to New York and visiting collections and agricultural institutes, European entomologists got a glimpse of the future, one in which public opinion increasingly influenced how scientists would spend their time. They learned of the wealth of American institutions devoted to economic entomology, from the Cereal and Forage Insects Field Laboratory of the Bureau of Entomology to the Chemical Warfare Service. The usual sections on systematic entomology and zoogeography, nomenclature and bibliography, and morphology-physiology-embryology certainly appeared, as did sections on genetics, ecology, apiculture, forest entomology, and medical and veterinary entomology. But in contrast to previous congresses, where just one set of sections met on "economic entomology," the Ithaca congress held five different sections devoted to applied concerns, including the use and toxicity of insecticides. This was quite a different program than the systematics and faunistic-filled sessions of the 1910 and 1912 congresses. And no doubt the day-long field trip to the New York State Agricultural Experiment Station seemed a far cry from the Oxford congress's leisurely field trip to Tring.

While the studies on insecticides may have steered far from the traditional purview of naturalists, it is important to point out that many of the "applied" papers addressed standard "natural history" research problems, including organisms' relation to the environment, the causes of distribution, and the study of life history. The applied entomologist (or, as the Hungarian entomologist Josef Jablonowski called them, "field entomologists") studied some of the same questions that had inspired Jordan to study natural history in the first place.[54] What animals occurred where and why? What environmental factors influenced distribution and population size? What interactions with other organisms determined the make up of a community? Given these questions, some of the results coming out of agricultural experiment stations surely would have been of interest to those entomologists, such as J. W. Tutt, who had long since campaigned for more attention to life history and behavior. But answering these questions now aimed at controlling insects, rather than understanding the biology of insects or evolution for its own sake. Under the new rubric of "ecology," for example, the study of what animals occurred where became the study of what factors influenced the occurrence of infestations and what would stop them. For many attendees at the 1928 congress, "the control of injurious insects" formed the primary goal, an aim that at least one contributor insisted (in stark contrast to Poulton) provided "the most beautiful and noblest object in the life of a scientific man."[55]

In addition, some of the methodological trends apparent at the 1928 congress differed starkly from the strongly observational basis of natural history defended by Lameere in 1910. It was a trend noted, and countered, by Royal Chapman in his paper on "The Measurement of the Effects of Ecological Factors." He urged members to remember that, as ecology inevitably became quantitative, observational natural history might be brushed aside "by the cold dry calculations of a mechanistic mathematics." Surely, he insisted, the results of purely observational natural history must always form the basis of quantitative work, and the fixed and controlled conditions of the laboratory be balanced by an awareness of the complexity of ecological phenomena in nature.[56] That Chapman had to insist on such a point shows that the voice of Jordan's young mineralogist friend urging the methods of physics and chemistry as a basic prerequisite for doing science in the 1880s, could still be heard loud and clear.

In opening addresses and closing farewells, some of the economic entomologists tried to assuage any existing anxieties regarding their dominant presence with generous overtures to the "pure entomologists" attending the congress. L. O. Howard used his presidential address to insist that the great work done in economic entomology in the United States had also shown "in a very forceful way the basic value of the labors of those ardent Entomologists who have been carried away by the fascinating scientific interest of other aspects of the science." Thus the members of the section on applied entomology would continue to look upon the members of the other sections "with deep respect, perhaps tinged with awe." He hoped that, conversely, the members of sections dealing with "pure" science would look to the applied men as useful, since they were helping "to reform the old ideas of Entomology and are bringing public appreciation and public funds to its support."[57]

In comments reminiscent of Lameere's first address of 1910, Howard spoke of the importance of the study of insects in general. He wondered why educational institutions did not give more importance to entomology (a complaint that must have bemused some European entomologists after reading the list of members from agricultural colleges). Why, Howard asked, must the teacher of zoology in inland institutions demand that his laboratory students devote hours and hours to pickled sea urchins when insects are available all around? He defended entomology's status as an independent science, and he "quite agreed" with Lameere's insistence that "Entomologists are almost as distinct from the Zoologists as the Zoologists are from the Botanists." This justified "establishing three categories, giving Entomology an importance at least equal to that of Botany or to the rest of Zoology."[58] Obviously, a "zoological" rival still existed that could be called upon in the name of emphasizing the distinction of one's own army, but in spite of Howard's diplomatic praise of systematic workers, it was quite clear who he thought entomology's new generals should be.

Once again, taxonomists at the congress expressed tremendous concern regarding how they were going to be able to get any work done, whoever defined the ultimate goals. As J. P. Kryger, of Denmark, explained: "The zoological museums are year by year becoming more handicapped. People have discovered that the museums are not only collections of dead animals, but also collections of living men who know all about animals which play a part in man's welfare." The number of objects sent for identification and remedies requested overwhelmed the modern museum man (a man Kryger facetiously defined as "a person with a typewriter who can do nothing else but answer inquiries"). No one, he lamented, asks for science.[59]

Walther Horn, intent once again on solving entomologists' problems through better organization, prepared the 1928 congress's section on taxonomy, distribution, and nomenclature with such complaints in mind. Opening with a long address on "The Future of Insect Taxonomy," he looked at the situation from the perspective of an entomologist struggling to maintain his own institution, now known as the Deutsches Entomologisches Institut. He began by recalling a visit to the Prussian Ministry of Education two years earlier when he had been met by the words: "You are coming to save a dying science." Reviewing the totally inadequate workforce, methods, and organization available to the field, Horn himself concluded that "Taxonomy is going to the bad." "The flourishing time of taxonomy is over," he announced bluntly, "In the struggle with evolutionary theories and other branches of biology, taxonomy sinks to a kind of Cinderella. There are exceptions, which, however, only prove the rule."[60] Leaving these words and a string of complaints regarding how taxonomists worked hanging in the air, he opened the "Forum on Problems of Taxonomy" with urgent flair. "Vigorously expanding research in ecology, life histories, morphology, genetics, and applied entomology has not been accompanied," he warned, "by a corresponding increase in taxonomists, upon whom these researchers are dependent for determinations."

Horn then proceeded to list the critical questions facing taxonomists as they tried to carve out a role for themselves amid new priorities: Did the general public have a right to insist government officers determine material? Must governments pay for this service? To what extent were specialists "morally obligated" to determine insects for other entomologists? What recompense could they expect for their time? Horn wished his colleagues to consider even whether insect collections were economically justified, and if so whether they should be centralized and supported by public money. Finally, in an effort to deal with the legacy of hundreds of private collection owners, he wondered whether the entomological public had a right to demand eventual public ownership of private collections "which have become the basis of important published work?" (Applied to bird collections, this would soon prove a question all-too-close-to-home for British ornithologists.) Fred Muir, who was present during the discussion of Horn's paper, must have been pleased with this detailed attempt to build the "international understanding" for which Muir had pleaded in 1921.

As Jordan had hoped, entomologists took up Horn's challenge in the interests of building better organization as the means of ensuring good work. René Jeannel, of the Muséum National d'Histoire Naturelle in Paris, and Gordon Floyd Ferris, of Stanford University, debated the advantages and drawbacks of building centralized, national collections.[61] Holland spoke on "The Mutual Relations of Museums and Expert Specialists" in an effort to clarify the unspoken rules governing compensation of experts who determined material for others.[62] Muir pleaded for some system of remuneration to be established for those completing identifications and determinations. "The payment of such work will become absolutely necessary," he argued, "as the museums cannot keep enough specialists on their staff to do all that is needful and they have no right to expect private specialists to work at their material for nothing. They would not expect it of a lawyer or an engineer!" J. B. Corporaal agreed, and pointed out that those working on groups of "wide economic importance, such as mosquitoes or weevils for instance, may often be asked to determinate large numbers of specimens, or worse still, single and often defective examples without precise indication of locality, the study of which is for himself as a rule of comparatively little interest, but frequently demanding considerable time." He insisted that the pay must be per day, rather than per specimen or species, in order to avoid shoddy work.[63] Clearly, the shadow of the fabled "species-maker"—a man who dreamed up species boundaries for either pay or prestige, rather than truth—loomed over taxonomists' conscientious efforts at reform.

Jordan's early letters suggesting an international congress had emphasized the need to deal with "the huge mass of insects as was necessary for Entomology as a *branch of science*." Now, both the 1925 "Horn-Escherich-Nuttall Resolution" and the 1928 Forum on Taxonomy explicitly placed taxonomic entomologists' wish list (which had, of course, existed for a long time) within the context of the rise of applied entomology rather than science for its own sake. For a time, Jordan proved willing to draw on the rise of applied entomology as a justification for more support for taxonomists. He announced, for example, in a statement to the *New York Times* regarding the congress, "It is absolutely necessary to arouse the interest of the public in natural history. The study of entomology is for the welfare of the public, on account of the role insects play in hygiene, in transmission of disease, and in destruction and protection of agricultural and horticultural crops, and forestry."[64]

Like a good diplomat, Jordan hid anxieties over territory (exemplified in his confession to Holland that he feared applied entomology would swamp pure entomology if the congress's "nurses" did not exert strict control) behind gestures of peace and cooperation. Yet in the midst of his own contribution to the congress entitled "On Some Problems of Distribution, Variability, and Variation in North American Siphonaptera," he permitted himself a gentle lament regarding the grounds upon which entomologists increasingly established research priorities. Bemoaning that "in Louisiana...some years ago over 4 million muskrats were killed for the sake of their pelts and not a single flea from this rodent is in the collections at Washington and Tring," he noted, "It looks to me as if insects must first become dangerous before they arouse an interest."[65] Now, indeed, the flea specimens were arriving in droves, and as the acknowledged "dean of siphonapterists," Jordan was constantly asked to provide quick identifications to medical workers. Such demands, while proof of taxonomy's importance, contrasted with the meticulously slow methods he believed necessary to make systematics scientific. Thus, although the dismissive attitude of Charles Rothschild's Cambridge professor toward the point of classifying fleas now seemed aptly rebuked, it was not quite clear what this new context would mean for responding to those who saw systematics as unscientific.

VARIOUS UTOPIAS II: THE INTERNATIONAL ENTOMOLOGICAL INSTITUTE

When Jordan presented the Zurich resolutions to the Association of Economic Biologists, he had been reprimanded for indulging in utopian schemes. But the incentives to reform proved too strong to back down. Intent on turning ideas into action, Jordan joined his friend Walther Horn in an effort to channel the rather vague concerns and resolutions expressed by the Zurich congress into plans to create an Entomological Institute for International Service. Aimed at providing a formal, institutional presence for the long list of entomologists' reforms, the fate of this particular utopia highlights some of the challenges that faced Jordan and his colleagues' efforts to maintain the entomological wing of the tradition via better, international organization.

Horn's letters to Jordan and others during the post-war period reflected his despair over the state of entomology in Germany.[66] In 1924, he confessed to Jordan that he knew of no one willing to carry on his work for the Deutsches Entomologisches Institut, since the position required one always to stand at the center of "the fight."[67] Horn's situation hints at the highly personal incentives driving some individuals to international organization, albeit incentives no less important than the usual internationalist rhetoric or practical calls for standardization. No doubt Horn longed for a supportive community that would provide respite from having to fight alone. "Cooperation and organization of the entomologists of the world would be the basis of the institute," Horn announced in Ithaca in 1928.[68]

Like the "Horn-Escherich-Nuttall Resolution" approved by the Zurich congress, Horn's prospectus for the Entomological Institute for International Service read like a Christmas wish list for systematic entomologists. To do taxonomy well, given all the problems canvassed during the 1928 Forum on Taxonomy, entomologists required a publication office that would issue catalogs and monographs, nomenclature lists, bibliographies, and rare books; an enquiry office to provide a "clearing house of information for pure Entomology"; a historical section that would collect biographies of entomologists, photographs, and histories of important collections; and answers to a myriad of wants lumped under "Museology and Systematics." In addition, the institute would provide space for the secretariat of the International Entomological Congresses, with the task of running publicity campaigns for the resolutions and rules emanating from entomological and zoological congresses and "drawing the attention of offending authors to them." Such a clearing house could organize specimen loans and the preservation of types, standardize literature citations, and distribute propaganda on the importance of entomological research to the press. The latter was very explicitly placed in the context of inspiring public pressure on both governments and academic departments to provide openings and adequate salaries. "If the public is interested," the circular concluded, "departments will listen."[69] Entomologists certainly knew who filled the coffers now.

Committed to better organization as the route to efficiency and success, Jordan campaigned for Horn's idea with such commitment that entomologists began referring to "Jordan and Horn's Institute." Upon his return to Tring after the 1928 congress, he printed dozens of copies of the congress's resolution in support of the institute and identified a list of associations to approach for support.[70] He also scheduled an informal meeting of some of his British entomological friends.[71] In his appeals to colleagues to support the international institute, Jordan addressed each and every time-consuming drudgery that prevented entomologists from carefully studying the range of geographical variation in a set of butterflies or beetles. He repeated the old exhortations for better organization so that entomologists could do research instead of hunting down references or "laboriously collecting separata and copying original descriptions from books he has on loan only, etc. etc. All such clerical work should be taken off the shoulders of the taxonomist and carried out by one Central Institute."

Again taking advantage of his wide network of correspondents, Jordan canvassed his most influential entomological friends to support the institute. He suspected that W. J. Holland, in Pittsburgh, would "laugh at me for plunging at once into a further matter which entails much work and thinking over, instead of putting names to the various kinds of 'bugs' which await my attention." But, perhaps chastened into consulting Holland and others in order to ensure success, he emphasized that the project could be realized only with the extensive help and guidance of "scientists of your standing, of international renown and great personal influence." Notably, Jordan now placed the call for such funds firmly in the context of the good of mankind. He admitted that the two million dollars he and Horn estimated the institute required seemed a large sum, but thought it "a small one for work undertaken for the great benefit of all humanity." He described to Holland how

> nations are spending untold millions on defense against one species of creation, their own species, and the immense losses in goods and lives yearly incurred through the actions of insects are passed over by ignorant or callous statesmen of Europe. It is a situation bordering on the ridiculous. Should not we Entomologists do our utmost to set things right in our domain for the benefit of mankind?

Jordan also wondered if Holland would talk it over with his friends at the Carnegie Foundation, adding, with close attention to the standards of those prospective patrons, "If you will take a lead in this international humanitarian project, the prospect of success is great."[72]

Jordan had to write again three months later to remind Holland of the undertaking,[73] and it is not clear whether Holland ever replied. Perhaps Jordan's pragmatic appeal to the good of mankind conceded too much to Howard's "class of workers." Jordan operated on the premise that those concerned could be convinced that applied entomology depended on good, scientific systematics yet so far the experience of at least some American taxonomists subject to the directives of their economic brethren did not justify such optimism. One worker complained that the chief of the U.S. Bureau for Entomology had a "poor idea of systematists," an opinion that resulted in very tangible results when only the "field men" received raises in pay.[74] Carrying out Jordanian systematics on a low salary, much less without the cosmopolitan network required, proved a utopian suggestion indeed.

Despite Holland's silence, Jordan found ready allies in his colleagues at the Entomological Society of London. After a lengthy address reviewing the totally inadequate "machinery" available for entomological research given the "gradual commercialization of the science," for example, the president of the society for 1928, James Edward Collin, urged his fellows to consider Horn's proposals for the new international institute in detail in preparation for the next international congress.[75] But, as Jordan waited for advice and suitable patrons, his and Horn's plans for an Entomological Institute for International Service floundered in the face of two crucial variables determining the success or failure of any scientific organization.

Jordan had been optimistic, in his letter to Holland in July 1929, regarding both the state of entomology in Great Britain and the role of taxonomy within the discipline, writing, "The Imperial authorities have grasped at last that Entomology is of fundamental importance for the overseas countries and that applied entomology must be based on sound systematics." He even permitted himself the hope that "our progressive government, so-called Labour... will succeed in reducing the construction of machinery for the reduction of mankind and employ the money saved for the welfare of mankind instead. We are all responsible."[76]

Four months later, the stock market crashed and the Great Depression began. Thus, within a little more than a year of the Ithaca congress, potential sources of financial support dried up in the face of a global economic depression, and those searching for "nice berths" had a difficult time indeed. "It is practically impossible for any one to get into a scientific position in the United States," reported John Merton Aldrich, an entomology curator at the Smithsonian, "they are reducing the forces everywhere, as well as reducing salaries."[77] By 1933 all printing and all travel from the Smithsonian was suspended, "heroic measures, but the only alternative was for the staff to take some leave without pay in addition to the 15 per cent cut."[78] Jordan noted that depreciation was hitting collectors particularly hard. He had to turn down a proposal from Miles Moss for a new expedition on the grounds of the financial crisis. "Matters are rather bleak in Europe," he explained, "and we have to retrench as much as possible."[79] Doing Jordanian systematics and organizing entomologists, much less finding $2 million to fund an international institute, was becoming more quixotic than ever.

Entomologists had little control over such global economic trends, but they had only themselves to blame for the second variable that interfered with their utopian plans. For decades, Jordan had urged the need for "international collaboration," building better relations between public and private entomologists, and countering the animosity that prevailed when entomologists met at the zoological congresses. Horn hoped an international institute would formalize the congress's efforts and turn its various resolutions into concrete action. But in April 1930, Jordan received a letter from U.S. entomologist Fred Muir warning him that Horn had "cut off all hope" of assistance from American entomologists for the international institute due to his role in a recent controversy in the International Commission on Zoological Nomenclature (ICZN). American entomologists, Muir explained, "did not want to help an Institute which would be influenced by those that voted with him."[80]

Jordan had had plenty of warning that the seas of nomenclature would be rough as entomologists entered the 1930s. With a rising number of applied entomologists on the scene, the problems of nomenclature became even more pressing. Some urged nomenclatural reform "without delay if we are to retain the confidence of the numerous workers in the many subjects that now touch the fringes of modern Economic Entomology."[81] Indeed, Tring's central reforms had become the target of several critics at the 1928 congress. Roger Verity, of Florence, Italy, protested, for example, against the British National Committee on Entomological Nomenclature's emphasis on the use of the term *subspecies* for geographical variation on the grounds that trinomials formalized the questionable assumption that such forms represented incipient species.[82] While C. H. Kennedy insisted that trinomials represented an inadequate and superficial patch-up job by evolutionary naturalists still trying to work within a system of nomenclature founded by the creationist Linnaeus.[83]

Jordan had hoped, of course, that the international congresses would facilitate consensus on all aspects of entomologists' work—from the details of nomenclatural rules to the theoretical assumptions behind the use of subspecies. Like other scientists who took existing models of international governance as their model, Jordan relied on the basic premise that rational procedures could be established upon which all could either agree or at least compromise. Yet he acknowledged the challenges facing such an enterprise, even with the seemingly simple task of agreeing on names for organisms. "Human nomenclature sits on the neck of divine nature," he would write in 1933, "and, there being an emotional force behind it, has assumed an astounding importance, like the emotional political forces under which humanity suffers."[84] Indeed, the particular shoals of nomenclatural dissent upon which Jordan's and Horn's institute was wrecked were composed of a heavy dose of those "emotional political forces"; namely, nationalism.

Zoologists had long recognized that the scientific traditions of a host country inevitably influenced the priorities and methods of the membership of a congress asked to vote on proposed rules. As a result, rules established by a congress meeting in one nation might be completely reversed by the next congress's members. To avoid such instability, the officers of the 1898 and 1901 international zoology congresses had insisted that only unanimous decisions of the commission would come before the full congress for a vote, a rule subsequently known as the Berlin Agreement. This so-called gentleman's agreement (since it had no concrete means of enforcement) had been roundly challenged at the 1927 International Congress of Zoology in Budapest by Austrian zoologist Franz Poche. Constantly hampered by the Berlin Agreement when proposing changes to the ICZN rules, Poche proposed that the agreement simply be abandoned. Thus, any nomenclatorial proposition given a majority (rather than unanimity) of votes in the commission could be brought before the congress for a vote.[85] As president of the commission, Jordan soon began receiving complaints regarding the measure on the grounds that "rules will be passed according to the locality of the Congress and when it is held in America you may have a total reversal of the whole set of rules!"[86]

When the International Congress of Zoology met again in Padua in 1930, matters came to a head with what Jordan understatedly described as "a little upheaval in the Section on Nomenclature."[87] For Poche had abandoned all pretense to following the ICZN's "gentlemanly" rules of parliamentary procedure by bringing his proposals directly to the floor of the entire congress. Unfortunately, it had been Horn who announced the proposals to the congress, giving tacit support to Poche's breach of parliamentary procedure. And the American opponents to Poche's tactics took note. The American secretary of the ICZN, Charles Stiles, threatened to resign from the commission (and take American zoologists with him) due to the influence of "reactionary" Austrian and German zoologists.[88]

The implications of Stiles's contrast between the diplomacy of aggression and that of international arbitration must not have been lost on his readers, as threats to parliamentary democracy once more loomed in Europe. Indeed, Stiles attributed his enemies' efforts "not to bad faith" but to national differences in "parliamentary technique and psychology between Continental Europe and North America." It was not the first time he acknowledged such "non-scientific" factors in zoologists' interactions. In an address at the Ithaca congress on "The Future of Zoological Nomenclature," Stiles had acknowledged that despite the theoretical ideal that science is objective and international, zoologists faced "practical as well as idealistic problems" in trying to form international rules. "Legislative and administrative ideas and experiences vary in different parts of the world." He recalled how recently a zoologist had insisted that the rules of nomenclature must be "democratic," noting that while he agreed wholeheartedly, "his ideas of democratic legislation gained in a country which up to about 8 years ago was an empire do not happen to square with my ideas on that subject gained in a country which has been a republic for more than 130 years." Even parliamentary procedure varied between countries, Stiles concluded, adding to the difficulty of reaching agreements.[89] Now, as newspapers chronicled the rise of Hitler in Germany and Mussolini in Italy, Stiles confessed that "from our American view-point people in Continental Europe are not so punctilious in parliamentary procedure as is customary in Great Britain and North America." Given such differences, he had come to doubt zoologists' ability to reach international agreement on matters of nomenclature. What must he have thought of the ill-fated efforts of politicians to save the League of Nations!

Although Jordan and his commission colleagues engineered a carefully regulated tabling of Poche's proposals at the next International Congress of Zoology in Lisbon in 1935, some effects of the controversy could not be smoothed over by any amount of parliamentary procedure. Muir reported to Jordan that the Americans, sure that Horn had been partly to blame for the Poche fiasco, would decline any further involvement in his proposed international institute—and by the 1930s everyone knew there was not much point in establishing such an organization without the support of American entomologists. The "Christmas list," in other words, had to be temporarily abandoned.

Although when the "nation of Entomologists" next met, in Paris in 1932, J. Harold Matteson urged that Horn's effort for an international institute be continued,[90] the coming decade justified little optimism that such taxonomic utopias could be fulfilled. Worldwide economic depression had destroyed the basis for the postwar reparations settlements in Europe, dealing a severe blow to the foundations required for cooperation and peace. Germany had collapsed economically, creating political instability into which extremism from both right and left battled for power. In the face of widespread discontent, Hitler's National Socialists made gains in elections in 1930 and 1932, his criticisms of the injustice of Versailles and the "futility of parliamentary democracy" inspiring increasingly desperate allegiance from a weary and humiliated nation.[91] Soon, the diplomats and statesmen to whom naturalists had looked for their models of international exchange and arbitration were again struggling to maintain peace and stability amid profoundly different visions of how society should proceed.

A LAD'S LAST MARBLE

Horn and Jordan had campaigned for an international institute on the grounds that insect taxonomists were continually overwhelmed by the lack of organization, especially given the increasing demands of applied biology. For a time, taxonomists had watched that rise with hope that, finally, entomology would take its proper place in society. Jordan had, for example, appealed to the Rockefeller Foundation's International Education Board to cover the expenses of the Zurich congress proceedings on the grounds of the "widely known importance of entomology to hygiene, agriculture, horticulture and forestry."[92] But with the pressure of the Great Depression descending on the discipline, Jordan's tone changed. When, after being unanimously elected president of the Entomological Society of London, he gave his presidential address for the end of 1929, he warned that "the very importance which Entomology has for the economic and hygienic welfare of humanity constitutes a danger which must not be overlooked." Jordan urged his fellows to remember that

> in a period of great economic pressure the clamour of the uninitiated public opinion for applied science might easily warp the judgment of a government and might lead to decisions which could cripple pure science, forgetting under the stress of the circumstances of the moment that pure research is the indispensable basis of all applied science and that to press for quick results in research is most disastrous, quick spelling nearly the same as quack.[93]

A few months later, and directly following the speech by Christopher Addison, parliamentary secretary to the Ministry of Agriculture, extolling the great importance of entomology to the Empire—since "the insect is the real enemy which has to be fought"—Jordan begged to disagree with the "general premise that Entomology must devote itself entirely to the study of those arthropods which we now know to be useful or injurious." Surely, he said, as a lover of insects the entomologist did not wish to increase the public's fear of insects, a horror that inspired them to stamp on any "innocent beetle that happens to run across his path."[94] The increasing workload of taxonomists made it increasingly apparent that, as the naturalist tradition had adjusted to new priorities, they may have taken hold of a double-edged sword. In some sense, the Tring naturalists proceeded with their daily work relatively isolated from many of these trends, or so it seemed to other naturalists. "It may seem preposterous to you," wrote one to Hartert in 1925, "but I daresay you are the most privileged Continental-born man I have met—because you can sit quietly in your Museum in peaceful Tring, with the silent mummies of our winged friends around you."[95] Rothschild and his curators certainly had the luxury of working on what they (or at least Rothschild) wished.

But even sitting quietly in peaceful Tring, delivering good systematics had been difficult. A few months after his presidential address to the Entomological Society of London, Jordan received a letter from William Forbes, of Cornell, inquiring whether he had been able to fit together the Old World, New World, and African Papilios into a unified whole. Forbes wished to rearrange the Cornell collection according to the latest knowledge, and Rothschild and Jordan were the acknowledged experts on the group. Jordan replied that although a classification of all the Papilionidae had been his and Rothschild's intention, the task had ultimately proved beyond their means, due to lack of material and time. The American material of Papilionidae had increased so much since he had written the revision (it seems that now Rothschild's work was subsumed in Jordan's) that he would have to "rearrange the lot" that winter. In addition, somehow their African material was "not so good as it ought to be for the purpose," and he had thus postponed analysis pending more collections. "It is now very doubtful," he concluded, "whether I shall find the time for it." He excused himself for not going very deeply into the various phylogenetic questions Forbes had posed on the grounds his head was full of many other problems.[96]

Within months, those "many other problems" became an avalanche of troubles. Holland may have wondered why Jordan, who had the loyal ear of a Rothschild, had begged him to go to the Carnegie family to help fund Horn's international institute. By 1932, a small group of American naturalists in New York could have explained to him that Jordan's and Hartert's "privilege" came with risks; namely, dependence on the foibles of a single patron. For the Americans had received the astounding news, from Rothschild himself, that he needed to sell his entire bird collection. Poor Hartert, who through gentle cajoling, endless hours of correspondence, and his own trips to the field had helped build this unprecedented collection, knew nothing of Rothschild's plan. He had retired from the museum in 1930 and moved permanently to Berlin at the behest of his wife, who had never recovered from her only son's death during the war and longed for her homeland.

Ornithologists had feared "some kind of collapse" would occur when Hartert left Tring since he had always countered, as the ornithologist F. C. R. Jourdain put it, Rothschild's "headstrong" and "irresponsible" disposition.[97] But no one had expected Rothschild to sell the bird collection. His initial letter to

L. C. Sanford, of Yale, written in mid-September 1931, offered the collection for $224,000, a price that Sanford estimated to be nearly one-tenth of the real value. Rothschild wrote he would need $25,000 at once.[98] Astonished, Sanford leapt into action to secure the collection, for example urging the American Museum of Natural History's president, Henry Fairfield Osborn, to use all his influence to obtain pledges from their patrons. Sanford explained how the offer of the collection, "although hitherto regarded as priceless," at such a bargain price "is largely due to the fact I have promised Lord Rothschild we will keep the collection intact and develop it according to his ideas."[99] Specifically, he had ensured Rothschild that the collection would be kept together in one place, no labels would be removed, and no specimen be considered a "duplicate."

The American beneficiaries of Rothschild's misfortune attributed Rothschild's having decided to sell the birds to the worldwide economic depression. "The crack in Rothschild's affairs was brought on by the S.A. collapse," Sanford explained, "He was in need of quick money."[100] In his initial letter to Sanford, Rothschild himself blamed the sale on "our enormously heavy new taxation," a factor that would have directly tied the loss of the largest private collection of birds in Europe to the profound changes urged by socialist reformers such as Alfred Russel Wallace. Wallace had insisted that the money required for the "salvation of the destitute" must be raised by halting the "excessive and harmful accumulations of the very rich" and obtaining every penny of taxation from the "superfluously wealthy."[101] Certainly, some contemporaries laid the blame squarely on those who had reformed British tax laws in the face of calls for revolution. "Confidentially when I received the letter from Lord Rothschild offering the collection I could not help feeling very sad," Osborn wrote to a journalist, "This is how the inheritance tax of England is leading that great country and robbing it of all its treasures."[102]

Jordan had indeed often complained in letters at the end of 1931 that "the country's finances being in a very bad way and taxation excessive, we cannot appoint another assistant to take over the bird-room, and so I have to struggle along as best I can."[103] When, in April, Jordan informed Moss that the birds had been sold, he blamed the move on a combination of the depression, the depreciation of the pound, and arrears of income tax. "The Nation got the money but lost the birds," he wrote.[104] Had the tax reforms called for by Wallace led directly to the loss of the Rothschild birds, this would have been a heavy indictment indeed of the economic and political winds of the century. But in fact, any role that the nation's dire economic situation and the new tax laws actually played in leading to the sale were exacerbated, if not entirely trumped, by a more direct call on Rothschild's funds—the demands for money, noted earlier, of the unnamed peeress with whom Rothschild had had an adulterous affair.[105]

In the end, Gertrude Vanderbilt Whitney donated the funds to secure the collection in honor of her late husband, Harry Payne Whitney, a great patron of the American Museum of Natural History and its expeditions. The deal was completed in extreme secrecy. When American Museum of Natural History ornithologist Robert Cushman Murphy arrived in London at the beginning of February 1932 to supervise the packing, Rothschild warned him not to say anything "to my mother about our business till everything is finally settled, when I wish to tell her in my own way, as I do not wish her to know of my money difficulties."[106] By contrast, Jordan had been privy to Rothschild's plans for some time. A month earlier, he had confessed to Miles Moss, safely incommunicado in the Amazon, that since Arthur Goodson had died and they could appoint no one new in the present circumstances, "Lord R. may decide to give up the birds altogether or the greater part of the collection."[107]

In explaining the situation to Moss, Jordan defended the sale of the bird collection rather than the insects on the grounds it had been readily salable at a good price, while the Lepidoptera, under existing conditions, would have fetched only a small sum.[108] He constantly tried to put the best spin on the matter, noting that, while he would be sorry to see the birds go, it would no doubt be the best policy; "That would leave us a little more money for Lepidoptera."[109] One senses in his rationalizing the gentle advice he must have given to Walter. "The amputation is very painful, but this institute is the healthier for it," Jordan insisted to Holland.[110] (At least one Tring associate confided that she thought Lord Rothschild quite right to send the birds to the United States. The collection would be safer across the Atlantic since war or revolution would doubtlessly descend on Europe soon. Besides, "How can Lord Rothschild manage them alone—and I fancy if the truth be known, old man Jordan is at heart a bit anti-bird. He grudges the time spent on them.")[111] But whatever advantages Jordan and others saw in sending the birds across the Atlantic, they were completely lost on Rothschild. Murphy recounted how the loss of his birds of paradise "pretty well busted him up. In fact, Doc Sanford was close enough to weeping himself."[112]

Given the secrecy required and the sensitive circumstances of the sale, Robert Cushman Murphy had his work cut out for him. He had to pack up all the birds surreptitiously, as British ornithologists visited the collection as usual.[113] And he had a devil of a time dealing with Rothschild. At one point Rothschild tried to insist kiwis belonged to the ostrich family (which he had not included in the sale), when in fact they belonged to a different order. Murphy came to sympathize enormously with Jordan, who had patiently managed Rothschild for four decades. "I am fond of old Lord R.," Murphy reported back home, "but when I return I can tell you some amusing stories of his attempts to slip one or two—no four or five—little tricks over me. Several times I had to use all my tact, and once I had to prime Dr. Jordan to help me out." Murphy concluded: "It's hard for a lad to give up his last marble."[114]

The man to whom most contemporaries attributed any science that came out of the ornithological department of Rothschild's museum did not learn of its sale until the birds had gone. Rothschild obviously felt guilty about not having told Hartert; he cabled Sanford that he felt obliged to do so, and would be making a public statement regarding the disposal of his collection.[115] It was, of course, Jordan who actually stood, alone, at the annual dinner of the British Ornithologists Union and formally announced the sale that would shift the balance of ornithological power across the Atlantic. "It was exceedingly brief and simple," Murphy reported, "The news came as an inevitable shock, and Dr. Jordan was afterwards besieged by questioners." He carried out his task well, for Murphy recounted that when Walter arrived later in the evening, the members made a point to greet Rothschild and talk as though nothing had happened. Murphy noted that although Rothschild initially "looked a bit hang-dog," he was soon "bellowing in his wonted manner with his table companions."[116]

The news of the sale astounded British ornithologists. Given that Rothschild was a trustee of the British Museum, they had assumed the collection would be bequeathed to South Kensington.[117] The ornithologist Percy Lowe encapsulated the feeling of many as he vacillated between dismay, blame, and pity in a letter to Murphy. Rothschild had told him personally that the Australian collection would stay in the British Museum, and he pointed out that this seemed "to have been the almost universal experience of members of our staff in regard to all Zoological branches." Lowe thought such promises explained Rothschild's having "occupied for years a very exceptional position as regards the Zoological collections of this Museum." Still, he did not think a single man or woman on the staff of the museum "did not sympathise with him in having had to sell a collection which must have meant something like parting with his Zoological soul." But no one could understand why Rothschild had not confided his desperate position to his fellow trustees and thus given the country a chance to acquire some of the material. "Frankly," he concluded, "we did not like the way it was done."[118]

The American ornithologist Frank M. Chapman echoed Lowe's sentiment that Rothschild had not behaved rightly. "If he had only had the foresight and the courage to give the British Museum the refusal of his collection before he offered it to us," he wrote, "his position would have been practically above criticism." But Chapman also agreed that, as Murphy had stated, Rothschild was "neither English nor American," and was "doubtless very much of a spoiled child and accustomed to having his own way."[119] Neither Chapman nor Murphy offered an explanation of what, precisely, Rothschild was, but presumably this was a veiled reference to the fact that, as a Jew, he was not expected to subscribe to the rules of gentlemanly behavior.

Eventually, disgruntled protests appeared in the newspapers. "True, even a Rothschild may well find the pressure of taxation such that he can no longer afford the luxury or fulfill the duty of maintaining his great collections," wrote one, "But not even a Rothschild should take such a step as this without offering the refusal to his fellow Trustees of the National Museum."[120] But such protests referred to Rothschild's duty in a time in which naturalists still debated the rights and obligations of owners of natural history collections. Some thought the rules, though implicit, were quite clear, insisting in the pages of the *Times* that Rothschild "had no moral right to sell specimens which they had given him on the tacit understanding they would eventually become the property of the nation."[121] As an article in the *Morning Post* explained, it was based on such an understanding that the museum had avoided competing with Rothschild when negotiating purchases.[122]

Trying not to blame New York, Jordan had carefully worded his own explanations of the sale to newspapermen. In firm agreement that science had nothing to do with personal property, and that duty to future researchers should indeed guide decisions regarding the *concreta* of the tradition, he emphasized the needs of scientific research rather than appealing to Rothschild's right to dispose of the collection as he saw fit. He told a reporter from the *Daily Mail* that the retirement or death of two of its caretakers meant the important collection could not be adequately cared for, while the offer from New York had provided an opportunity to carry on the rest of the museum.[123] The *New York Times* reprinted his explanations that the maintenance of a great private museum was a heavy burden and, under pressure from prevailing economic conditions, the choice had lain between selling either the birds or a more unusual collection of insects.[124] But, as reporters who could get no information from the British Museum began to press him as to why the nation had not received first bidding, some frustration escaped; "The British Museum," Jordan was quoted as saying, "has not got the money, so that it was no good offering the collection to that institution. The Government has reduced the Museum grant."[125]

The response from the British Museum was swift. Secretary G. F. Herbert Smith called Jordan's statement "misleading, inaccurate, and unauthorized," and though he conceded the Treasury had reduced the grant available to meet the normal purchase of specimens for the collections, he insisted that additional funds might have been procured to meet an emergency.[126] Jordan defended himself on the grounds the newspaperman had reduced a long conversation to "very bare statements," but he also pointed out that the matter had been urgent, and appeal to the general public, specialists, or rich men for subscriptions "was quite out of the question."[127] Miriam Rothschild once mused that there must have been times when Jordan felt that he was expected to carry the can, and "Walter should have learned to do his own dissections."[128] One wonders if Jordan similarly felt his boss should have made his own explanations to both newspapermen and the British Museum authorities.

One observer who lamented the loss of Rothschild's two thousand type specimens to a buyer across the seas concluded that this terrible development "closes what has probably been the most progressive chapter British ornithology has ever seen."[129] But deciding what this sale meant for the future of British ornithology depended on one's vision of the path natural history should be taking amid changes in biology. Increasingly influential, generally younger ornithologists most certainly did not believe that the loss of Rothschild's collection closed the "most progressive" chapter in British ornithology. David Lack and R. E. Moreau, for example, who would soon take over the *Ibis*, the British Ornithologists' Union's journal, caricatured the period of ornithology that focused on amassing collections as "hidebound" and unscientific. Phyllis Barclay-Smith gave the most colorful description of the generation thus replaced when she wrote that most British Ornithologists' Union members saw birds as a "source of amusement at which to blaze off a gun. For the more highly specialized naturalists the criterion might be, 'There is a rare bird—shoot it.'"[130] Some, including Moreau, acknowledged that Hartert and Rothschild had been different from the average "hidebound" ornithologist in their emphasis on geographical variation and a more evolutionary taxonomy. But to many of the self-proclaimed "biological" generation, a new "scientific ornithology" entailed close attention to both living birds and explicit theoretical context, rather than the endless gathering of facts represented by rows of specimens.

All sides of the old "systematics" versus "biology" debate recognized, however, that it would be impossible to amass such a collection ever again. As Murphy pointed out, Rothschild had made expeditions to regions where avifauna was undergoing rapid changes or was threatened with extinction.[131] Indeed, as a primary argument for securing the collection for the American Museum, Sanford had written that "islands that have completely lost their fauna are represented in entirety."[132]

Other elements of Rothschild's collection were becoming extinct as well; namely, an entire network of private collection owners. British natural history's strong reliance on private individuals had, for a time, constituted a strength—one that bolstered calls for entomology for its own sake. Although the Cambridge entomologist David Sharp, his sights set on government patronage and centralized collections, had argued against private collections in the 1890s, Jordan had used his authority as president of the Entomological Society of London to argue that private collectors be encouraged as much as possible. "The efficient amateur," he noted in 1929, "has always been one of the greatest assets in systematics."[133] The following year he repeated the point: "The importance of well-labelled collections cannot be stressed enough, and as an intensive biological survey at public expense is still far off, the private collector should be encouraged as much as possible, the lack of collections impeding investigation."[134]

Yet, as his occasional tributes to applied entomology showed, even Jordan found something useful in American entomologist Austin H. Clark's insistence that they must begin to see entomology as "a commodity subject to sale and purchase." Clark thought American entomology's need to prove "its mettle" had been a good thing. "Recognition won by an uphill fight is always on a solid basis. . . ." he wrote, "If science relies entirely or chiefly on the social prestige of a limited group of aristocrats it will rise or fall with those aristocrats."[135] Taking "aristocrats" to refer in general to the wealthy classes, the Tring naturalists—and British ornithologists in general—had experienced this fact dramatically in the winter of 1932.

Six hundred miles away from Tring, the ornithologist Erwin Stresemann, at the forefront of the move to a more biological ornithology, was working down the hall from Hartert's office in Berlin when Hartert staggered out of his office after hearing of the sale with tears in his eyes, mumbling "My collection! My collection!"[136] The possessive pronoun is a telling example of the amorphous lines concerning the ownership of the museum among the Tring Triumvirate. Rothschild (or, more accurately, the bank) paid the bills. But Jordan and Hartert spent their scientific lives building and working in the collection. In some sense their knowledge conferred a scientific and intellectual ownership on the thousands of specimens. Yet in the end, the museum was as much an object of commerce as of science, evidenced by the thousands of letters listing price per specimen from natural history agents, dealers, and collectors. And as an object of commerce, the collection belonged solely to Walter Rothschild, to dispense with to pay his debts if he so chose.

Although Murphy dismissed any British or German claims to Rothschild's collection, he did acknowledge the possessive claims of his fellow ornithologist Ernst Hartert. After writing "the friendliest letter I can draw up,"[137] he was relieved when Hartert's reply arrived full of patient resignation. "Of course it is a very hard blow to me," Hartert began, "to see 'my' collection disappear to America." And the secrecy, he wrote, was indeed regrettable. "But," he explained from long experience, "Lord R. sometimes likes to keep things clandestine, even if there is no reason for it, except reasons which he imagines. In a way, however, I am glad I was not obliged to discuss the matter!" He only hoped that the Americans would not give away, or exchange too many examples "for often the large series are important."[138]

That plea formed an almost feeble reprisal of Hartert's firm, authoritative exhortations to collectors of the 1890s, when he and Jordan had mobilized the natural history network along new lines in order to create a "scientific, worldwide collection." But the world had changed since the days when Rothschild had called his *Schaun Sie! Schaun Sie!* through the corridors of the museum almost half a century earlier. Following a trend that had influenced the work of their American fellows for years, British naturalists had to increasingly adjust their wish lists to the priorities of new networks of professionals in the field. And their collecting efforts were now often directed at highly focused research programs that had little to do with the study of geographical variation or even evolution.

Trying to adjust to this shift in context, H. A. Baylis, of the Zoological Department of the British Museum, conceded during the 1925 discussion on "The Place of the Systematist in Applied Biological Work" that museum workers must serve the interests of applied biology. But he urged that, in return for this service, the applied biologist send as many specimens as possible and allow the museum entomologist to keep part of the material when desired. As Jordan and Hartert had once exhorted Tring's collectors and natural history agents to collect in a certain manner and amass long series, museum workers such as Baylis now exhorted economic biologists to aid systematists in their taxonomic decisions by sending sufficient material. "Longer series of specimens would cost the sender no extra trouble," Baylis explained, "and it would help enormously in increasing the value of the national collection."[139]

Clearly, the positions of authority from which Jordan and Hartert had exhorted collectors to "do better" were not strengthened by the institution-changing winds of the twentieth century. William Brierley had placed systematists in the position of working toward their own elimination; the Prussian minister of education had insisted they pursued a "dying science." Surely both were premature obituaries, but certainly the tables had turned.

SEVEN

The Ruin of War and the Synthesis of Biology

The turbulent first decades of the twentieth century raised fundamental questions for entomologists. If the old scientific and social infrastructure of the taxonomic wing of the naturalist tradition was to be abandoned, could the tradition adapt? Could the tradition be reformed to fit new concepts of science, and could new justifications successfully garner support amid new priorities? Could reforms actually be implemented in this rapidly changing century, in which so much of the original context of the naturalist tradition was being either recklessly destroyed or conscientiously abandoned?

Jordan and his fellow entomologists alternatively waxed optimistic and pessimistic on these questions. Some saw enormous opportunity in the rise of applied entomology, while others grew wary of justifying work on insects solely on economic grounds. But whether anxious or optimistic, Jordan and his friends worked hard to adjust the tradition to new techniques and justifications, even as they built on the infrastructure established in another age. Jordan's congresses, now meeting every three or four years, provided one place to canvass the various answers posed, although by the end of the decade at least one congress member wondered whether, given the expansion in both the diversity and number of papers, the congress could indeed act as the "International Parliament for Entomology."[1]

As entomologists tried to get their bearings in the interwar period, they were met with ambiguous signals as to how much of the old world could survive. In going on an expedition to South-West Africa (modern-day Namibia), Jordan safely traveled the infrastructure of European empire to inventory the animals and plants of far-off regions, even as calls for the end of imperial rule grew louder. As he hiked the Alps to study subspecies of fleas, researchers breeding *Drosophila* flies in laboratories linked geographical variation directly to changes in chromosomes. As he worked to secure the future of the Tring Museum as a center of scientific research following Walter Rothschild's death in 1937,

Victor Rothschild (Charles's son and heir to the title) campaigned for a strongly reductionist, experimentalist biology firmly focused on applied science. Finally, even as the evolution synthesis of the 1930s and 1940s vindicated Jordan's methods of systematics, the research programs thus established—and another world war—highlighted the enormous challenges taxonomists would face in maintaining and reforming the tradition.

THE EDGES OF EMPIRE

By the time entomologists convened in Paris for the fifth International Congress of Entomology in 1932, talk about an international institute aimed at dealing with the organizational challenges facing the varied "nation" of entomologists had ended abruptly. A brief notice appeared in the congress proceedings that, given current economic circumstances, the committee organized to develop the institute would be dissolved.[2]

Intent on solving taxonomists' troubles from some angle, the ever-persistent Walther Horn delivered a paper entitled "Thoughts about Entomological Systematics, Mathematics, Genetics, Phylogeny and Metaphysics" to the Paris congress. He urged increased cooperation between "modern, scientific biology," which emphasized mathematics and experiment, and his own tradition, comparative, morphological, descriptive, and historical entomology. In marked contrast to Auguste Lameere's opening address of 1910, Horn directed his fellow "descriptive" entomologists to read the experimental work of Sturtevant, Goldschmidt, Fischer, Lenz, and Standfuss. But like Jordan, Horn expressed ambivalence toward claims that all science must be mathematical or experimental. While he acknowledged the draw of mathematics as exact, he warned that such efforts belied the complexity that every entomologist knew plagued neat conclusions.

Naturalists often emphasized the extraordinary complexity of the living world, but Horn also had something to say about mathematicians' ignorance regarding the complexity of actually doing natural history. Suppose, he suggested, a taxonomist tells the mathematician of his constant struggle to describe the endless species of insects. Eager to help, the mathematician at once asks how many types of insects have been described and how many remain unnamed? The systematist replies that perhaps two million species and about two million races exist, of which about one million have been described. The mathematicians proceeds: Can one assume that it requires about thirty minutes to identify a particular form, and that there are about two hundred working entomological systematists in the world? The systematist agrees to these assumptions. Satisfied, the mathematician sits down and calculates. Assuming three hundred working days a year of eight fifty-minute hours each, the project of describing all these unknown species should take, for those two hundred systematists, exactly three years, seven months, and fifteen days. "The result is clear, simple, and very gratifying," Horn notes. Unfortunately, he warned, it was also "completely false." The calculation assumed, first of all, that "no mistakes are made, that a 'complete' collection is at hand, that 30 minutes is all that is required for each form, and that a new species is delivered promptly every half hour!"[3]

Delivering new species at a constant rate had long depended on being able to move entomologists, collectors, cash, and specimens, around the world, and that movement was dampened considerably by World War I and the Great Depression. The backlog of undescribed species sitting in the halls of large museums provided enough to keep curators and specialists busy, but, at least by Tring standards, the description of new species ideally proceeded in concert with the directed accumulation of new material by which conclusions could be tested. Horn's imaginary mathematician's calculation assumed that specimens arrived from every possible place according to taxonomists' wishes, and that taxonomists—or someone—could pay for those specimens.

Both these assumptions proved tenuous after the war. At Tring, for example, Jordan wrote Miles Moss that he could not in good conscience encourage those collecting natural history specimens as a business.[4] "Conditions are unfortunately bad everywhere," he warned Orazio Querci, who was trying to raise money for an expedition to Serra de Estrela in 1933, and "the number of Lepidopterists who can afford to purchase specimens dwindles more and more."[5]

An optimist might have noted that amid the decline of the old network, a new one had appeared. A constant stream of specimens arrived from regions where insect pests and carriers of disease threatened European colonization. Applied entomologists sent out to fight insects in Africa and elsewhere sent specimens home for determination in droves. Such work drew on the old context of Empire even as it emphasized a new awareness of the ties between insects and disease as the reason that entomologists should be on the imperial payroll. Many of the papers at the Paris congress, for example, recounted the efforts of entomologists to catalog and control the various "enemies of mankind" in their nations' colonial possessions. The language in these papers could at times read like military briefings. One entomologist spoke on the locust problem in Egypt as a war against invasion complete with scouts, campaigns, officers, and a range of combat methods.[6] Back home in Britain, even the content of the traditionally purist Entomological Society of London had turned the nation of entomologists into an army fighting its objects of study. In the papers of Charles Swynnerton, director of the Tsetse Research Department in Tanganyika, members read of the insect ecologist's primary goal as the study of the "weak points in (insects') defence" in order to "deprive them of one or other of their vital requirements." The entomologist became a general on the front lines of a warfare against a terrible enemy, devising "plans of defence," "attacks," and conducting "campaigns." In fact, a summary of the "The Present Position" of flies in one region read just as smoothly if one substituted "the Germans of 1915" for "the tsetse of 1936."[7]

As the new territories obtained by the war's victors provided an outlet of attention and energies as the depression dragged on, naturalists had plenty of evidence that the exploration of colonial possessions would continue to provide a context in which they could collect, explore, inventory, and serve. Of course, amassing worldwide collections of specimens had long depended upon both unlimited access to far-off lands and the compliance of local populations. Both relied in turn on distinct assumptions about the relative authority of nations and peoples. Rothschild's collectors, for example, operated on the assumption that Europeans were justified in exerting authority in far-off lands and thus collected specimens wherever they wanted with freedom. When Captain Angus Buchanan led an expedition, funded by Rothschild, to the isolated Aïr Mountains (in modern-day Niger), these assumptions were on full display, even as Buchanan expressed dismay that values might be changing.

Buchanan's travels resulted in the discovery of dozens of new species and subspecies, duly described in Tring's journal by Rothschild, Hartert, and naturalists at the British Museum. Rothschild composed a preface to Buchanan's account of the expedition, praising him for having worked "absolutely single-handed" to collect more than eleven hundred birds and mammals and more than two thousand Lepidoptera. A few pages later in the book, Buchanan described his relief when ammunition finally arrived, in boxes on the "heads of carriers that were groaning under their loads." Rothschild meant, of course, that the "single-handed" Buchanan was the sole European on the expedition, and therefore, in his view, he might as well have been completely alone.

Yet Buchanan's account contained the anxious laments of a man who saw the world in which such statements could be composed with confidence disappearing. In this world, Buchanan could speak of his "boy," even if the hired man was middle-aged, praise him as "ever faithful as a dog to his master," and note that he was "certainly not of a race or rank to claim intimate acquaintance" since "he was as black as the ace of spades." Combining naturalists' attention to geographical variation with long-standing notions of hierarchy, Buchanan could soliloquize on what he saw as the terrible consequences of recent trends in colonial policies—"from what sane source is past understanding"—that abandoned the "hard plain fact" of the inequality of the races.[8] That Buchanan thought it important to emphasize such a point hints that there were alternative views about native ability and Europeans' right of authority over them, and that the alternative was increasingly influencing colonial policy.

Indeed, the imperial endeavor upon which these naturalists depended had never been without its critics. As Europe navigated a fragile peace, the supposed acquiescence of the colonized to the paternalistic assumptions of colonialism and the mandate system showed signs of fraying. Colonial troops from colonies in Africa and Asia who had fought for Britain and France during the war had received for their pains a stiff resistance to the principle of racial equality at the Versailles conference. By 1930, opposition to imperialism had led to concerted action. After a decade of relative quiet in British India, Gandhi had started a new campaign of civil disobedience. In the Middle East, uprisings in Iraq forced the end of Britain's mandate there. Men such as Ferhat Abbas in French Algeria and Ho Chi Minh in Vietnam fought against the control of colonial powers.

In an expression of solidarity with such efforts, Alfred Russel Wallace once wrote: "With all my heart and soul, I protest against and condemn the doctrine that we have any right to force our rule upon people who do not want it, under the pretence of better government."[9] But even the principled Wallace failed to file his protest by disengaging from the study of biogeography, one of the "most obviously imperial sciences in an age of increasing imperialism."[10] These ties were certainly noticed by others as, during the first decades of the twentieth century, opposition to naturalists from imperial powers appeared, especially in areas of "informal Empire." Zoologists in the United States, for example, had been having trouble getting specimens out of South America, a collecting ground they had dominated after the Great War. Political instability provided one source of difficulty; rising nationalism and attendant concerns over sovereignty produced another. Charles Townsend complained in 1931 that he could not get word from his contacts in Peru, where a coup d'etat in 1930 had resulted in a series of interim presidents and political chaos. Pressed by either ideology or their opponents, all sides had taken a staunch anti-imperialist line. (One of the candidates in the 1931 election led a delegation to the International

Congress Against Imperialism and Colonial Oppression in 1927.)[11] Townsend eventually concluded that his correspondents feared writing because the new government frowned on Americans getting anything more out of Peru.[12]

References to similar barriers to collecting had been appearing in correspondence regarding Mexico as well. "There is a strong nationalistic movement in the land," the chief entomologist at the Mexican Department of Agriculture, Alfons Dampf, explained to U.S. National Museum entomologist J. M. Aldrich in 1925, "and all the scientific work, performed by a foreigner, is hampered in an incredible manner." He warned Aldrich that now permission to collect "is issued for a limited time, for a determined zone and for determined species of plants and animals."[13] Making sure new species arrived at the taxonomist's desk at a constant rate obviously relied on more than economic and political stability. Access to and centralization of specimens in European (and American) capitals relied on a certain contingent, and increasingly contested, balance of economic and political power.

Certainly, medical entomologists could extol their work's importance for the health of all inhabitants, even as they insisted that controlling insects would advance the march of European civilization. But to men such as the director of the Dirección de Estudios Biológicos in Mexico, Alfonso L. Herrera, whom Dampf blamed for the new permit laws in Mexico, the simple act of removing specimens from Mexico to Washington, DC, or New York served as yet another expression of the political and economic hegemony of the United States. The naturalist tradition in which naturalists worked and the infrastructure through which they moved, although focused on the diversity of animal and plant life, was often inseparable from the aim of establishing authority, if not outright possession, over other regions. Even if packing up butterflies and beetles seemed less insidious, extracting specimens had become part and parcel of the same assumptions that had permitted the imperial powers to extract natural wealth from far-off lands.

For decades Jordan had been primarily on the receiving end of both the conceptual and practical results of this imperial infrastructure. Yet in the winter of 1933, at the age of seventy-two, he organized an expedition to South-West Africa, then under a League of Nations mandate after the Treaty of Versailles stripped Germany of colonies. He thus traveled to a land where the task of inventory and control now fell primarily to British and South African entomologists. "I go under the auspices of the British Museum," he explained, "i.e. have the moral support of the Foreign Office and therefore various facilities."[14] Jordan undertook the expedition, in part, to "ascertain for Museum's purposes how much a traveler without European collecting staff might be expected to accomplish,"[15] an aim inspired by the lack of expedition money.

Following the example of anecdotal accounts of expeditions that had filled natural history journals for a century, Jordan peppered his own narrative with adventurous mishaps, commentary on colonial politics, and observations of local peoples. And like many of his fellow naturalist-travelers, Jordan described the landscape through the eyes of a European who was certain it could be civilized only by colonists' efforts to improve the productivity, and, indeed, the morality of these lands. He thought it quite proper that the most productive lands be in the hands of Europeans.[16] Jordan certainly was more generous toward native ability than Captain Buchanan (after trying the experiment of teaching his "boys" how to collect for a few days, Buchanan had ridiculed the idea that natives could prepare specimens). But he accepted the assumed hierarchy of ability that inspired the original question. At one point, having proceeded to Portuguese-controlled Angola, Jordan recounted seeing an old native sitting in a primitive shelter aside a field of maize. "He did not seem to be troubled by the deep thoughts of a Gandhi," he commented.[17] Nor, it seems, were naturalists.

Ultimately, natural history's strong historical ties to the infrastructures and assumptions of Empire would one day make collecting specimens in the field seem just another example of imperial exploitation, rather than scientific research. And this, too, would have to be taken into account in calculations of the rate at which specimens would arrive at taxonomists' desks.

WHERE SUBSPECIES MEET

In his effort to ensure that the museum specimens captured important knowledge from the field, Jordan's principles of scientific systematics expanded the amount of time required to produce a good description of the specimens coming from far-off lands. It took him, he said, about sixteen hours to describe a new flea or beetle.[18] The descriptions that resulted contained constant caveats that taxonomists needed more specimens and breeding experiments (in other words, more hours) to test his conclusions. Even when specimens arrived in London, they might not necessarily add to naturalists' stock of knowledge. When the zoological results of Jordan's expedition to South-West Africa appeared in *Novitates Zoologicae* in 1935, the author assigned to process Jordan's fishes noted that the high percentage of undescribed species from the expedition only emphasized zoologists' ignorance of the region.[19] The description of Jordan's specimens depended entirely on the presence of taxonomists with both expertise and time to work on certain groups, and Jordan himself warned that "it is not to be expected that all the insects and other evertebrates will be worked out in the near future."[20] Certainly, processing even this small sample of the Earth's species would take longer than the three years calculated by Horn's imaginary mathematician.

Still, the daunting timeline did not matter much if one denied that finishing the inventory formed the taxonomist's primary job. The mathematician's equation had been premised on the goal of completing the catalog of all insects on the globe. Yet Jordan dismissed any suggestion that taxonomists must focus on completing the catalog. Rather, he insisted that the mere description of new species did not "help the least to solve the all-governing questions of evolution, but adds simply more 'species' to the hundreds of thousands of 'species' already made known."[21] Although he defended the careful description of species and varieties as necessary to the "higher aims" of biology—namely, the study of relationships and the origin of species—inventory for its own sake made no sense to him. Thus even if an army of taxonomists had been available to process his undescribed species at the rate of one every thirty minutes, it is not clear he would have wished to hand over the specimens.

Even as he had planned the expedition to Africa with "great anticipation," Jordan pursued research that betrayed a fundamental tension between naturalists' practice of rushing across a landscape collecting a few specimens—which, Jordan readily admitted, his expedition had done—and the aims and methods that he believed were necessary to make systematics both scientific and important to biology. During summertime field trips to Continental Europe in the 1930s, he spent an inordinate amount of time studying a single species, the widely distributed mouse flea, *Ctenophthalmus agyrtes*, rather than collecting many different species. He took as his guide one of the basic research questions established by Darwin's *On the Origin of Species*. Given the importance of geographical variation to evolution, what happened where subspecies meet? Like other biologists who had seen the evolution of sterility as an important factor in the origin of species, Jordan believed it necessary to examine the points at which aversion to breeding was developing. And this meant hours and hours collecting and studying the distribution and morphological variation of just a few subspecies.

Jordan knew subspecies of the mouse flea demonstrated gradation in morphology, an expected fact, he insisted, once one assumed that subspecies represented incipient species. Having accepted a distinction between species and subspecies based on the ability to interbreed, he wanted to find evidence of gradation from complete interbreeding to noninterbreeding in areas where subspecies came into contact. "Do they peaceably amalgamate when they come into contact," he asked, "or have they acquired such an aversion towards each other that they clash and keep strictly apart though living side by side, as in many human cases?" The territory in question, he noted in 1930, lay before their door, "but where is the collector who alone can help us?"[22] In an effort to provide that missing collector, Jordan had returned to his naturalist roots, crossing the Channel whenever he could in order to hike the landscape and trap small mammals himself. With the natural history agents and professional collectors disappearing, Jordan advised his colleagues at the 1935 International Congress of Entomology in Madrid to do the same. "Study the variation of a single common species," he urged, "and spend your holidays in the districts where its subspecies meet."[23]

In focusing naturalists' attention on geographical variation, the project to ensure that species the taxonomists "made" reflected realities in nature had thus morphed into a project to understand those species at an unprecedented level of detail. And what Jordan found astonished even him, despite his extensive experience with the complexity of nature. He had long insisted that he could say some things with certainty. Subspecific characters and territory, he wrote, went hand in hand. Now, after many field trips, he found that where certain subspecies of the mouse flea met around the town of Bagnoles, in Normandy, each kept not just to a certain territory but a "definite biotope," one staying in the wood and the other in the open country. Suddenly even the detailed locality labels of Tring had become insufficient. "If the specimens had not been studied on the spot," he explained, "but been labeled in the usual way with locality, date and host, and then examined at home, the fact that the two insects are separate in space at Bagnoles would have been concealed. The areas of distribution do not overlap, but dovetail."[24] He had reached this conclusion regarding the variation in a single species after many field trips, continuously narrowing down the potential places where the subspecies met, followed by meticulous examinations with a microscope. Taking thirty minutes to describe the boundaries of each species seemed, by comparison, laughably inadequate.

About the time Jordan began his detailed attack on the question of what happened where subspecies of the mouse flea meet, in the late 1920s, a young Ukrainian entomologist named Theodosius Dobzhansky began tackling the very same question. The similarities and contrasts between the two naturalists highlight both the challenges and possibilities of traditional natural history collections. Dobzhansky had attended the 1928 International Congress of

Entomology at Ithaca, where he presented his work on variation in Coccinellidae, or ladybird beetles. Dobzhansky arrived from a country where, in contrast to the anxiety-inspiring battles noted by Auguste Lameere in 1910, a stark division between the zoology of the laboratory and the zoology of the field did not exist. Indeed, his Ithaca paper on "The Origin of Geographical Varieties in Coccinellidae" called into question some of the basic assumptions on which both Jordan and William Bateson had distinguished the conclusions of geneticists from those of naturalists. Based on a statistical study of the variability of color patterns in ladybird beetles, Dobzhansky concluded that all kinds of intermediate conditions occurred between what Jordan, Bateson, and others held to be the quite distinct phenomenon of geographical versus nongeographical variation. "The study of those intermediate conditions is perhaps the most important problem," he concluded pointedly, "because it shows clearly that the nongeographical and geographical variation are not such different phenomena as is often considered to be the case."[25]

Historian of biology William B. Provine once noted that Dobzhansky's research on ladybird beetles soon entrapped him in the limitations of morphological systematics. With plenty of information about ladybird beetles' natural populations and systematics—the familiar working material of naturalists such as Jordan—Dobzhansky soon became frustrated by his lack of knowledge of the group's genetics. He had concluded that a species must be thought of as a population that had "genetically differentiated to such an extent that it became reproductively isolated from other populations." The next step in analyzing the mechanism of speciation, then, would be to examine the genetics of races and species. But no one knew anything about the genetics of ladybird beetles.[26]

Meanwhile, in 1927, Dobzhansky had begun studying genetics in Thomas Hunt Morgan's famous "Fly Room" at Columbia. This laboratory reflected everything William Bateson had envisioned for the future of the study of heredity. Bateson had called for zoologists to organize their subject by problems, rather than taxa. In doing so the animals under investigations would become models of organisms in general, rather than objects of interest for their own sake. Certainly, some entomologists had highlighted the role of their organisms of choice in elucidating general biological problems. Henry Walter Bates had cited "the facility with which very copious series of specimens could be collected and placed side by side for comparison" as a reason to use them to study particular problems.[27] But Bates was imagining long series of specimens in museum drawers that could then be meticulously compared. As Jordan had seen during Morgan's visit to Tring in 1922, the Fly Room operated according to entirely different methods; namely, breeding, manipulating, and mutating stocks by the thousands.

The successes of the Morgan lab in the study of heredity had already completely transformed biologists' views of both heredity and biology. E. O. Essig recounted in 1936 how "Thomas Hunt Morgan, in his researches on heredity, has elevated the lowly vinegar fly, *Drosophila melanogaster*... 'of which it has been said that it has apparently been created by God solely as an object of hereditary research,' to the pedestal of world importance."[28] The fact that the organisms involved were insects proved superfluous to any attempts to raise the status of entomology per se. No one in Morgan's lab considered themselves a member of the "nation of Entomologists."

Except, perhaps, Dobzhansky. In contrast to his American fellows, Dobzhansky came from a tradition that combined the study of the genetic variability of wild populations with the study of laboratory stocks—a tradition largely isolated from the British and American schools since many important articles from the 1920s and 1930s were in Russian and unavailable in the West.[29] In contrast to most American and British geneticists, Dobzhansky worked among a network of natural history collectors who studied long series of specimens and saw the determination of distinct species and subspecies as a central practical and theoretical problem.[30] Jordan would have felt quite at home with such a research program.

But Dobzhansky combined his own field studies with the huge amounts of data being amassed by the Morgan group in order to address problems initially raised by his work with ladybird beetles. Morgan's lab had accumulated a vast amount of information that allowed Dobzhansky to examine differential sterility in an organism with few chromosomes, and in this way the genetics could be relatively easily analyzed.[31] By contrast to Jordan, Dobzhansky could actually observe, through chromosome inversions, what happened "when subspecies meet" at a genetic level. Ultimately, he inferred that although single gene changes often caused variation within a population, geographical races generally differed by complexes of genes. In doing so, he demonstrated the fallacy of both Bateson's claim that evolution occurred solely by nongeographical variation and Jordan's insistence that evolution depended solely on small-scale, continuous, geographical variation. Indeed, in a 1933 paper in *American Naturalist*, Dobzhansky cited Jordan as having wrongly overemphasized continuous, geographical variation at the expense of discontinuous variation.

Unlike Bateson, Dobzhansky did not thereby dismiss the importance of geographical variation. After all, he had been led to his conclusions by the careful study of "where subspecies meet." But, unthreatened by the heady claims of young geneticists extolling large mutations, he denied any grounds for assuming that the behavior of continuous characters differed from that of discontinuous characters.[32] Mendel and Darwin, it seemed, could indeed both be the guiding heroes of modern biology.

Both Dobzhansky and Jordan based their research programs on the assumption that the study of the relation between two subspecies was the study of the origin of species. Dobzhansky wrote that a comparative study of different species possessing different degrees of geographical divergence could provide "insight into the process of the evolutionary divergence in time and in space."[33] But by the early 1930s, Dobzhansky was amassing long series of specimens of a single species and its subspecies in numbers Jordan could only imagine. Flies could easily be collected in the wild, raised in milk bottles, bred quickly, and required no expensive cabinets for storage. The requisite data could thus be obtained at an extremely low cost, compared with the expensive effort to amass large collections of butterflies and beetles, and much faster than the painstaking setting of mammal traps in the hopes of catching a few fleas. While Dobzhansky collected stocks of the various races of *Drosophila* and reared them in the lab, producing hundreds of generations, and repeatedly back-crossing them, Jordan was, as often as he could get away from Tring, packing his bags and making the trip across the Channel, hiking about the Alps, and setting his traps for the mice upon which his species of choice fed. And while Dobzhansky observed chromosome inversions, Jordan relied strictly upon morphological comparison.

Over time, Jordan conceded the importance of the recent findings of genetics. In a 1938 contribution to a discussion of species concepts at the Linnaean Society of London, he noted how "we know now that specific differences exist down to the germ-cells. What fifty years ago was a mere suggestion has since been proved to be fact."[34] But he worked within a tradition that only in the final decade of his life took up such tools in earnest. Meanwhile, collecting and breeding flies could be done fast, cheaply, and it certainly did not look like a sport for rich people. Aware of the tentative nature of conclusions based on even large series of specimens in the museum, Jordan acknowledged that "field researches and experimental breeding will have to supplement the investigations of the museum's systematist in order to make the results conclusive."[35] But Jordan envisioned taxonomists doing the main research, and experimentalists providing "supplementary" data. Workers such as Dobzhansky turned to field researches and experimental breeding in genetics laboratories, supplemented by museum systematics, rather than the other way around. That is, assuming they saw museums and taxonomy as supplemental at all. Dobzhansky certainly did, but those in Morgan's lab thought him somewhat odd.[36]

The increasing primacy of experimental work and genetics emerged unmistakably when entomologists convened at the Fifth International Congress of Entomology in 1932 in Paris. In 1910 and 1928, Lameere and Howard had bemoaned biologists' attention to sea urchins rather than insects. Now, during his presidential address, French economic entomologist Paul Marchal praised biologists' increasing use of insects as material for study. Marchal cited the work on *Drosophila* as having provided an unprecedented foundation for continued scientific research. Morgan's work on these insects, he announced, allowed workers to establish the laws that govern heredity and unravel the mechanisms behind the evolution of new forms. He did warn that entomologists must not thus abandon the fields, woods, and the passion for observing the outdoors; and he paid tribute to the systematists' study of geographical subspecies, which had opened up new perspectives to both entomologists and biologists more generally.[37] But he made quite clear to what organism and which workers "first honors" must go. (Inspired by such tales, at least one entomologist, Johannes Zopp, enquired of Jordan whether he thought the new chromosome research could be applied to moths.[38])

The commonality of Dobzhansky's and Jordan's questions and general approach illustrates how the problems and data of a reformed taxonomy, focused on geographical variation, had provided a foundation from which to proceed with the study of speciation. But it was not clear in the 1930s that museum taxonomists, despite the overwhelming amount of data on variation at their fingertips, were still the best ones to continue that job, much less lead it. Jordan's argument regarding the importance of geographical variation had been vindicated, even as Dobzhansky called him to task for dismissing nongeographical variation. But this did not necessarily lead biologists to dismiss as misguided Bateson's insistence that the experimental study of heredity should dominate biology. Despite Bateson's claim that systematic experiments in breeding called for more patience and more resources than any other form of biological inquiry,[39] the capital involved proved far less than that needed to amass long series of specimens, the results less tentative, and the conclusions not so long in coming. And perhaps most importantly, patrons willing to provide capital, such as the Rockefeller Foundation, were waiting in the wings of laboratories, in stark contrast to the halls of many natural history museums.

"THE END OF TRING AS WE HAVE KNOWN AND CHERISHED IT"

When Jordan advised his colleagues to spend their holidays "where subspecies meet" in order to help solve the mystery of the origin of species, he was at a podium during the sixth International Congress of Entomology in Madrid. It was the summer of 1935. For a congress aimed at ending the isolation among entomologists in the interest of unity, the meeting was unfortunately not a large one. Reduced travel budgets and, as one entomologist explained, Spanish entomologists being "not very much in the front," produced a limited attendance.[40]

Yet what a congress the Spaniards organized! Accounts of the meeting described endless dances, banquets, and excursions, including an audience with the president of the new leftist Second Spanish Republic.[41] Richard Goldschmidt, who had been invited to speak at one of the general sessions, recalled that the republican government was anxious to promote science and had launched a well-organized campaign to encourage international gatherings to meet in Madrid.[42] In the midst of at least one nation's program of leveling society, natural history seemed safe.

But the concluding banquet proved to be one of the last evenings of gaiety for most of the Spanish entomologists. Jordan had once described the differences between two populations of subspecies as comparable "to two constituencies, the one conservative, the other liberal, which distinction does not mean that all the voters in the one constituency are conservatives and those of the other liberals."[43] Soon, the analogy hit somewhat too close to home for his Spanish colleagues. For little more than a year after the congress, Spain had descended into civil war. A conservative military coup d'etat led by General Franco besieged the republican government, and for three years, conservatives and liberals tried to force unity, rather than rest with a majority, with disastrous consequences for the populace. As a result, the proceedings of the 1935 congress would not appear until 1940, and then only in condensed form. The immediate problem for natural history, it seems, was not a certain ideology but the interminable twentieth-century conflicts between ideologies.

Jordan voluntarily brought his correspondence with Spain to a halt as war descended on the country, a wrenching move when knowledge about European flea distribution depended on being able to gently hound his colleagues with questions and requests for material. As General Franco's troops surrounded Madrid, Jordan informed Walther Horn that he had stopped writing to anyone in the city since letters from England were distrusted by the Bolsheviks in charge of the Spanish capital, "all too willing to express such distrust with a gun shot."[44] He became anxious when the secretary of the congress, the wealthy Candido Bolivar, failed to send him manuscripts.[45] "I wonder what is happening to the members of this family under present conditions," wrote the Russian refugee Andrey Avinoff from that other bastion of upper-class patronage, the Carnegie Museum in Pittsburgh. He hoped they were all safe and the scientific possessions of the museum secure.[46] Ultimately, some of Spain's most prominent entomologists, and natural history's most ardent patrons, had to flee the country.[47]

The Spanish civil war showed how the old world could disappear in a sudden conflagration. Elsewhere it passed away more slowly. By 1936, Walter Rothschild was working at his collection less and less due to increasing fatigue.[48] His beloved mother's death in January 1935 had deprived him of his most ardent champion, and he had injured his left leg in a fall, succumbing for months to gout.[49] Soon he was working just one hour each day, moving about the halls of the museum with two walking sticks. By the fall of 1936, diagnosed with pelvic cancer, he had moved from Tring Park to the more manageable and modest Home Farm, just three hundred yards from the museum.[50]

In view of Walter's declining health, Jordan and Rothschild's niece, Miriam, tried to secure the museum's future by clarifying the language of Rothschild's will. With the family's agreement, Jordan wrote an addition to the will outlining how the buildings and collections would be bequeathed to the British Museum, passing to Miriam if the bequest could not be accepted.[51] Having driven to Germany in July 1937 for the jubilee year of the University of Göttingen, Jordan received word via telegram that "the end seemed near," and he dutifully rushed back to Tring.[52] Finding Rothschild too ill to answer letters, Jordan took up the task of processing his employer's accumulated correspondence.[53] Rothschild passed away peacefully a few days later.

The ornithologist Ernst Mayr, working through Rothschild's bird collection at the American Museum of Natural History in New York, concluded his letter of condolence to Jordan with the lament: "This means the end of Tring as we have known and cherished it."[54] Indeed, the loss of Rothschild symbolized much more than the end of the rarified way of life at the Tring Museum. To many zoologists, his passing symbolized the loss of a whole system of patronage. For years, entomologists had blamed the paucity of governmental support for natural history on the fact that, in the words of the South African entomologist A. J. T. Janse, "much individual assistance seems to be expected from wealthy men."[55] Now, many urged that the discipline could no longer rely on devoted, wealthy amateurs, since, as one president of the Entomological Society of London announced, "it is unlikely that in the days to come there will be so many men of the necessary leisure available as has been the case in the past."[56]

The change had been described to Rothschild in his capacity as a British Museum trustee a few years before he died, in a letter from the Cambridge professor of zoology, John Stanley Gardiner. "There seems to me to have been a very great change in Zoology in the last 40 years," Gardiner wrote. As a young man he had known many wealthy collectors who had helped and encouraged him, but practically all had passed away. "Unfortunately, there does not appear to be any succession of similar men looming on our horizon," Gardiner lamented. His concern had very specific causes, which he proceeded to outline. Members of the public were becoming increasingly interested in animals, he explained, and the museum had to react to the "demands of this entirely new 'electorate.'" In particular, the museum must stay "one step ahead" of the "loose unconsidered advocacy from outside, accompanied perhaps by press articles and some times comments that sound rather ill-natured."[57] Museums filled with dead specimens had always had their critics, even within the diverse naturalist community. Now they came from members of an electorate with the power to shut institutions down.

A shift in power had occurred within science, too. Natural history had been formed within a society dominated by the rule of the landed and capitalistic classes. It persisted in a world of science ruled by middle-class professionals.[58] As a result, some seemed quite content that the age within which Rothschild had counted as a zoologist (Jordan eulogized him as the last nonprofessional zoologist) was passing. A few years prior to Rothschild's death, one entomologist noted with satisfaction that in the past thirty years "Entomology, which used to be regarded as the amusement of rather eccentric persons, has taken its proper and important place in the field of Science." Now, new developments in applied entomology and biological theory had thankfully brought the study of insects "to the front rank among the Sciences which claim the recognition of all serious persons."[59]

Some of the older generation felt less than impressed by such changes. Richard Meinertzhagen complained in 1939 that Britain was losing her older ornithologists at a very fast rate; "We have few private collectors coming on," he wrote, and "there are but half a dozen youngsters in the country who have collected and scarce one who can skin a bird up to the best standard."[60] But to many, Rothschild, Meinertzhagen, and their friends embodied a tradition of wealthy naturalists studying birds and insects for their own amusement.

"Serious persons" followed new rules, studied insects and birds for different reasons, and worked for new institutions.

When Jordan had arrived at Tring, the two worlds had worked together well enough, even if at times Jordan and Rothschild had wrestled over the proper methods of natural history work. In his memorial, Jordan hinted that Rothschild tended to apply ornithologists' criteria of specific distinctness, such as size, to butterflies and moths. This was despite the fact that such characteristics often depended on the quantity of the caterpillar's food and thus were not useful as a criterion of distinctness "unless corroborated by other differences."[61] Jordan often admonished Rothschild to pay more attention to those "other differences," particularly less conspicuous ones. He occasionally warned correspondents when sending reprints of Rothschild's work that he doubted very much whether his employer's conclusions were always right, especially since many of the groups he worked on required more detailed study than Rothschild gave to them.[62] Although, Jordan explained in his memorial, Rothschild had studied at Bonn and Cambridge during "a period when the acquisition of knowledge in internal and external morphology was one of the main objects of the Zoological curriculum," he had never taken kindly to the microscope and microtome; "Structural details easily escaped him. Like an artist, he perceived the animal as a whole and not the details which made up the picture."[63]

Although Jordan gave Rothschild credit for letting his "secretary" dissect specimens, in contrast to collection owners who wanted their specimens in pristine condition for display,[64] the account of Rothschild's "artistic eye" clearly distinguished him from those in possession of a practiced "scientific" eye. Later in life, Jordan's "scientific" eye—and the professional identity it represented—even seemed to influence what animals the Tring curator noticed when in the field. He rarely, for example, took note of birds or large mammals during his expedition to Africa. "The only large carnivore killed was a hyaena," he reported, "which fortunately yielded some fleas."[65] The geneticist Richard Goldschmidt later recounted how an old biology teacher had ridiculed his dream of being a zoologist with the response, "What, a zoologist? Zoology is no science but a sport for rich people!"[66] Was Jordan's lack of attention to big game a passive rebellion against natural history's strong association with wealthy sportsmen?

At times it seems Jordan even regretted the years spent building an extraordinary knowledge of notoriously popular butterflies and moths. For although Rothschild's ability to amass an enormous collection had given Jordan access to the material required for "scientific systematics," he did not purchase a single Lepidoptera for the museum after Rothschild's death.[67] By 1938, he was informing friends that he was abandoning the insect order "to which I had formerly devoted so much time as Lord Rothschild's assistant."⁶⁸ Sensing the lapse in "Jordanian" papers on butterflies and moths, John M. Geddes wrote in 1949: "I have always understood that you were personally interested in the family Sphingidae, and rather specialized in this group. Have you given this up in recent years?"⁶⁹ Jordan, it turns out, may have been a bit antibutterfly as well as antibird. But perhaps the less popular fleas and Anthribidae avoided the implication of entomology and zoology as a "sport for rich people."

Jordan had tried to set the record straight regarding his "personal interests" during his 1930 address to the Entomological Society of London. He knew that anyone who had seen the collections at Tring would certainly think that "when I dream an entomological adventure, my unrestrained mind would wander in tropical forests in chase of beautiful Lepidoptera, whose glorious colours are ever under my eyes at Tring and to the study of which I devote more time than to any other subject." But no!, he announced. "The dream is always the same." He came upon "unknown species and varieties of *Carabus*, on whose structure I gaze enraptured." He confessed he did not know how a Joseph, Daniel, or Freud would interpret the dream. "To me its meaning is that I am at heart a Coleopterist, as in my youth."⁷⁰ Entomologists studying fruit flies for departments of agriculture were not the only ones who had to give up their groups of choice in the interest of securing patronage.

Abandoning his own studies of Lepidoptera did not, however, equate to forsaking the collection entirely. No matter how dazzling to the artist's eye, Rothschild's Lepidoptera also represented an unsurpassed source of scientific information on geographical variation and distribution. As part of the Rothschild bequest, and with the help of Miriam Rothschild and the British Museum's keeper of entomology, Norman Denbigh Riley, Jordan put forth a plan to the trustees to transfer all of the British Museum's Lepidoptera to Tring. Defending his proposition, Jordan pointed out that Rothschild's two million specimens could not possibly be accommodated in London, and the scientific staff there "is so much interrupted by visitors that their scientific work suffers a great deal." He insisted that the man left in London to care for the reference collection and deal with visitors be a technical assistant, not a scientist.⁷¹ Jordan thus proposed a new, state-supported institution in which the collections of London and Tring would be combined for the benefit of scientific research on pure taxonomy, far from the distractions of the tax-paying public.

Riley backed Jordan's plan to the hilt. He, too, knew the dangers of moving Tring's entomological collection within the purview of the British Museum and other biological priorities. In his memorandum on the subject, he explained how the proper study of the collection would "occupy the whole time of several generations of lepidopterists." The staff at South Kensington, though competent, could not justifiably undertake such work "since it would alienate them from the critical taxonomic work necessary to the formation of reliable reference collections there." Riley urged that the researchers at Tring be given the resources necessary to complete monographs for which the London entomologists had no time.

In full recognition that their scheme was impossibly ambitious, Riley concluded his memorandum by insisting that, at the very least, the Rothschild collection must be kept intact. It had been amassed, he explained, in order to provide "material adequate to the detailed study of the local, geographical and individual variation of species, not as an end in itself, but as a means of shedding light upon the problem of the evolution of species."[72] Thus, Riley emphasized that under the terms of the bequest, "no specimen formerly in the Rothschild collection is to be considered duplicate."[73] The last thing he and Jordan wanted was for some museum administrator to view Rothschild's long series of specimens of the same species as a source of revenue or exchange.

To seal their case, Riley drew attention to the inability of the modern age to amass such a collection ever again:

> Taking a broad view of the whole, the dispersal of these great collections would be a zoological disaster; for that destruction which now accompanies "civilization" throughout the world has already rendered it impossible to bring the like together again.[74]

Had Riley defined "civilization" according to nineteenth-century criteria, he might also have mentioned the imminent decline of imperial control of far-off collecting grounds. Or the demise of definitions of science that emphasized the importance of both collections and taxonomy. Or the impact of Wallace's vision on the amount of both private and public revenue available for resource-intensive natural history pursuits such as specimen collecting. (Jordan, at least, blamed the limited amount of funds available for the museum following Rothschild's death on the high amount of death duties that had to be paid, the result of Wallaceian tax reforms in the 1890s.[75])

Committed to ensuring that Rothschild's unique collection remained available for scientific research, by late 1938 Jordan was canvassing wealthy friends for anyone willing "to help the nation" buy a large portion of the grounds of Tring Park in order to build a number of "biological institutes for research."[76]

He took advantage of a Linnean Society gathering on the question of species to advertise the scheme. While his fellow speakers not only argued over "what is a species?" but questioned the importance of systematics and large collections, Jordan tried once more to focus attendees' attention on building the institutions and organizations required to amass the *concreta* necessary to answer such *abstracta*. He tied his proposal regarding Tring explicitly to the crucial role of systematics in biology, describing, for example, the important lessons taught by systematists' careful investigation of the distinctions between subspecies, "the significance of which is not yet understood by every biologist."[77]

Although Jordan's and Riley's vision of a research institute at Tring had land, a collection, and in Miriam Rothschild an ardent guide, that vision had to compete with other sciences seeking patronage as well as new reasons for doing science. The new Lord Rothschild, Charles's son Victor, may have inherited the family title, but, although he would become a biologist, he had none of his uncle's or his father's enthusiasm for natural history. Those at the museum rarely saw him, for, as Jordan explained, Victor preferred Cambridge "on account of his scientific work."[78] That scientific work departed completely from the tradition of the Tring Museum. Victor patronized biochemists such as John Tyler Bonner, a champion of an increasingly reductionist biology, not the meticulous friends and seemingly archaic disciplines of his naturalist uncle and father.[79] Ideologically and politically, he self-consciously separated himself from the world in which his eccentric uncle could pursue his hobbies and call them science. He described himself as having a "scientific albeit anti-entomological bent" and avoided using his vast wealth in support of his own pursuits.[80]

Perhaps, as the son of a man who the 1939 edition of Ripley's *Believe It or Not* claimed had paid £10,000 for a flea from a grizzly bear,[81] it is not surprising that Victor Rothschild wished to distance himself from Tring's natural history empire. Strongly influenced by Cambridge professors whose "political sympathies were inimical to capitalism *per se*," Victor turned away from the bank in even more profound ways than had his Uncle Walter, sitting as a Labour Party peer in the House of Lords.[82] In stark contrast to calls for the independence of pure science, the new Lord Rothschild insisted that scientists must earn their keep in society by establishing closer ties between science and industry.[83] Such calls would have profound implications for the ability of different traditions and disciplines to garner legitimacy and support in coming decades. While geneticists could call on their potential usefulness to both agriculture and medicine (at least until they had established a sufficient institutional base to dispense with such overtures to utility),[84] naturalists often had a difficult time explaining the point of accumulating huge collections of butterflies, moths, and birds. Victor Rothschild's interests, values, and absence from Tring served as a constant reminder that the criteria by which natural history would be judged had shifted dramatically.

By February 1938, Jordan, unable to quickly secure a plan for the future of the museum in this changed world, described the situation as "nightmarish": he did not know what would happen in the future, there was so much to do, and time simply flew. "I wish matters would straighten soon," he lamented to his old friend Moss, "and allow me to go on with research work."[85] By then entomologists widely knew Jordan as the world's expert on both fleas and Anthribidae. With this honor came the dubious distinction of having hundreds of specimens sent to him for identification from all over the globe, especially from economic entomologists trying to fight insect pests. "I _must_ clear away the accumulation of small collections sent to me for determination and description," he lamented in July 1939.[86] He had hoped after 1936 to devote himself "entirely to those insects in which I am specially interested," the Anthribidae.[87] Correspondents such as J. B. Poncelet, who had sent Jordan specimens for determination in 1935, had long been complaining that they still awaited "with impatience the descriptions."[88] But his letters were peppered by apologies that he had not been able to complete much work on the group yet. "Other matters," he explained, had claimed his entire attention.[89]

Although Miriam Rothschild tried to help, within a year of Rothschild's death much of her attention, too, was forced elsewhere. By the end of 1938 she warned Jordan that she was "snowed under" with refugee problems as the Nazi regime descended on Europe (though she wondered with determined prodding if his "big paper on fleas was ready").[90] A month later, she confessed that her own papers for _Novitates Zoologicae_ would have to wait.[91] Soon, a whole new series of "other matters" would claim Jordan and his colleagues' attention, again halting correspondence, the revision of manuscripts, and, ultimately, the ability to even obtain paper on which to publish accounts of expeditions, descriptions of new species, or detailed analyses of where subspecies meet.

"PROVIDED EUROPE DOES NOT GET QUITE MAD"

Jordan and his friends had kept an anxious watch on political developments on the Continent for some time. Walther Horn warned him in 1932 that if Hitler came to power, German science's internationalism would end, since nationalism and internationalism by their very nature, he explained, contradict each other. Horn believed the rising nationalism in Europe had already caused many of his compatriots to lose interest in the activities and endeavors of foreigners, and he did not think many Germans would attend the entomology congress in Paris.[92] In the face of rising nationalism, Jordan soon suggested removing the word *international* from congress titles altogether, a tacit recognition that convening entomologists as representatives of nations somehow undermined the endeavor of unifying them as scientists. Horn frankly replied that he knew Jordan's wish to use the words *Cosmopolitan Congress of Entomology* entailed a silent protest against *national* since, as "everyone knows," the cosmopolitan "knows no nations."[93] But he suspected, given current political trends, that future entomological congresses would simply be a series of national meetings, name change or no.[94]

Meanwhile, political internationalists were wrestling with rising nationalism on a far wider scale. As the ambitions of Italy, Germany, and Japan tested both the resolve and the power of the League of Nations to direct the interests of diverse nations toward a peaceful future, the League's self-appointed mandate of settling international disputes through negotiation and arbitration was drowning in a wave of inquiries, commissions, and reports. Mired in parliamentary procedure, it had become an easy target of derision for those who did not take the virtues of liberal models of governance for granted. In 1933, ten months after taking over as chancellor, Hitler had not only withdrawn from the final disarmament conference but removed Germany from the League of Nations altogether. Japan followed his lead after the League condemned its invasion of Manchuria. Italy would withdraw in 1937, following ineffective sanctions of its invasion of Ethiopia in 1935.

As Jordan's homeland explicitly renounced parliamentary democracy and all it stood for, Jordan confessed he began to find the "old country" uncomfortable: "One can not speak out freely, and everyone is afraid of spies and informers."[95] He did not record the "confidential outpourings"[96] heard while visiting friends and relatives. But British entomologist J. C. R. Fryer noted in a report to the Ministry of Agriculture that Walther Horn could be "frank to the point of indiscretion" in expressing his dissatisfaction with the present political regime.[97] Jordan's friends at the German Entomological Society were also desperately warding off Nazi regulations, and would soon be investigated by the authorities for their refusal to commence and close meetings with the required toast to Hitler.[98] Years later, Jordan's friend Erich Martin Hering would note proudly that nowhere in the entire run of the society's journal during the Third Reich, from 1933 to 1944, would one find the words *Hitler, Führer,* or *Nationalsozialismus.*[99]

By August 1937, Jordan began refusing to loan specimens of importance to Berlin, though he denied this had anything to do with German politics.[100] He and his friends tried to keep an eye on Horn. They worried that his sensitive temperament would not bear the Nazi regime for much longer. Horn's anxieties for his family and institute following the Great War had already brought on a breakdown from which he never completely recovered. Though he had diligently campaigned to hold the 1938 International Congress of Entomology in Berlin, he now abandoned his fellow organizers. "Our friend Horn seems to be in the dumps," Jordan reported to H. J. Turner. "He has resigned as General Secretary of next year's Entomological Congress (Berlin) and talks of retiring from his Museum."[101] Turner thought it best if the Berlin congress found another secretary. "He might break down after what he has had."[102]

Both Horn and Jordan had known that the congress organizers would have to make concessions to the demands of the Nazi Party, such as in the choice of honorary presidents.[103] After Jordan urged him to keep the number of speeches during the opening ceremony to a minimum, Horn warned that, in accepting government support, they would not be able to restrict government representatives from speaking.[104] When the draft congress invitation arrived in Jordan's mailbox, he read that the congress would be held under the patronage of "General Göring, Minister-President, Commissioner for the Four-Year Plan and Chief Forester of the Reich," and a whole series of Nazi officials. These names disappeared from the final invitation, perhaps another example of the passive resistance to Nazi interference of which Hering was so proud. In any case, almost a decade later Hering still smarted from the troubles he had had with the propaganda ministry in planning the congress. He later attributed the choice of Stockholm for the next congress to the committee's wish to meet in a country not governed by a dictatorship.[105]

Although the idea of not holding the congress in Berlin did not arise in Jordan's correspondence, some individuals certainly questioned whether they should attend. A friend warned the American entomologist G. P. Engelhardt that "things don't look so good over there now," and he should spend his vacation in the United States.[106] But despite such misgivings, the Berlin congress hosted a "gigantic assembly" of more than 1,100 members from 350 foreign countries. The German ornithologist Erwin Stresemann boasted of the entomological gathering to the American ornithologist Alexander Wetmore, in the midst of planning the next International Congress of Ornithology for Washington, DC. "Go and try to beat this record in 1942!" Stresemann gloated, "There were 13 sections meeting at the same time, in the Berlin University. The concluding dinner was followed by a ball, which lasted till 4 in the morning."[107] Andrey Avinoff, of the Carnegie Museum, seemed nothing but impressed. "Let me tell you how enjoyable was the Congress in Berlin," he wrote to his friend William Schaus in DC.[108] But others, such as Riley, Jordan's colleague at the British Museum, noted vaguely that although the meeting had been highly efficient, it had "somehow been an uncomfortable congress."[109] J. C. R. Fryer thought that amid the lavish scale of entertaining "it was not easy to differentiate between that carried out for the purpose of propaganda and that dictated by other motives."[110]

International politics aside, some entomologists found the congress uncomfortable on other grounds. It was after returning home from Berlin that Fryer expressed doubts during his presidential address to the Entomological Society of London whether, given the number and diversity of the papers—and a membership of more than one thousand—the congress could continue to act as the "International Parliament for Entomology."[111] If it did, surely the balance of power portended by the Berlin congress made some nervous. For compared to the congress at Madrid, applied entomology dominated the congress in Berlin to a dizzying degree. Jordan and Horn stood to deliver their manifestos for studying geographical variation and for addressing the crisis facing systematic entomology, respectively, but the majority of the papers highlighted entomologists' role in improving agricultural and industrial productivity, forestry, and human health. The program's introduction set the tone with impressive descriptions of the damage done by insects to forests and food, and as germ carriers. "National frontiers have no meaning for such enemies of mankind," it declared. "The problem of their destruction is of the same great importance for the well being of all countries and the necessity of controlling them leads all nations to cooperate."[112] It seemed, perhaps, an ironic claim in a country that had withdrawn from the League of Nations and whose troops would soon storm over the national frontiers of Czechoslovakia and then Poland.

Like many Germans traumatized by their nation's postwar decline, some naturalists saw in the rise of the National Socialists enormous prospects for not only stability and prosperity but a return to the scientific glory of the prewar period. In the aftermath of the Great War, German naturalists, including Ernst Mayr, had often sought patronage elsewhere, since institutions such as the Berlin Zoological Museum had limited funds for specimens, much less expeditions.[113] Now, a few months after the congress, Jordan received a letter from Rudolf Hans Braun, with whom he had traveled in South-West Africa, waxing enthusiastic over the prospect of Germany regaining her colonies in the near future. Braun had long collected in Africa independently since bad market prices prevented natural history societies from sending anyone overseas, but he now held out for the approaching day when Germany would have colonies once more.[114] Jordan replied to Braun that indeed the British Museum would be interested in purchasing specimens. He even thought Braun would find plenty of private collectors willing and able to make purchases, hinting that the collecting network had recovered to some extent. He warned, however, that things were not as easy as when Lord Rothschild had been alive, since requests for advances must be submitted to the British Museum. Jordan ignored the comments regarding German colonies.[115]

Perhaps Braun had thought he was writing to a fellow patriot, but Jordan had long since adopted a new homeland, to the point that he insisted that he did not possess the "strong German accent" that his friends found so endearing.[116] He had written just a month earlier to Horn, trying to cheer him and insisting that the looming clouds threatening war would disperse. "It makes me happy that your optimism has proved to be true!" Horn wrote after the Munich Agreement.[117] While Jordan and Horn hoped that the meeting in Munich would bring an end to their homeland's expansionist ambitions, Braun saw Britain and France's concession to Hitler as the first step toward Germany's return as an imperial power, and German naturalists' return to the field on their own terms.

As Hitler proved Chamberlain the fool by marching into the rest of Czechoslovakia in March 1939, and as Miriam Rothschild scrambled to help Jewish refugees escape the expansion of the Nazi regime, Jordan tried to help members of the entomological community. When Austria fell under the jurisdiction of the Nuremberg Laws, the Jewish entomologist Adolf Hoffman, his fortune "wholly ruined," appealed to Jordan for aid. Jordan dashed off a series of letters to his British colleagues, in the end raising a subscription of £6 a month so Hoffman and his wife could emigrate to England.[118] Meanwhile, he began advising correspondents on the Continent that it would be better not to mention the name Rothschild when dealing with the German government.[119]

All the letters in the world could not help some of their friends. In June, Jordan received word that his most ardent partner in the effort to organize entomologists internationally, Walther Horn, had committed suicide. In coming to terms with his friend's death, Jordan later explained that Horn had strongly opposed the Hitler system and ultimately "the fear of being put into a concentration camp and maltreated drove him to make an end of his life. He was a thorough internationalist in science and very outspoken."[120] As the network of systematic entomologists lost one of their most vociferous reformers, many looked to Jordan to continue the campaign. Martin Schwartz urgently requested Jordan's advice in choosing Horn's successor at the Deutsches Entomologisches Institut in order to protect and continue Horn's vision for German entomology. He thought E. M. Hering might do, but requested Jordan to confidentially sound out the executive committee of the congresses for suggestions. Surely, Schwartz concluded, the institute should not be given over to the sole control of an applied entomologist.[121] Like W. J. Holland a decade earlier, Schwartz recognized that an applied man at the helm would not necessarily be in the best interest of all kinds of entomology.

Committed to improving naturalists' knowledge of diversity and geographical distribution in spite of the looming clouds on the horizon, Jordan and his daughters had continued plans for a trip to South-West Africa in November 1939. They had hoped to stay for half a year, and Jordan had been drilling his correspondents in the region, seeking information on good collecting sites. "I shall leave for South-West Africa the middle of November," he informed V. G. L. van Someren optimistically in mid-August, "provided Europe does not get quite mad and starts a war."[122] After Germany invaded Poland in September and Britain responded with a declaration of war, Jordan canceled the trip. "It is a great disappointment," he wrote, "which has to be borne with a smile, like the income-tax."[123]

By September 1939, all letters went through censors again, and Jordan began writing only in English. "It is a great pity that the war interferes with life in so many ways," he lamented, "so many innocent people suffering and losing their livelihood because trade cannot go on in the present deplorable circumstances."[124] The official accounts of the proceedings of the International Congress of Entomology from the previous year in Berlin had still to be dispersed to foreign members, and Jordan dispatched only a few at a time "in order to avoid the risk that a whole lot goes down in a torpedoed boat." "It is a sad time, unbelievable madness," he wrote to Bror Yngve Sjöstedt in Stockholm, "We cannot correspond, naturally, with any one in Germany, but our other correspondence should be continued." Convinced that the German people would not be able to stand another period of hunger, he thought the next congress would have to be postponed until 1942.[125] Meanwhile, the proceedings of the 1935 Madrid congress finally appeared. A brief apology for the delay introduced the volume, accompanied by a nationalist tribute to the triumph of General Franco's "Glorious National Army" over the Communists.

The insect trade, like so many things, ground to a halt for the second time in three decades. No collections could be bought during the war, since, as Jordan told the German ornithologist Oscar Neumann (who had fled Berlin), no money could be sent abroad. "I personally would buy Siphonaptera and Anthribid beetles, if I were allowed to pay for them," he wrote, "but that is not possible during the war, and whether I have the means to do so after the war is doubtful."[126] As entomologists tried to carry on some semblance of work, petrol rationing interfered with travel, even the thirty miles or so between Tring and London. As Riley noted, "a trip to Tring and back would use up two thirds of my monthly ration!"[127]

The time permitted for entomological matters was dispensed by Jordan and his colleagues no less sparingly than victuals, gasoline, and paper. By 1940, he had decided to work at insects only after dark, devoting daylight hours to "digging for victory" in his garden.[128] The Jordan household took in three evacuees from London and two Polish refugees, and Jordan lost almost thirty pounds working in his garden.[129] He laid aside his work on the Anthribidae classification; "Diodyrrhunchus or whatever its correct name is and allies are probably beyond my reach at present; well, I must wait till after the war."[130] Correspondence, specimen determinations, and manuscripts fell by the wayside, as a paper shortage and other restrictions shut down printing. Although he lamented that the next issue of *Novitates* would be delayed, since he had many new fleas and Anthribidae that awaited publication, he conceded to friends that "under the present conditions, that seems to be a very small matter."[131] Meanwhile, he tried to encourage the specimen network in countries outside the war zone, asking entomologist J. D. Sherman if he knew of any North American collectors who could send him small Anthribidae. "It seems almost funny to speak of wanting minute insects at a time when an heroic struggle is going on before our door so to speak," he wrote, "but it is advisable to go [on] with the work one do[es], and keep cool."[132]

Keeping cool would be a tall order amid the disruption. "If the war had not broken out," Jordan explained to Ernst Mayr, hard at work studying Rothschild's collection of birds at the American Museum of Natural History, "Tring Park might be by now property of the British Museum and gradually be developed into a centre for biological research; the Park would be ideal for new institutes. Well, that was my hope."[133] As mobilization proceeded, the government instead earmarked Tring Park for use by the military authorities, although after being cleared out the buildings remained unused for the duration of the war.[134]

Jordan accepted all he could from those trying to find a refuge for their collections. "Bombs are rare," he informed one entomologist, and in any case the Germans aimed at the railway tracks, which they missed.[135] An incendiary bomb hit one of the Tring cottages, but it was removed quickly and the museum escaped harm.[136] "We have so many collections of importance stored that biological science of all countries would suffer if such collections were destroyed," Jordan wrote.[137] Although he went regularly to the museum after gardening, he could not work there late into the night, as he had formerly been accustomed to, because a man on fire watch slept in his working room.[138]

While "digging for victory" and working in the museum, Jordan joined many of his colleagues in trying to maintain some semblance of the natural history network's access to information. He worked to ensure those in Madrid, who could not get periodicals outside of Spain due to financial regulations, still received the *Zoological Record*, even offering to pay their subscription himself. "I am much in favor of the unity of science," he wrote to the Secretary of the British Council in explaining his scheme, "As permanent secretary of the Entomological Congresses and president of the International Commission on Zoological Nomenclature I emphasize the adjective International and am aware of some of the difficulties."[139] He campaigned for the Czech entomological publisher Wilhelm Junk and his wife to move safely to England from Berlin through enlisting the support of the Entomological Society of London. "The Home Office," he wrote, "should welcome the immigration of Dr. & Mrs. Junk, as Junk's world-wide experience would be at the service of British publishers during and after the war and would help to make London the centre of the publication and sale of Entomological literature."[140]

But whether the Home Office, or any other governmental body, would care about the publication and sale of entomological literature depended, once again, on whether some tie could be made between entomology and the war effort. Even some entomologists argued that work conducted by the London societies that had no material bearing on the country's war effort should be stopped entirely. "Until we have won the war nothing matters," wrote one.[141] Entomologists who could make the case that their work had a material bearing on the war quickly stepped to the fore in such a climate. Work continued on the British Museum's handbook of insects of medical importance, for example, as Jordan found when pressed into adding a section on fleas.[142] He had always had a difficult time getting collectors in exotic places interested in fleas and Anthribidae rather than just Lepidoptera.[143] Now, as he wrote to J. B. Corporaal, he found it fortunate that both insect groups proved of some economic and medical importance since this meant collections continued to arrive.[144] The traditionally "purist" Entomological Society of London had long bowed to the dominant wind, but now even more papers appeared with explicit notes that their work had been inspired by medical or agricultural problems.

Despite the fact that the research program he had outlined for entomological systematics had been called to a halt once again, Jordan urged his friends not to despair. "The power of humanity to recuperate is immense," he wrote to Miles Moss.[145] Unable to agree with Jordan that the war was one of a recurrent series that would soon be over, Moss chided him for his optimism. "In view of the modern high-explosive bomb which can blow everything into smithereens in a second and burn up the residue," Moss wrote, "surely we have spent our entire lives in vain in storing up the treasures of science and art and literature in Museums, Galleries and Libraries! A hideous reflection! . . . And _now_—how much has been wantonly and brutally destroyed!!"[146]

They would not learn the answer to Moss's query for three long years. In the meantime, Moss stubbornly defended the "treasures" stored up in museums, not as tools in the fight against disease or insect pests, but as worthy of preservation in their own right. He even gestured to the old defense of the natural theologians, that such collections demonstrated the existence and benevolence of God. Ultimately, six years of another world war dampened Jordan's optimism as well. By 1947, he would confess that scientific conditions in the world were such that, with all free communication and exchange halted, somewhere in Paradise old father Linnaeus must be repenting that he had baptized the first species in "Systemae Naturae" *sapiens*.[147]

"WITHOUT THE COLLECTION I AM HOPELESS"

Jordan once conceded in a letter to Moss that no one could possibly return to the former state of society once Hitler was beaten. "Our civilization will certainly be affected by this struggle," he wrote, "and many who are now comfortably off or rich, will be poor, while others will have amassed fortunes on which the State will very likely lay its hands. . . . When peace has come, the war against ignorance, the perpetual and greatest foe of humanity, will absorb all our energies and probably level the incomes as much as the present war expenditure does."[148] Wallace would have grieved the cost but commended the result. He had long insisted that nationalization of wealth and the leveling of incomes provided the only route out of the "social quagmire" that had gripped British society for so long.

Jordan blamed both the war and death duties for his inability to turn Tring into a research center for biology. Changing views of what counted as science and how it should be justified, and a consequent lack of interest among potential patrons, had not helped. In addition, in the middle of the war, Jordan had received a cryptically worded admonishment from the secretary of the British

Museum that the plans rumored for Tring, including the purchase of the Tring estate, would be "outside the function of a museum."[149] But what exactly were the functions of a museum and could these institutions survive without reevaluating those functions?

In one of his rare asides on the debates over natural selection, Jordan once wrote that "the difference in number of genera and species in two districts equally accessible to the original stock may be due to the dying-out of some species in one district." Jordan cited the American zoologist Edward Drinker Cope's claim that the disappearance of species from natural causes arose from restricted variability. "It is indeed easily conceivable that a species or race well adapted to a certain definite condition of life," he wrote, "has much less chance to survive under changed conditions than a morphologically and physiologically variable species which offers a broader basis for new adaptations. The old nobility gradually dies out and a new one arises from the promiscuous non-specialized lower strata of life."[150]

Jordan was talking about insects. But the soliloquy could stand in for the fate of both natural history museums and their patrons as well. At the beginning of the twentieth century, and despite the rhetoric of geneticists such as William Bateson, the territory of modern biology could reasonably be viewed as equally accessible to the various trajectories of natural history and biology. The question then became, would a naturalist tradition tightly bound to certain characteristics of the nineteenth century be able to adapt to changing conditions, or would a "new nobility" less hemmed in by tradition arise to take its place?

The naturalist tradition was not the only field of science facing this question, of course. Jordan often expressed fear for the future of science in general within the post-war world. "What will happen after the war," Jordan wrote to a fellow entomologist in 1943, "depends most likely on the attitude towards SCIENCE in general by the government which will then be in power."[151] Science required funding, and Jordan believed that the income of scientific societies would go down considerably during and after the war. Institutions dependent upon voluntary contributions, he wrote, would not be able to carry on as before.[152] Since the state would then remain the sole patron for science, the criteria by which scientists were judged would depend on the nature of the government in charge. But Jordan also knew that systematics and natural history collections—long supported by "voluntary" contributions—would inevitably suffer more than many other sciences during the recovery period.

Jordan knew, for example, that the universities, providing a haven for some sciences, would not fill the gap in patronage for natural history collections and taxonomy. Cautioning one entomologist against leaving his collection to Birmingham University, he explained that the zoological departments of universities were "out of sympathy with systematics and will remain unsympathetic during the next one or two generations."[153] Taxonomy, he explained to Charles Radford, a mite specialist at the British Museum, had left the universities and been relegated to Museums. "A reputation in Taxonomy therefore carries little weight in the mind of the professor of Zoology or Botany at a University."[154] As for the British Museum, a traditional center of taxonomic research, he noted that it preferred to hire young men to whom they did not have to pay too much. "This sort of administration of a museum for RESEARCH," Jordan confessed, "appears to me very queer."[155]

The lack of adequate sites for systematics as a science obviously concerned Jordan, and had been one of the main reasons for his plans for Tring. Who precisely he considered the primary culprit in nudging taxonomy out of university zoology became clear in a letter written to Moss in 1942. After describing his hopes that Tring Park would appeal to the various keepers of the British Museum as an ideal place for biological research stations as opposed to museums for exhibitions of specimens, he concluded, "Zoology requires research institutes now, since the Zoology of the Universities has practically been replaced by genetics."[156]

Riley had expressed optimism during their campaign for Rothschild's collection to form the centerpiece of a new research institute that those genetics-obsessed universities were actually "rediscovering" the importance of taxonomy. Indeed, he thought it necessary to look principally to the universities for students qualified to turn Tring into a research institute on the grounds geneticists and cytologists were recently paying more attention to taxonomy.[157] Riley had reasons for being hopeful. Throughout the 1930s and 40s a group of biologists were developing the conceptual, institutional, and disciplinary framework within which a choice might not need to be made between genetics and systematics, or between experimental work and natural history.

In response to the enthusiastic claims of Mendelian geneticists, Jordan had insisted in 1916 that "Mendelism cannot account for the geographical phenomenon embodied in the problem" of what factors led to the appearance of structural differences between subspecies and species.[158] He had been joined by other naturalists such as E. B. Poulton, who used Tring's data on geographical variation as well as Jordan's conclusions to counter "the extravagant claims of Mendelians."[159] But by 1931 Poulton could announce triumphantly during his address to the Zoological Section of the British Association for the

Advancement of Science that the work of geneticists W. E. Castle, T. H. Morgan, and others had proved discontinuous variation could indeed create continuous, gradual variation. This work would be lauded by Julian Huxley as a "modern synthesis" of seemingly contradictory mechanisms of evolution.

In the course of his 1931 address, Poulton had reminded his audience of the "splendid work of the Tring Zoological Museum under the guidance of Lord Rothschild, Dr. Hartert, and Dr. Jordan, and the conclusions published in their journal in 1903." He repeated yet again Rothschild and Jordan's words that "whoever studies the distinctions of geographical varieties closely and extensively will smile at the conception of the origin of species *per saltum*."[160] The naturalists, in other words, had been right all along. In again drawing attention to these trends during the 1938 discussion on species concepts at the Linnean Society of London, Poulton noted how his and Jordan's conception of species as interbreeding communities had recently been confirmed by Mendelian research, as summarized in Dobzhansky's *Genetics and the Origin of Species*.[161] Jordan himself took note of the change in geneticists' views regarding continuous variation after attending the 1936 meeting of the British Association for the Advancement of Science. "I was particularly pleased that natural selection is back in biology," he reported to Horn, "Just a few years ago, geneticists insisted they could produce new species in the laboratory." He excused such claims as the enthusiasm of youth, obviously pleased by recent developments.[162]

The man who would most clearly establish the importance of the "modern synthesis" for the work of a new generation of systematists—and was even then launching an unprecedented publicity campaign for systematics in the process—had once collected for the Tring Museum. As a young ornithologist Ernst Mayr first visited Tring in 1927, five years before the birds went to New York. Having been hired as a member of a joint expedition to Dutch New Guinea and the former German New Guinea (now under a League of Nations mandate) for the Berlin Museum, the Tring Museum, and the American Museum of Natural History, Mayr went to Tring to be, in his own words, "indoctrinated" by Hartert in how to collect birds scientifically.[163] In his account of the expedition for *Novitates Zoologicae*, Mayr noted that Hartert and Rothschild directed him to the mountains of the northwest peninsula of New Guinea, their rationale being the usual Tring complaint that, since series from this area were poor and known animals had been based on a few specimens with poor labels, available material proved inadequate for establishing the zoogeographical boundaries and relationships of the animals.[164]

There was some talk of Mayr replacing Hartert as curator of Tring's bird collection upon the latter's retirement in 1931. But the collection was soon being packed up, and Mayr soon found himself, in what he later described as a "twist of fate," in charge of Rothschild's birds when he was hired by the American Museum of Natural History in New York.[165] Over the next decade, Mayr worked on integrating the Rothschild collection into the rest of the museum, building an expertise in systematics and biogeography that he then channeled into a broader study of geographical variation and speciation.[166]

As war raged in Europe and Tring did its best to absorb military personnel and refugees from London and abroad, Mayr began sending Jordan inquiries about the systematics of both fleas and *Lymantria* moths. In justifying the queries, Mayr explained to Jordan that he was formulating arguments against Richard Goldschmidt's book *The Material Basis of Evolution*, certain, no doubt, that he would find in Jordan a ready ally. Throughout the book, Mayr explained, Goldschmidt argued against the idea that subspecies are incipient species, insisting instead that species arise from "revolutionary reorganizations of the germ plasm." (Indeed, Jordan's 1896 and 1905 papers had served as targets of Goldschmidt's critique.) Mayr wished to bring out the weakness of Goldschmidt's conclusions, he noted, using the renegade geneticists' own material. But the United States had no specialist in *Lymantria*, and he hoped Jordan could help with the group's systematics and geographical distribution.[167]

In taking a break from his wartime responsibilities to respond to Mayr's letter, Jordan began by apologizing that he could not answer the long list of questions, for the collections of Lymantriidae had all been moved to London. Besides, Jordan noted, the man working through the British Museum's collection of the family was engrossed in war work and had not yet studied the genitalia, "necessary for sound systematics." He had not seen Goldschmidt's book, but offered that, based on Mayr's description, Goldschmidt seemed to interpret his discoveries in the same way as Bateson, "being entirely on a wrong track." Jordan then recommended that Mayr read an article by K. Mather in *Nature*, quoting, to one of the deans of the "modern synthesis," Mather's words that geneticists now realized increasingly that small variations are the material of evolution.[168] "It is a pity that men like Goldschmidt have not studied the systematics of a family of mammals, birds, or insects before they entered on a discussion of the basis of evolution," he wrote. "The geographical variation is so general that it is quite unscientific to neglect the factors involved." He concluded that, if the material had still been at Tring, "I should certainly, for my own satisfaction, test Goldschmidt's systematics; but without the collection I am hopeless."[169]

As Jordan lamented his inability to establish a biological research institute at Tring due to the war, and tried to carry on work without access to the old network, Mayr used naturalists' long-standing emphasis on geographical variation—now becoming key to the "synthesis"—to raise the status of systematics. In acknowledging Jordan's letter, Mayr wrote that he agreed entirely with Jordan's view that locality played a crucial role in the differentiation of species. He planned, he noted, to set these ideas down fully in a forthcoming book.[170] The opening line of Mayr's *Systematics and the Origin of Species* made his aims explicit: "The rise of genetics during the first thirty years of this century had a rather unfortunate effect on the prestige of systematics."[171] He intended, in no uncertain terms, to take that prestige back. In doing so, Mayr readily insisted that not much in the book was original. It was merely, he wrote to the ornithologist Richard Meinertzhagen, "a well-organized presentation of the point of view of the modern taxonomist."[172] His summary of that point of view certainly impressed many naturalists. The famous biologist E. O. Wilson recalled that as a young entomologist he considered Mayr's book "his bible."[173]

Mayr began churning out books to maintain the momentum, including manuals of taxonomy that incorporated the approaches of what Julian Huxley had begun calling "the new systematics." The new systematist, Mayr explained, focuses on the study of infraspecific categories, or subspecies, using a species concept based on interbreeding and studying long series of specimens, while the old focused on species defined purely morphologically and based on a few specimens. The major problems of the "old systematist" were those of the "cataloguer or bibliographer," while "the new systematist tends to approach his material more as a biologist and less as a museum cataloguer." Most importantly, he tends to "consider the describing and naming of a species only as a preliminary step" to more synthetic and far-reaching generalizations.[174]

Jordan could wholeheartedly agree with such an aim, even as he called some of the resulting generalizations into question as too simplistic.[175] The "new systematics" affirmed the importance of careful attention to geographical variation and the analysis of "where subspecies meet." And it vindicated careful work on museum collections as central to the most pressing questions in biology. Mayr's understanding of speciation, for example, was bound to his experience as both a collector and curator of Rothschild's unprecedented collection. For his book, he had drawn extensively on conclusions produced by taxonomists working on collections of a whole range of organisms, insisting within the first pages that "it can be claimed without exaggeration that lasting generalizations can be based only on systematic groups that are well known."[176]

Yet how would the altered context in which naturalists worked influence their ability to do this "new systematics" (which, of course, was not so new, as Mayr frankly acknowledged)? For while geneticists and naturalists may have made their peace under the grand conceptual rubric of the modern synthesis, square footage had been allotted, budgets determined, and positions named. In the first chapter of *Systematics and the Origin of Species* Mayr lamented that specialists dealing with little-known groups (which included most groups of organisms) could barely keep up processing new material. "Only seldom," Mayr lamented, "does he have time to go beyond the purely descriptive phase of work and try his hand at putting some order into the growing 'heap' of species. Newly discovered species are likely to upset his ideas any day. He is forced by necessity to do 'old' systematics."

Mayr acknowledged that taxonomists' ability to apply the "new systematics" depended "as much on the available material and the degree of knowledge attained in his group as on his training and point of view."[177] Fully aware of the dependence of their manifesto on the continued availability and accumulation of collections, Julian Huxley and his fellow contributors to the 1940 volume *The New Systematics* called for increases in museum staffs to ease the burden of routine description and naming and urged museums to sponsor experimental laboratories and field stations. Huxley insisted that increases in the staff of entomological departments in particular were urgently needed "if they are to escape from the burden of routine description and naming and take full part in the activities which may be described under the head of the New Systematics."[178] It was a bold call in the middle of a world war that had inspired demands that everything peripheral to the war effort be halted.

Clearly, the modern synthesis was built on the careful work of a generation of taxonomists. But whether they could continue that detailed work in the postwar world was not at all certain, particularly for entomologists. John Smart, of the British Museum, stated frankly in his contribution to *The New Systematics*, entitled "Entomological Systematics Examined as a Practical Problem," that no other group of scientists "prostitutes its knowledge and efforts to the extent that systematic entomologists must in the maintenance of their routine work." He called for an end to the "enormous weight of curatorial and identificatory routine work that falls on the shoulders of those engaged in the professional study of insect systematics." All this drudgery resulted in a low output of "basic research." Premature ambitions to incorporate the conclusions of the new genetics would only increase their burden if the "exceedingly heavy curatorial duties" of those in charge of collections and the great confusion in the literature were not first addressed.[179]

Like Horn and Jordan before him, Smart turned his attention to the burden of the tradition and the panacea of organization. His words echoed both Jordan's optimistic manifestos for the international organization of entomologists and Horn's call for an international institute. But given that the congresses were now defunct and the international institute canceled, even the basic informational index that Smart called for seemed a distant utopia. Meanwhile, Julian Huxley conceded that in view of Smart's description of the "overburdening of the entomological systematist by the mere number of kinds of insects, of which approximately 10,000 new species are still being described every year," the ideal work of the "new systematics" "must wait for many years before being adequately discussed or put into practice in this, by far the largest group of organisms."[180]

As the cries for organization continued and entomologists faced both a huge backlog of specimens and constant calls to determine incoming collections, working entomologists who tried to apply the principles of the new systematics to their specimens looked at what Mayr had been able to do with the data of ornithologists—and Rothschild's collection—with envy. Elwood Zimmerman wrote to Mayr at the end of 1942; "I wish we knew as much about some of the groups of insects as you know about the birds.... How few authors ever indicate relationships derivation and geographical affinities when they describe new things."[181]

The lament would have sounded heartbreakingly familiar to the reforming entomologists Jordan had joined in the 1890s. Mayr certainly acknowledged his luck at having "what is probably the largest and most complete bird collection in the world" at his fingertips. But Mayr insisted that, short of access to such a collection, "if a specialist wants to examine a genus critically, he can draw on the resources of sister institutions and borrow their material."[182] To war-torn Europe, where constant calls by international congresses to establish easy exchange routes and cooperation had been silenced by blitzkriegs, this optimistic comment by a naturalist working peacefully in New York must have seemed somewhat exasperating.

EIGHT

Naturalists in a New Landscape

The events and trends of the decade and a half after the Ithaca congress of 1928 held ambiguous hints of what the future would be like for taxonomists and entomologists. Britain's colonies served as a source of specimens, but taxonomists could not keep up with the influx of unnamed forms. The loss of private collectors sealed the authority of a new professional class of scientists working in state-funded institutions yet deprived natural history collections of traditional sources of patronage, specimens, and cultural prestige. The conceptual developments of the "modern synthesis" seemed to vindicate the claims of systematists who had emphasized the importance of geographical variation, but fulfillment of the research program of the "new systematics" required a level of organization that had eluded entomologists for decades. As state patronage provided new positions and funding, entomologists became subject to public oversight whose attitudes toward scientific research could shift at the turn of an election. Finally, even as the war opened up government coffers for applied entomology, that worldwide conflict and its aftermath shut down the routes of open communication and travel necessary for international congresses, visiting other collections, and obtaining specimens.

Jordan had always insisted that "a small army" of entomologists was really required to study insects adequately,[1] particularly since the meticulous methods he insisted taxonomists use in order to complete scientific systematics slowed down the description of the living world, not to mention classification and explanation. Convinced that existing organizations could not solve entomologists' problems, he had hoped that entomologists' own international congresses could ultimately raise and direct such an "army." When the army did arrive, in the form of increased support for entomology, its loyalty to new priorities created tensions between Jordan's justifications for doing scientific systematics and new patrons and priorities that heightened demand for quick, efficient inventory and identification.

Within these altered circumstances, pursuing the collection of long series of specimens and doing good systematics in peace must often have seemed, as Rothschild and Jordan once said of sorting through zoological nomenclature, like a quixotic undertaking, particularly since the windmills could come in so many varied shapes and sizes. By the final decade of Jordan's life in systematics, a new generation of systematists looked back on the work he had been able to complete in the 1890s and 1910s with awe. They held him up as exemplary—a scientific taxonomist whose work demonstrated that the jibe "species-making" did not reflect what good taxonomists do. But given the direction of scientific and social change, could a hero from such a different world be of much use to a new generation?

RECOVERING FRIENDS, COMMITTEES, AND CONGRESSES II

When Ernst Mayr's *Systematics and the Origin of Species* appeared in 1942, Jordan was eighty-one years old. Britain was struggling through the third year of a war that disrupted work at the museum and destroyed once again what entomologists had rebuilt of the natural history network. Obtaining access to both taxonomic literature and groundbreaking theoretical syntheses had become virtually impossible. Miriam Rothschild tried to get a copy of Mayr's book, for example, but months passed and the book had not arrived.[2] In any case, the time available to peruse the manifestos of up-and-coming biologists proved limited, for Jordan was swamped by requests to determine Anthribidae and the promise to complete the flea catalog.

Some legacies of the past could not survive this straitened period. After half a century in print, *Novitates Zoologicae* was shut down in the name of wartime economy.[3] Meanwhile, with an eye toward keeping open those avenues that remained for obtaining more *concreta*, Jordan found time to describe some fleas sent to him by a "kind donor," who, as he explained to Riley, "it is advisable not to disappoint."[4] Entomologists now needed every last ally in the constant search for specimens.

At the war's end, Jordan's focus shifted, for the second time in three decades, to determining what had been lost. News of Continental entomologists and their collections slowly began arriving as, "the shooting and bombing being over," Jordan's and other entomologists' "peaceful correspondence" came "to life again."[5] Erich Martin Hering's wife, having heard a rumor that Jordan had died, greeted the first letter from Tring with the words "Dr. Jordan lives, he has written!"[6] From René Jeannel, director of the Muséum national d'histoire naturelle in Paris, Jordan learned that two of the Paris museum's entomologists, Eugène Bouvier and Ferdinand LeCerf, had died during the war and that their colleague Pierre Lesne's health was failing.[7] In Hering's letters he read how, although Hering and friends had "tried from all sides" to save their colleague Walther Arndt (a member of the International Commission on Zoological Nomenclature) after he was denounced as an enemy of the Third Reich, Arndt had been executed. Hering himself had been accused for his "anarchist attitude," but his denouncer had shot himself before the Gestapo could do their worst.[8]

As correspondence resumed and the statesmen tried to put the world back together, Jordan pressed his correspondents for news of the fate of collections. Did Hans Eggers know whether Erich Martini's flea collection had been saved? Had the Wagner collection survived?[9] Jordan soon learned that the Paris Museum collections and the Congo Museum appeared to be safe,[10] and that Otto Bang-Haas's natural history agency in Dresden was intact.[11]

But for each collection that survived, word came of one that had been lost. Bombing had destroyed the Hamburg library, and Witold Eichler wrote from Stuttgart that his entire collection had been lost.[12] The homes of Italian entomologists Orazio Querci and Roger Verity had been destroyed, and Walter Roepke had lost nearly all his belongings, his laboratory badly damaged and looted.[13] Adolph Kricheldorff lost his entire collection, and the 1944 edition of the *Deutsch Entomologische Zeitschrift* had burned at the printers.[14] When Jordan wrote with relief that the Berlin Museum insect wing had not suffered damage, Hering informed him that in fact the Lepidoptera hall had sustained two direct hits, completely destroying the Satyriden, Nymphaliden, Danaididen, and parts of the Lycaeniden collections, including many type specimens. One-fifth of the Lepidoptera and ornithological collections were wholly destroyed. Hering reported that he had been able to save some collections by storing them in a cellar, including Otto Staudinger's microlepidoptera collection and the Papilios. The beetle hall had been destroyed by a bomb, but fortunately the specimens had been evacuated, with the exception of the Brenthiden.[15] After the list of lost collections arrived from his friend, Jordan questioned anxiously whether the destruction at the Berlin Museum included the Anthribidae; Hering replied that they were safe.[16] Given the destruction wrought by the war throughout Europe, in the end Jordan was surprised to find how many collections had survived.[17]

As they traced the fate of butterflies and beetles, Jordan and his friends waited anxiously for the authorities to let them mail food packages to friends in

Germany. By mid-1946, a family friend in Germany, surnamed Nolte, described to Jordan how his family had been given about four grams of fat per day; "The fact that humans can not exist long on such rations in the long run and continue to work," he wrote, "you as a biologist may understand." Nolte wondered how a true democracy could develop in such conditions. He suspected—although he had hopes for the Marshall Plan—that Communism would seem to be the only answer to workers given so little to eat. In the meantime, he recommended that Jordan not visit Germany. "Keep your beautiful memory of the old Lower Saxony country intact."[18] Jordan had learned soon enough that "Old Hildesheim has gone up in smoke."[19]

Whether the taxonomic wing of the naturalist tradition had also gone up in smoke also had to be determined, of course. A postwar report described how in Imperial Japan a thriving natural history community's libraries and natural history collections had been obliterated by the war. Since the 1930s, following the example of their Western colleagues, Japanese naturalists had fully joined the endeavor to catalog the spoils of conquered lands. From 1941 to 1945, a new institute, the Research Institute for Natural Resources, had been set up in Tokyo "to act as a repository and clearinghouse for all the scientific material sent to Japan from invaded territories." Expeditions to China obtained collections "long needed for comparative work in Japan," with reports on subjects such as "The Geography of the Greater East-Asia Co-Prosperity Sphere." Now, naturalists learned that the fire raids of 1944 and 1945 not only "destroyed some of the most valuable collections and libraries in Japan but also eliminated the reserve stocks of recent periodicals and private publications," including the buildings of the Tokyo Research Institute for Natural Resources and the homes, aviaries, libraries, and collections of the naturalists Prince Taka-Tsukasa and Marquis Kuroda.[20]

For many, the naturalist tradition itself seemed irreparably damaged, and it proved difficult to convince the authorities to both lament the fact and authorize remedies. German ornithologists had hoped that the military government would grant a special license to the major ornithological journals to commence printing by 1946, but the U.S. Military Government Publication Control Office in Stuttgart refused to allocate the paper needed on the grounds that such publications were luxuries.[21] Not surprisingly, the pressures of securing scarce resources in the postwar environment channeled entomologists' justifications of their work along distinctly applied lines. Even Jordan, who increasingly urged that pure entomology be pursued for its own sake, again stressed the field's economic and medical importance. In working with N. D. Riley to help E. M. Hering restart the activities of the German Entomological Society in the Soviet-controlled section of Berlin, Jordan advised Riley, for example, to emphasize how Berlin had been a center of applied entomology. Applied science "is much to the fore under the regime of the Soviet-system," he explained; such an argument might influence the Russians.[22]

The Berlin entomologists were not alone in having to adjust to new masters. In becoming an employee of the British Museum after Rothschild's death, Jordan had joined colleagues subject to the opinion of government officials and, ultimately, the public. His shift in employment was emblematic of the end of an age in British science. Due to the naturalist tradition's strong ties to the British social and economic infrastructure of the nineteenth century, the changes reached into the depths of its network. As Jordan reported to Hering, the war had dramatically altered British society. The owners of large country houses, for example, could no longer afford to inhabit luxurious buildings due to the rise in income taxes.[23] Nor, of course, could they afford to build collections of natural history. Almost three decades earlier, after the First World War, Stanley Gardiner had bemoaned the loss of large-scale, wealthy collectors. Now the Second World War made the transition complete. As Jordan warned Walther Horn's widow while helping her sell Horn's collection of Cicindelidae to the British Museum, it was the Treasury, not the museum trustees, who would have the final say about the purchase.[24]

Much of the collecting infrastructure upon which Jordan, Hartert, and Rothschild had built the Tring Museum had gone the way of the old patrons. As Jordan warned his colleagues at war's end, the collecting network in the tropics was ruined. Indeed, he thought new attempts to collect would be interrupted for some years after the war.[25] In trying to explain the situation to those in the field, he described how the purchasing power of the British Museum had been severely curtailed and would remain so "for the indefinite future."[26] Almost a decade later, Jordan still warned correspondents who wished to amass large collections of butterflies and beetles that he did not know how they could come by them since the war had destroyed the natural history trade.[27]

Jordan exaggerated slightly, for although in Britain they lacked workers, food, and goods of all kinds, his correspondence with America was quite lively. The Americans, he noted to Horn's widow, seemed to be getting on with work faster and with better resources than anyone.[28] British Museum entomologist Riley obviously envied his American colleagues: "Your people seem to have been getting all the pickings in the Pacific," he wrote to a colleague in Washington, DC. But while he lamented that the war had not produced the "crop of collections" that the more "static" conflict of 1914–1918 did, Riley admitted it was perhaps just as well, given "our huge arrears and all our other troubles to overcome."[29]

The problems created by the loss of almost half a decade of correspondence, publications, and some collections soon became apparent yet again. "From the scientific point of view," Jordan wrote to one correspondent, "we are greatly hampered by the shortage of paper for scientific publications and by the lack of publications issued in the enemy countries during the war."[30] Olov Lundblad, writing from Stockholm, asked Jordan when it would be possible to correspond with German entomologists. "At present I am working out a water mite collection for the Brussels Museum," he explained, "and I can hardly finish my paper without getting some types for comparison from German museums."[31] Examining type specimens to sort out synonymy had been one of the primary ways taxonomists had responded to the legacy handed to them by the diversity of the naturalist tradition. But the types were often off limits now, if not destroyed. "You know as well as I do," A. J. T. Janse wrote to Jordan, "that many species have been described insufficiently and studied most superficially, with the result that a large number cannot be recognised without access to the original material." They would have to come to some sort of ruling at the next congress, Janse urged, regarding whether such species could be ignored, or whether "we are to ballast our taxonomy with such useless descriptions."[32]

In firm agreement that the "next congress" should indeed provide the route out of chaos, Jordan began making plans to reconvene the "nation of Entomologists" as soon as his international correspondence resumed. Again he insisted scientists must lead the fight toward recovery by their internationalist example. The memory of his friend Walther Horn served as encouragement. "I honour his memory," Jordan wrote to Frau Horn, "Science knows no national borders. He fought for this idea, an idea that one now defends eagerly in England."[33] He began, in January 1946, with the members of the executive committee, "as far as they are reachable," arguing that the congress should meet as soon as possible, despite the obstacles that "the politicians—i.e. governments" would no doubt put in their way. "The invitation should be general," Jordan insisted. Whether the congress was large or small did not matter; "the main point for Entomology is to preserve continuity."[34]

Jordan received encouragement from the knowledge that Ivar Trägardh and Bror Yngve Sjöstedt in Sweden wished to host the 1947 congress and from the French entomologist René Jeannel's support for an early congress. "He sees no reason against it," Jordan noted triumphantly to Riley.[35] The news that a recent, successful International P.E.N. (Poets, Essayists, and Novelists) congress had met in Sweden in May 1946 also bode well, he thought, despite the absence of the Russians.[36] Hering, working in Eastern Berlin, replied with enthusiasm, particularly at the choice of Stockholm, since "it was not under a dictatorship."[37]

Jordan placed entomologists' continued efforts to restart their meetings firmly within the old internationalist ideals of the prewar period. In advising Riley, who was trying to convince the authorities to let preparations for the congress move forward, Jordan wrote, "According to current news scientists in general are in favour of unity of science irrespective of frontiers and race difference." Riley, he advised, should certainly point out that systematists had fought for this ideal for centuries. "In point of fact Systematics have achieved this unity," he wrote, "whether a species is described and named in China, Germany or Peru, for the purpose of science, a reference as to the description enters into the bibliography of that species throughout the globe."[38]

Yet once again the ideal belied the complex realities—and new challenges—facing the old project of establishing "one name for each creature." Riley would later praise Jordan for his ability to hold "unfalteringly to his course" when meeting the turmoil and changes of the first decades of the twentieth century,

> untrammeled by personalities or politics, gently turning aside the unwelcome, blandly ignoring what was best ignored, oblivious of rules, regulations or formalities that hindered progress, avoiding disputes in favour of sweet reason and the main issue, always courteous, friendly and infinitely patient.[39]

But now, pressed by the turmoil of the postwar years, Riley replied to Jordan's idealistic statement with a confession that he felt "not by any means satisfied that the human heart shows any sign of improvement yet, at any rate in the international sphere."[40]

Certainly the prospects of convening a truly international congress proved dim as the cold war descended on Europe. Jordan had been having trouble communicating with Soviet entomologists since the 1930s: he received not a single answer to a 1937 letter to the directors of Russian biological institutes inviting them to the Berlin congress. As nations that had once eagerly participated in building the naturalist tradition's networks of information and specimen exchange fell behind the Iron Curtain, the all-important paths of communication between naturalists were severed. Soon, entomologists in the West began complaining of being "handicapped" from completing necessary revisions since they were unable to obtain specimens from the USSR and countries under its influence.[41]

"So the door is actually closed," Jordan exclaimed in disbelief after packages of journals he had sent to Russia were returned in March 1948. "The papers on Fleas were not particularly dangerous politically!"[42] Requests for loans of specimens from the director of the Moscow Museum were met with silence.[43] Meanwhile, Stalin's anti-Western cultural campaigns of the late 1940s attacked internationalism in academia, maintaining that Soviet culture and science must proceed uncontaminated by capitalist, bourgeois civilization. The possibility of amassing "scientific, world-wide collections" was thus firmly, and officially, ended. As Russian entomologists began publishing new species in Russian,[44] the division of the globe into two ideological spheres of influence destroyed the ideal of eventually having "one name for one species." For if Jordan was right that international exchange and communication represented a basic prerequisite for the standardization required for scientific systematics and entomology, then the natural history work which, ironically, both Walter Rothschild and Alfred Russel Wallace had loved so much, could only be damaged by the fragile truce between their two worlds. The battle between the opposing ideologies behind Rothschild's world and Wallace's vision thus ended in a dangerous draw, as Europe was divided into two blocs. The legacy of the nineteenth century was profound indeed.

THE QUEST TO "CLEAR UP THE CHAOS" IN WEEVILS AND FLEAS

Jordan continued his efforts to complete scientific systematics in this new environment. He had presented his collection of 21,420 specimens of Anthribidae (including 1,292 types) to the British Museum in 1940 on the condition that he could incorporate it into the museum's material himself at Tring, using his own specimens for exchanges.[45] He had received very little from abroad during the war, apart from fleas amassed during an investigation of the plague in Argentina. But he assured correspondents that dealing with the Anthribidae and the classification of the fleas gave him enough to do—probably "more than I shall finish."[46] "The number of undescribed species is rather appalling," he exclaimed to J. Vinson as he combined his own collection with the British Museum's.[47]

In thanking correspondents for the few specimens of Anthribidae sent during the war, he repeated the now familiar Tring line that no specimens would be useless in determining the all-important question of the extent of individual variability.[48] The familiar laments regarding poor work appeared as well. One worker who had failed to look underneath his specimens of Anthribidae had given different sexes of the same species distinct names.⁴⁹ Others had described new forms based on purely individual differences in size and coloring despite the fact, as Jordan wrote in 1929, that work on structural differences had shown that distinctions presented by the "outerwear" were as stable and reliable as the "old diplomacy" of prewar Europe.⁵⁰ Fortunately, he wrote, his eyes were still good and his hands steady and he could continue to "dive beneath the surface" of his specimens, although he had stopped going to the Entomological Society of London meetings due to hearing loss.⁵¹

Intent on harnessing the knowledge obtained by Jordan's good eyes and steady hands for the sake of scientific systematics, his coleopterist friends urged him (forty years after Jordan had first written to Poulton that he hoped to work out a generic revision and classification of the Anthribidae) to publish his conclusions. "We need such a foundation!" fellow specialist Elwood Zimmerman urged, "And you are the only person in the world qualified to do it."⁵² Zimmerman had been pestering him for almost a decade to "clear up the chaos" in the "muddled" group.⁵³

But Jordan refused to rush into publication. His firm belief that the ultimate goal of systematics was to provide a solid foundation for biology justified, in his view, extreme caution in putting forth any conclusions regarding relationship. He once wrote that he had focused on the Anthribidae because the study of such a small family, in which naturalists had described only a couple of thousand species, had "the great advantage that one is familiar with and remembers the general appearance of practically all the species, and, if the degree of relationship between the genera has been ascertained by oneself," theoretical questions could be tested with ease.⁵⁴ But the system he had developed for fulfilling this goal, amassing long series of specimens and the comparison of many characters, meant the work must be slow. "More extensive study on the spot is needed!" Jordan urged, in order to determine the extent of individual variation necessary for a thorough revision of the group.⁵⁵

Getting anyone to "the spot"—namely, the field—during wartime had often proved impossible, of course, but the situation after the war demonstrated how difficult completing monographs and revisions could be even when consignments of unknown forms arrived in droves. As Jordan's correspondence picked up, he repeatedly warned colleagues that inadequate material prevented him from making firm conclusions regarding the individual variability of each species. He excused his slowness in determining species on the grounds that beetles did not lend themselves to breeding experiments as well as Lepidoptera. (And, of course, Lepidoptera did not lend themselves to breeding experiments as well as Drosophila, by this time the primary tool of an army of researchers studying speciation.) Jordan explained that this made the process of determining the basic units of classification, the species, difficult, and dependent on the comparison of long series of specimens (although he admitted that "copulating individuals help to determine what goes with what").[56] The study of more specimens, species, and characters was needed, at all times guided by the recognition that "without material the systematist is helpless, and in the absence of sound systematics the biologist also works in a fog."[57]

Given Jordan's constant effort to convince systematists that the "material can never be enough," Elwood Zimmerman must have been astonished when, in 1949, Jordan turned down his offer to send 1,500 specimens of Figian Anthribidae to Tring. In explanation, Jordan offered a common taxonomist's excuse that he already had extensive collections waiting to be "worked up."[58] After the war, as naturalists set to work processing a backlog of undetermined collections, Anthribidae had arrived from the museums of Paris, the Museum of the Belgian Congo in Brussels, and even amateur collectors in the Indies, with requests for him to describe any new species. In 1949, Belgian entomologists flooded him with so many cases of Anthribidae to determine that he began holding on to one of the cases while sending back the others "so they don't send any more for a while."[59] The masses of material proved no help if one did not have the time to process them. And time had to be prioritized.

Pressed by Zimmerman to return to the coleopterists, Jordan eventually confessed the main reason why he wished to slow the number of Anthribidae specimens. By 1946, he was immersed in the task of completing a flea classification at the behest of Miriam Rothschild. As he explained to Zimmerman, with limited time at his disposal he had decided to focus primarily on "the study of FLEAS from many countries."[60] Throughout 1947 he constantly informed correspondents that the next year "will have to be principally devoted to the study of the taxonomy of FLEAS."[61] One wonders whether Zimmerman silently cursed the siphonapterists as, in this "age of specialism," another specialty competed for Jordan's time. To generously wish the flea work well may have been more than a strapped coleopterist could manage, just as those working in museums often found it hard to wish drosophilists well. (Zimmerman was not known for patience in the face of taxonomists' troubles. Laments filled his letters to taxonomic colleagues: "What a mess!" "It's maddening!" "No body knows nothin' about these critters," "Woe is me," "Bother, what chaos!" Jordan's and Horn's laments had been slightly more academic, but Zimmerman captured with eloquent succinctness taxonomists' increasing frustration with their assigned task.)

Meanwhile, specialists on fleas exhorted Jordan to synthesize his experience into a comprehensive survey of the Siphonaptera. Jordan had hoped to publish a classification of fleas in 1937, and all concerned with the group had eagerly awaited the result. University of California at Davis siphonapterist M. A. Stewart encouraged Jordan to finish with the words, "It will certainly be received with a great deal of enthusiasm, particularly since it will have been prepared by the Master."[62] But years passed and the promised classification did not appear. Nor did the catalog of Charles Rothschild's collection. After the war, Miriam Rothschild took matters in hand and spoke with Riley about the need to get someone to assist Jordan with the catalog. Well aware that the British Museum had few assistants to spare, she offered to "look after the financial aspects of the matter."[63] The wishes of a Rothschild with access to purse strings, the obligations of state patronage (the terms of Charles Rothschild's original bequest demanded that Riley have someone complete a catalog),[64] and loyalty to a lost friend all now combined to determine Jordan's remaining hours at his desk. To make sure the work got done, Riley and Rothschild arranged for an expert on bird lice who had extensive experience at the microscope, G. H. E. Hopkins, to assist Jordan.[65]

Intent on (or harassed into) finally completing the catalog of fleas, yet unwilling to do so without attention to the "deeper" aims of natural history, Jordan organized a trip to Switzerland with his daughters to collect fleas as soon as possible after the war's end, "in order to be able to answer some contentious questions."[66] In the midst of postwar travel restrictions Jordan had confessed that he often dreamed of sitting on a rock at 7,000 feet among alpine flowers, looking across the valleys at peaks and glaciers and consuming a piece of good cheese.[67] Now, he, Ada, and Hilda finally crossed the Channel in the summer of 1946 to spend their holiday in Pontresina, Jordan choosing the place because he wished to collect a series of a species of flea known from only a single specimen.[68] As often as they could in subsequent years, the Jordan family traveled to Switzerland, their routes determined by questions of variation and distribution in Siphonaptera.

His travels showed him how much Europe had changed. Parts of the Continent were closed off entirely. Jordan had been invited to spend a month at the Instituto Espanol de Entomologia, but had to decline on account of the problems "excitable northern neighbours" might cause for a stranger rambling through the countryside.[69] He had hoped that the Balkan countries would be open to collecting after the war so he could solve certain problems of flea distribution, but the region remained largely inaccessible to entomologists. "It is a rather depressing thought," he wrote, that "it is more difficult and more dangerous to walk about in Eastern Europe with a butterfly net than in Central Africa."[70]

Meanwhile, inflation devoured the funds required to keep even a famously frugal entomologist in the field. In 1952, the money once sufficient for a trip of five to six weeks on the Continent now lasted at most sixteen days—hardly enough time, Jordan wrote, for a collecting journey to "fill in the gaps" in distribution.[71] Returning home, he tried to remedy the gaps in his knowledge by enlisting what remained of the old correspondence network, writing to friends and acquaintances with enquiries about areas he had not visited. Unable to devote much money to the endeavor, he would gladly, he wrote, send English beetles in exchange for large series.[72]

Meanwhile, the coleopterists continued to hope that Jordan would return to his "beloved beetles." "I hope that your correspondents are not flooding you with such a mass of specimens that you are being kept away from your all-important task of writing your generic revision," Zimmerman wrote in 1955. "You are the only one in the world with the proper understanding of the Anthribidae."[73] Jordan had long since patiently replied to such demands by explaining that he now knew of three times as many species as when he first became interested in the family in 1893. "I have dealt," he wrote to one correspondent, "with Anthribidae only as a side-line of my work and even now most of my time is occupied by the study of fleas."[74] But as one of the few experts on the group, he constantly received requests for assistance. The Commonwealth Institute of Entomology relied on him for determinations, although in 1952 Jordan warned them, "The determination will have to wait, as I am at present struggling to deal with two collections of tropical fleas and four collections (no, five besides the two you sent) of Anthribidae."[75] At the age of ninety-three, he was still warning coleopterist colleagues that he could not determine collections sent just yet, for he needed to compile the results of a lengthy investigation on the highly variable fleas of the Mediterranean.[76]

When, in the mid 1950s, Jordan finally returning to his beloved Anthribidae, he was well aware that his time was short and the number of undescribed species immense. He thus decided to concentrate on putting together a "reasonable classification and arrange the specimens at least in their genera if I have no specific names for them." He confessed to Riley that he would be "much relieved if an Angel came from heaven and described and published all our new species of Anthribidae," thus permitting him "to get on with the study of their relationship to each other (and subsequently to bother you to give me sufficient drawers for putting them in order)."[77] Yet despite his constant complaints regarding the woeful lack of specimens and species, he again astonished fellow workers with his choice of characters and detailed study. With great anticipation, Zimmerman wrote: "I note with much interest that you are not only making a study of the genitalia, but you are also studying the hind wings of the Anthribidae in preparation for your generic revision. Is not this study of the wings quite new?"[78] It was no wonder Zimmerman begrudged the time this honored expert spent on fleas.

The division of labor inspired by natural diversity meant losing a specialist could be disastrous to work on a particular group or region; thus, experts on other groups pressed Jordan to return to their fold as well. Afraid to lose Jordan's expertise, his lepidopterist friends urged him to work on butterflies and moths again. No doubt they found his brief replies refusing to describe their new species discouraging. After Jordan had laboriously compared two specimens sent by the lepidopterist Ernest Rutimeyer to the series in both Tring and the British Museum and found no match for them, Rutimeyer implored him to describe the two new species on the grounds that he was the only one alive with the required knowledge and necessary comparative material at hand.[79] But Jordan would not be moved.

Jordan's friend E. M. Hering, in the midst of attempting to mend the natural history museum in Berlin and limited by the restrictions placed on East Berlin by the Soviets, also begged Jordan to return to the group he had studied so extensively with Rothschild. Hering appealed to Jordan's sense of duty as an entomologist. He conceded that if one had been restricted to a certain order during one's official life, one would of course wish in his free time to turn exclusively to his favorites. But surely lepidopterists still had a great claim on his time and expertise by virtue of the extensive knowledge he had built over the years. He should, at least, Hering urged, publish a phylogenetic table of the Macrofrenaten, "which would make it possible to indicate the family affiliation of the butterfly in each case with absolute certainty." Jordan was the only one, he insisted, who could correct the available tables, which all failed at the difficult border lines and with exotics. To clench his argument, he added the warning that surely it would henceforth prove impossible for anyone to acquire the comprehensive knowledge required.[80] Hering, no doubt, felt the old world slipping away—both its access to specimens and the presence of experts who could complete good classifications.

But Jordan would not be moved, although in apology to Hering he sent a long explanation of the circumstances under which, as a "private secretary to Rothschild," he had come to work in so many different areas of entomology. Upon reading it, Hering quickly forgave Jordan his desertion; "Quite natural that you should not now return to Lepidopterists," he wrote.[81] To others, Jordan stated briefly that he had not had the time nor the inclination to describe new species of Lepidoptera he had himself collected in Africa, much less the specimens of anyone else.[82] He confessed to a Belgian friend in 1947 that he had not looked at literature on Lepidoptera since 1940, nor collected butterflies or moths since he gave up their study.[83] The loss was hard to swallow for those, such as W. J. Holland at the Carnegie Museum of Natural History, who had regarded Jordan as "the ablest lepidopterist now working in Great Britain."[84]

Well aware of the importance of ensuring entomologists studied the *concreta*, Jordan did take up a few specimens of butterflies or moths when colleagues wrote to him about specific specimens in the Tring collections.[85] Obtaining access to collections had been difficult enough in the tumultuous twentieth century, and he did what he could to help other workers use the material available. As Riley had explained when describing the importance of the Rothschild collection to the trustees of the British Museum, there was no other collection like Tring's on the planet, and the possibility of amassing such a collection in one place ever again was slim.

Even apart from the loss of wealthy patrons, the relative ease with which European naturalists could explore any land they pleased had ended. With the increasing power of independence movements throughout the British Empire after the war, conventions that naturalists had simply taken for granted came under scrutiny. Natural history collections had been built on assumptions regarding hegemony and the superiority of Western culture that meant European naturalists—though in competition with each other for specimens—pursued a common project of inventory. But in a postcolonial world, it was not clear that the interests of all would be served by specimens amassed in European capitals.

As we have seen, novel ideas regarding sovereignty over natural resources had occasionally (in the form of "socialist governors" and "nationalists" such as A. L. Herrera) interfered with collecting in the past. Now, after India gained independence in 1947, men such as Richard Meinertzhagen could do nothing without a permit.[86] France was soon fighting wars to maintain her colonies in Madagascar, French Indochina, and Algeria. Independence movements in the Dutch East Indies resulted in the Netherlands giving up that territory in 1949. When, in 1951, the Swiss lepidopterist Rutimeyer Fry asked Jordan regarding the British Museum's rules governing exchange, he explained that he found it impossible for both financial and political reasons to obtain specimens from the former Dutch East Indies; chaos reigned there, and probably would for some time, so he did not know when they could renew collecting operations. They were thus dependent, he wrote, upon those with large collections such as the Tring Museum.[87] Meanwhile, as men and women called into question the basic assumptions of superiority upon which European authority had been based, some saw the act of "discovering" new species that native populations may have known and even used for centuries emblematic of the arrogant cultural superiority that had defined the previous age. In all their discussions at prewar international congresses regarding whether natural history collections should be centralized in European national capitals or be dispersed throughout each (European) country, naturalists had never envisioned this.

As naturalists adjusted to new sensibilities and assumptions, we have at least one hint that Jordan, like Meinertzhagen, could not adapt in time to avoid conflict. At the age of ninety-six, he submitted a manuscript on fleas of the genus *Stenoponia* to the British Museum for publication. Loyal to a new system of peer review, his assistant G. H. E. Hopkins sent the paper to Harry Hoogstraal, of the U.S. Navy's Department of Medical Zoology in Cairo for comment. Though expressing great respect for Jordan, Hoogstraal was "shocked and appalled by the inclusion of several political references that I feel have no place in a report such as this." Since the annotated manuscript cannot be found, precisely what Hoogstraal found offensive can only be guessed at by the strong remarks he felt "morally obligated" to make in condemning the paper: "Due to the friendly cooperation and substantial assistance of our Arab and Muslim hosts, both of whom are roundly damned in this paper, a good share of the data in this report were obtained."[88] In response to Hoogstraal's comments, Hopkins convinced Jordan to delete the "objectional, non-scientific, and non-relevant" sections of the paper,[89] and Hoogstraal, placated by the alterations in the manuscript, was in the end "sorry he blew up."[90] But the episode highlights the demise of two historically accepted components of the naturalist tradition—components with which Hoogstraal refused to be associated: first, the inclusion of long soliloquies on the politics and culture of the lands being surveyed; and second, explicit statements regarding the inferiority of those cultures.

Eurocentric anecdotes and political commentary were not the only things being expunged from scientific papers. In the interest of conveying the objective nature of results, reforming editors were even changing the structure of sentences in accounts of observations or experiments from active to passive voice. At least one American entomologist noticed such shifts with a long rant that editors who shoved authors along "in the direction of third-persons and passive voices, austere polysyllables, and chilling Germanic dependent clauses," made them "forget what induced us into entomology."[91] As editors purged first-person adventure stories from the pages of scientific journals, other elements—theory, hypothesis, experiments, graphs, statistics, and applied concerns—entered in full force. And it soon became apparent that the generation submitting papers in this new framework might not be concerned by the challenges facing those dependent on great collections of specimen-based "facts."

AVOIDING THE SNAKE IN THE GRASS

During the debates over "how to do entomology" of the 1890s, H. J. Elwes had insisted that if entomologists decided that one must try and discover explanatory laws in order to join the "highest ranks of science," a dangerous check would be placed on "the humble endeavors of persons like myself, who find their greatest pleasure in collecting and arranging the material which must form the only solid foundation for such work as has been done by Darwin, Wallace, Bates, Weismann, and others."[92] Elwes and others had argued that a first-rate naturalist—indeed, a first-rate scientist—could in good conscience continue the work of careful description based on natural history collections. But as the context, institutions, defenses, and research priorities of entomology had changed, what counted as a "scientific" contribution also shifted.

Specifically, increasing pressure on younger workers to make groundbreaking contributions to the *abstracta* altered the relative value placed on description versus theory.[93] This shift would have profound implications for the scientific status of natural history museums and collections, often perceived as monuments to the nineteenth century's demand (in the words of Charles Dickens's character Thomas Gradgrind) to "Stick to the Facts, Sir!" In contrast to Entomological Society of London presidents who took young fellows to task for speculating, entomologists now criticized their fellows on the grounds that their work did not pay enough attention to theoretical context.

The new journals that appeared on the scene after the war made the new methodological framework, in which old journals must adapt or be considered "unscientific," quite clear. In submitting a manuscript to Ernst Mayr's new journal, *Evolution*, the ornithologist R. E. Moreau suspected his paper dealt too much with details to be of interest to the journal, although it presented a case of unusual geographic variation.[94] Mayr replied that indeed there was too much ornithological detail for the journal, whose editors maintained that descriptive material should be included only to the extent needed for the evolutionary conclusions.[95] The tight ties between method and place involved in this shift became explicit when, as a member of the Entomological Society of London's publication committee, Jordan fielded complaints from entomologists who found their work turned down on the grounds that it was "based on Museum material."[96]

In recognition of these changes, and after experimenting with *Stylops, a Journal of Taxonomic Entomology* in 1931 for the descriptions of new insects, the Entomological Society of London had divided its *Transactions* into two series; Series A for general entomology and Series B for taxonomy, in 1936. The division in some sense formalized Albert Günther's old boundary between "biology" and "taxonomy," a boundary he had reprimanded Jordan for crossing. But now it was obvious that young entomologists should publish in Series A if they wanted to impress a new generation of mentors. And as scientists emphasized theory as the hallmark of science, much of the day-to-day work required for a good revision, indeed, the revisions themselves, no longer seemed to count. Indeed, Jordan could read in one of the primary texts of the modern synthesis that "Elaborate monographs," even those which included a phylogenetic and zoogeographical basis, were "not, strictly speaking, scientific research," since they did not "propound and then prove or disprove theories."[97] This was not a young, enthusiastic mineralogist touting the methods of physics and chemistry or a Cambridge professor of zoology ridiculing an interest in classifying fleas. Rather, these were the words of the thirty-three-year-old British Museum entomologist John Smart.

Watching this shift take place, Jordan had used his end-of-the-year address as president of the Entomological Society of London at the end of the 1920s to issue a gentle reprimand to those who defined the search for *abstracta* as the only thing that counted as science. He did so by interpreting a recurrent dream, during which his enraptured gaze on the structure of unknown species and varieties of *Carabus* showed him "that I am fond of hard facts as typified by the hard carapace of a *Carabus*, which shows variation in structure so beautifully." He interpreted the dream's frequent switch to an encounter with poisonous snakes, "from whose attacks I extricate myself with some difficulty," as a warning "to beware of the snake in the grass when searching for an explanation of facts discovered and phenomena observed. Our longing for an answer to the why and wherefore makes us indeed apt to theorize hastily." Using this cautionary tale, he urged his fellows "to distrust the obvious, in particular the outward appearance of things, and to investigate the distinctions of species more deeply."[98]

During his second address, the following year, and perhaps chastened by some young entomologists for his methodological conservatism, Jordan acknowledged the importance of having a theory to guide research. Even erroneous theories served as a guide to research, he explained. Though he agreed much remained to be discovered and "many a weary description" still had "to be penned," the very existence of an amazingly large number of species "which every capable collector sends home from the tropics and subtropics" cried out for explanations of how nature had produced this "mass of different and yet in many ways similar insects." He insisted that "one must look and think, not merely look," when processing the incoming specimens.[99] The differences between Jordan's first and his second address, the one emphasizing caution, the other the virtues of theory, reflect the tensions present as one generation's methodological ethos gave way to another's. As a museum taxonomist surrounded by cabinets of specimens that could mercilessly contradict generalizations, he knew the scientific status of the *abstracta* depended on the careful accumulation of the *concreta*. On a practical level, Jordan's methodological caution meant that even those who had captured him to work on their particular organisms of interest had a difficult time convincing him to publish any phylogenetic conclusions.

As Miriam Rothschild and Hopkins tried to help Jordan complete both the catalog of the British Museum's flea collection and compose a classification, they confronted the cautious methodological stance that had made Jordan an ally to both sides of the old "biologists vs. systematists" divide among entomologists of the 1890s. "You need more collecting and less phylogeny," he admonished them when they proposed their own classification. This reprimand was delivered, Miriam recounted in dismay, despite their having twenty thousand tubes of flea specimens at Tring.[100] She found Jordan's caution and his frequent joke that "I'll know everything there is to know about flea classification when I am dead" exasperating.[101] Yet Jordan would not back down. He persisted with his old exhortations for more specimens and more time at the microscope, sounding, no doubt, a bit like Albert Günther trying to bludgeon the younger generation into good science via gentle but firm disapproval.

In the end, the pair obtained Jordan's permission to publish his own phylogenetic tree only "under duress," a task made all the more difficult by Jordan's increasing deafness. As Miriam wrote, "All these importunings and entreaties had to be yelled at the victim, who refused a hearing aid on the grounds it only increased the noises in his head."[102] Though he still insisted they needed more material, the persistence of Rothschild, Hopkins, and Riley eventually paid off. In 1948, Jordan finally published a taxonomic key to Siphonaptera, since described as having ended the "literal chaos" that had plagued the group's classification.[103]

Jordan's "preliminary" (as he termed it) classification of fleas proved E. M. Hering had been right to lament the loss of Jordan's expertise. Those who used Jordan's key, accustomed to navigating a chaotic mass of questionable descriptions and classifications, often declared their astonishment at the quality and robustness of his work on the group. The American medical entomologist Robert Traub confessed that, while he had originally wagered that Jordan's and Charles Rothschild's names for Indian fleas had "represented over-diligence, or over-enthusiasm, because of their possible role in diseases," he soon found that the species fit the figures and descriptions so well "that all the differences Jordan saw and evaluated may be exactly as he said." When the conclusions of the old master seemed questionable, Traub said he would make no change of Jordan's names without seeing the topotypical material first since "certainly he proved to be absolutely correct in so many other instances."[104]

This accuracy absolutely depended, of course, on slow, meticulous, painstaking work and large amounts of material. Advising Miriam Rothschild in 1954 on flea systematics, Jordan had insisted: "The basis of systematics is the knowledge of the latitude of the individual variability of subspecific and specific populations. This knowledge can only be acquired by the study of large series of specimens."[105] Faced with two highly variable species in her own work on fleas, Rothschild acknowledged that "more collecting and a more critical examination of large series of specimens, together with breeding experiments, will be necessary before we can satisfy Dr. Jordan's criteria for deciding the status of these two fleas."[106] It was only based on such meticulous work, Jordan insisted, that those taunted as "species-makers" could ensure, amid the complexity of a nonstatic nature, that they provided a foundation for accurate generalization (and, furthermore, prove that they did not deserve the taunt).

Jordan knew entomologists' workloads had a direct effect on the ability of that foundation to serve the generalizer. "The harassed specialist," he once wrote, who knew that only a small percentage of existing insects had been described, was all too tempted to steer entirely clear of speculation as premature. He spoke from experience. In 1929 he had known of 80 species of Anthribidae described from Java. Two collections that arrived within the year had doubled the number, and 50 of those turned out to be new species.[107] In 1897, when the theory of the plague flea's role in disease transmission had first been proposed, 68 species of fleas were known. By 1954 that number had risen to 1,454 species and subspecies, 640 of which had been described by Charles Rothschild, Jordan, or both.[108]

The number of species entomologists had been able to describe, even without Jordan's detailed, meticulous methods, had always been only a fraction of the species they knew existed in nature. One of the main points of contention in the debate over whether entomologists should allow speculation in the 1890s was whether the available facts justified abandoning the long tradition of defining sound entomological work as descriptive. Although he argued firmly against making the description of new species the central goal of systematics, Jordan could not brush aside what Miriam Rothschild called his "no doubt right Darwinian complex" of never believing one had enough material, whether species or specimens. He consistently finished any references to evolutionary mechanisms with a precautionary warning that all he had said assumed that the classification used was correct in its main lines. "And so," he concluded, "after having thrown a stone among the giants, we come back to systematics."[109]

Inevitably, when Jordan occasionally did throw "a stone among the giants" he did so in order to convince the theorists to pay more heed to the conclusions of good systematics. Given the quality of these rare excursions, Riley once wondered why Jordan speculated on evolution so infrequently in the 420 papers he published between 1903 and 1958.[110] In reply to his own question, Riley suggested that the reason lay in every good entomological taxonomists' awareness that no matter how extensive his material, "it is hardly ever more than a small fraction of what he needs as a basis for satisfying work." He recalled that, while Jordan "deplored the amount of time which had to be devoted to descriptions of new genera and species," he knew this work was absolutely necessary "if a sound classificatory basis was to be provided for the study of evolution." The drudgery paid off in some instances; "The 'only truly satisfactory classification of fleas,'" Riley writes, "is said to be that published by Jordan when he was nearly ninety." Yet despite having described 150 new genera and 1900 new species of Anthribidae (nearly two-thirds those known at the time of his death) Riley noted that, in his studies of this group, Jordan "never completely extricated himself from the drudgery phase. He would discuss their infinitely bewildering variety by the hour; but he never achieved, in this group, a system of classification which satisfied him."[111]

Between the time Jordan captured his first Anthribidae and the day he penned that nineteen-hundredth-or-so name, entomologists' world had changed much. At Oxford in 1912, Jordan's friend Sjöstedt had compared the International Congress of Entomology to the Olympic Games recently held in his own country. Sjöstedt suggested that in the case of entomology the competition was between ideas, and the winners would be those ideas that best solved problems.[112] But the competition had not only been between ideas. Why and how to study insects was also at stake, as well as who would decide the answer to these questions. If entomology was, as Jordan and Holland's exchange leading up to the 1928 congress had implied, a nation with generals competing for power, then the stark contrast between the postwar congress proceedings and those of the earliest congresses demonstrated who had won.

When entomologists met at the first postwar congress in Stockholm in 1948, papers on the ecology and physiology of insects with an eye toward control and the biological effect of different insecticides dominated the meetings. The resolutions urging the need for increased support came from sections such as "Forest Entomology," "Tropical Agricultural Entomology," and "Insects in Stored Products." Five of the eleven sections covered applied entomology, and the campaigns for international cooperation came from the international committee for applied entomology.

In his opening address, congress president Ivar Trägådh urged entomologists to embrace this trend. He described how, in the years in which entomologists had first founded the international congresses, "zoologists looked down upon the entomologists" as "bug-hunters or worse." Economic entomologists "were in their turn slighted by the systematists who looked upon them as a kind of farmers, dabbling in entomology but with nothing like the same standing in science as the true entomologists."[113] Now, and in marked contrast to Jordan's insistence that entomologists avoid frightening the public into caring about insects, Trägådh called for entomologists to "rouse the interest of the laymen in our activities" by appealing "to their most fundamental trait, their egotism, and make them realize that the insects constitute a growing danger to their welfare and economy."[114] He offered the requisite gestures of cooperation to those who must name and describe all these enemies of mankind (indeed, Trägådh had determined his own share of new species in his position as a forest entomologist). But he clearly stated the grounds on which entomology must appeal for recognition and status. Given that the man in the street knew next to nothing about the purpose of entomological investigations, Trägådh insisted that "the only way to alter this state of things is of course to enlighten the people about economic entomology." "Mr. Middleswenson himself" must thereby be convinced of the "activities of the nefarious insects."[115]

Jordan, in noting that he would welcome applied entomology at the meetings of the Entomological Society of London, had placed his generosity toward the new priorities firmly within the context of its usefulness to pure entomology. He explained how applied work on fleas, for example, had demonstrated that different species of fleas had different abilities as vectors, supporting the pure entomologists' division of these forms into distinct units. But Jordan clearly distinguished between the aims of the systematist and applied workers. The help the systematist could give to the protection of man from insects was, to him,

> only a side-issue or by-product; he is a student of pure science, devoting his time to the discovery of new species, of new connections between them and of new facts bearing on the relation between the species and its surroundings, the driving force in this pursuit of knowledge being the irresistible attraction which the subject has for him.[116]

The different goals driving the different "classes" of entomologists, as L. O. Howard called them, sometimes escaped in frustrated asides in correspondence. One disgruntled taxonomist complained, for example, that "these modern economic workers seem to have the scientific curiosity of a plumber wiping a joint."[117] Of course, such jibes often obscured the rich interrelationships and cooperation between these classes of workers. Dividing entomologists into "applied" and "systematic" hid the presence of much cross-fertilization between the different institutions, journals, and societies in which these workers pursued the study of insects. Jordan's description of subspecies, not as types but as populations consisting of different proportions of certain characters, could be applied here, as well. Still, amid diversity of form and function, the environment had, indeed, determined the relative fitness of each population of the naturalist tradition amid constant change.

GLORIFIED OFFICE BOYS

When, at the age of ninety-two Jordan abandoned his "Darwinian complex" and briefly spoke in 1953 to the Lepidopterists' Society on various problems of evolution that could conveniently be studied using butterflies and moths, he concluded abruptly: "But enough of it. I must stop suggesting what one or the other of you might do. Talking of these problems unburdens my mind, for I should have liked to try to solve them myself if there had not been many other matters which required my attention and consumed my time."[118]

Those other matters had multiplied over the years. Even during Tring's heyday of the 1890s, Jordan and Hartert had been pressed by an immense amount of work dealing with visitors and managing and building the collection. With Walter Rothschild's death and the bequest of the Tring Museum to the British Museum, Jordan officially joined the publicly funded, professional civil service that had employed scientists in increasing numbers since the 1870s as the nation enlisted scientists in the service of both home and Empire.[119] This patronage had come with both costs and opportunities, of course, particularly as pressure mounted to justify natural history museums to a new electorate on the grounds of tangible applications to human welfare.

In considering the place of natural history collections and the taxonomist after World War II, Jordan left consideration of utopian visions of some "future state of things" to others. Having retired from committees, commissions, and congresses, the current state of things dominated his correspondence. He was less than impressed by the rhythms and rules that governed civil service life. Plagued by postwar restrictions and a "paper shortage everywhere except for the endless forms issued by the Civil Service," Jordan noted resignedly that he supposed "it is as difficult to govern efficiently as it is to revise a group of insects muddled up by previous authors."[120] Riley tried his best to smooth the usual bureaucratic paperwork for his friend, who famously abhorred red tape, sweeping regulations at congresses aside with "tolerant disdain."[121] Riley convinced the museum trustees to make exceptions to many of the rules placed on the activities of other department workers. He made sure, for example, that Jordan would not have to ask for special permission each time he wished to make an exchange of specimens during his work on the Anthribidae.[122]

It was a generous attempt on Riley's part to reduce Jordan's administrative duties so that he would have more time for scientific work. Jordan once confessed to Miles Moss in 1936 that he wished he could "give up the management of this institute and devote myself entirely to research on our beloved insects without being continually interrupted by routine work or demands for information."[123] Surely Riley had heard similar laments, and obviously did not want to make things worse. He knew the effect that the pressures of administrative work could have on an entomologist's time. In thanking a friend for congratulations on his promotion in the Entomological Department, Riley replied, "It has its drawbacks, however, I assure you. It is weeks since I last handled a 'bug' or did any work on the collections. I am a kind of glorified office boy."[124] One of Riley's memorialists defended his never having produced a "great Jordanian revision" on the grounds that as the head of the entomological department Riley was at the constant mercy of the demands of departmental administration.[125]

Indeed, plagued by mounting administrative duties and lowering status (and, as he lamented, a tendency for pay scales to be adjusted accordingly), Riley had to spend much of his time campaigning for more support for the museum. As part of his effort to drive home the importance of systematics to the powers that be, he urged Jordan to publish an account of how misidentifications of fleas had confused work on plague transmission. "It might be useful in our dealings with the Treasury in the near future," he wrote, since the episode was evidence of the entomology staff's importance. "The Treasury seem at the moment inclined to look upon us all as rather third-rate curators," he explained, "and to ignore the real work of the Museum, hence they are trying to thrust upon us some rather inferior salary scales, a move which we are doing our best to fight."[126] Jordan obliged his friend. "These cases of mis-identification make it abundantly clear that the study of systematics is of fundamental importance for applied Entomology," he wrote, "as it is, in fact, for all biological research." Once again turning to better order as the answer to systematists' problems, he blamed the entire episode on the "faulty organization of biology" and urged that the matter be forcefully brought before the newly formed UNESCO.[127]

But while Jordan used the defense when pressed, he knew the fact that both Siphonaptera and Anthribidae proved of some economic and medical importance was a mixed blessing as new priorities began driving the production and use of collections. Jordan observed that the occasional postwar visitors to Tring, for example, came primarily to study fleas for medical-hygiene research.[128] But his incoming correspondence was peppered by the laments of colleagues overwhelmed by the new roles of collections. His fellow weevil specialist Elwood Zimmerman, for example, complained of being swamped by "urgent requests" from agriculturalists in places such as New Guinea (then under Australian control) for "us to work up their stuff and supply them with names for a number of new things which are turning up as plant pests."[129]

As Trofim Lysenko's reign descended on Soviet biology, ending in the lethal persecution of "bourgeois science" in the name of "science for the people," Jordan gave voice to the anxieties felt by scientists trained by another century's values. In writing in 1946 to J. B. Corporaal, an entomologist working in Sumatra, he reflected on the state of affairs produced by two world wars and its implications for pure entomology. He thought that "unless the conditions in this world of ours (or does it entirely belong to the politicians and their menials in the armies?) improve very rapidly we enthusiasts for pure research in Entomology will have to be content with the small additions of specimens which dribble in from time to time."[130] He knew that the kind of world the politicians created in peacetime would have profound implications for the kind of science that could be done. "I hope there will be a place somewhere for you as a Zoologist," he wrote to another friend who was looking to immigrate, "In a socialistic world (socialistic in a political sense) pure science will have to fight for a position."[131] By the late 1940s, Jordan had joined the Society for Freedom in Science, an organization that prided itself on having defended the importance of "pure (fundamental) science in the dark days of the early forties," when calls for science to be of material use to mankind had threatened to swamp "freedom in science."[132]

Those who tried to sustain a role for natural history museums, Jordanian systematics, or both in this new century thus had to confront the potential mire of quite different goals driving science. As Riley struggled to deliver on his insistence that the museum's scientists could be useful in the postwar crusades to improve man's welfare, for example, he was forced to confront Jordan's notoriously meticulous attention to insects for their own sake. Under pressure to fulfill the terms of Charles Rothschild's bequest—namely, the catalog of the flea collection—and well aware of Jordan's scrupulous methods, Riley exhorted one of Jordan's assistants, F. G. A. M. Smit, not to let his respect for Jordan allow him to lose sight of the project's main goal. He reminded Smit that as a civil servant his first duty must be to the museum that had hired him.[133] No doubt he feared Smit would be influenced by Jordan's systematic prudence and the results thereby be delayed. As Riley's admonition made clear, British Museum entomologists served a state-supported institution with ultimate oversight resting in the hands of British government, rather than the discipline of entomology, an enthusiastic patron such as Rothschild, or some idealistic concept of scientific systematics.

Installed for almost six decades in one of natural history's "nice berths," Jordan could afford both to wait for more specimens and to spend endless hours at the microscope. He could also indulge in warnings that careful descriptive work must take precedence over theory and couch his defenses of the discipline in idealistic tributes to entomology for its own sake. But entomologists pressed by a new reward system and methodological ethos could not always adopt such ideals. The new generation of entomologists struggled to reconcile entomology's postwar context with Jordan's warning that "to press for quick results in research is most disastrous, quick spelling nearly the same as quack." In 1964, Jordan's successor as "dean of the Siphonapterists," Robert Traub, received a letter from a colleague admonishing him for spending so much time on the systematics of Egyptian fleas. "We are not kids playing at bug-collecting any longer," Traub was told. "We are both spending a lot of government money to get research data and material, and we owe it to the government to finalize the studies of these data and materials."[134] To a systematist who admired Karl Jordan, and who knew well his constant calls for "more specimens! more data!" such a letter must have been disconcerting.

As Elwood Zimmerman tried to continue Jordan's work on the Anthribidae, he had similar problems. He, too, took Jordan's work as a model, and grew increasingly frustrated by his inability to apply these methods in practice, much less receive official backing for the endeavor. He found the old network completely overwhelmed. He wished to send specimens to Washington, but found entomologists there "so swamped with routine work that they cannot accept these offers." He wondered what had become of a Samoan collection he had sent in 1940, a Fijian collection sent in 1937–1938, and other collections, but conceded that he did not see how the men there could be expected "to turn out sound and efficient work" under their present handicaps.[135] "It is difficult not to be overwhelmed by the seemingly endless numbers of insect genera and species," he wrote Ernst Mayr, "If conditions were such that I could do nothing but describe new species, it would be possible to describe 10,000 new beetles and still leave the field comparatively untouched!"[136]

But Zimmerman most certainly could not "do nothing but describe new species," even had he wanted to. Financial support for his work was at the mercy of the booms and busts of the sugar industry.[137] Eventually, with that industry mired in a depression, he had to leave his post at the Hawaiian Sugar Planters' Association Experiment Station when the authorities insisted he stop working on a planned "monumental" *Insects of Hawaii* and focus on lab routine, plantation inspections, light-trap surveys, and quarantine work.[138] The association seemed embarrassed, he explained to Mayr, to have a scientist on the staff doing basic research whose productivity could not be measured by "How much sugar is *he* putting in the bag?" One trustee had asked point blank what possible use his *Insects of Hawaii* could be. Zimmerman wrote to Mayr in despair:

> It often seems that certain research, especially taxonomy, is something almost to be ashamed of being caught doing on official time, or worthy of only inferior intellects, or in the realm of stamp collecting—something to be hidden from view when the big boss or the trustees come around.[139]

Four years later, Zimmerman applied for a job at Stanford University to replace taxonomist Gordon F. Ferris as professor of entomology, though he knew the university was replacing the "David Starr Jordan school," a generation that valued systematics, with experimentalists. Still, he was not without hope since "there is the Stanford Museum collections which must be curated."[140] (In fact, the administrators at Stanford dispersed the collection to other institutions, and eventually the University of Hawaii ended its support for Zimmerman's *Insects of Hawaii* because it "no longer wishes to become involved in 'old fashioned biology,' and they have no intention of becoming involved in 'museum work.'"[141]) By 1960, Zimmerman confessed to a friend that he now understood why a professor had once warned that he "could not in good conscience recommend the field of systematic entomology to his students. It is a sad business."[142]

Underfunded and unappreciated, Zimmerman looked at the British Museum with envy—even though that institution itself complained about lack of funds. "There are 113 persons (excluding cleaners, housemen and messengers) working on the insect collections," he reported to Mayr, "69 of these are full-time established staff paid by the Museum." He compared this to the state of affairs at the United States National Museum (USNM), where he had been admonished for being completely unrealistic for proposing they should have at least 100 entomologists, rather than six. "The USNM," he concluded, "simply is not doing its job.... They can do little more than keep moth balls in the cabinets."[143] Zimmerman's praise for conditions in London may have been in part strategic, as he tried to humiliate U.S. authorities into doing better. "We should compare our facilities with those of little England," he wrote, "and feel genuinely ashamed of the situation."[144] Meanwhile, British entomologists described those in London as "overworked in these days of intense research in applied or economic entomology."[145] Those entomologists who "stood at attention" as Jordan passed through the entomological department of the museum must have thought his life in the country, largely isolated from such pressures, the equivalent of an entomological fairy tale.

LATE FOR A KNIGHTHOOD

By the time Zimmerman composed his lament regarding the state of systematics to Ernst Mayr, that boundlessly energetic protégé of ornithologist Erwin Stresemann and Hartert had been campaigning to improve the status of the field for years. Mayr, having moved from the American Museum of Natural History in New York to a position at the Museum of Comparative Zoology at Harvard University in 1953, had his work cut out for him. Just down the hall, the codiscoverer of the structure of DNA, James Watson, was carrying out his own, none-too-subtle blitzkrieg of the naturalist tradition, refusing even to acknowledge the entomologist E. O. Wilson in the hall. Watson represented the triumph both intellectually and institutionally of Jordan's mineralogist friend's opinion that only those sciences based in the methods of physics and chemistry counted as science (although it is not clear that Watson would have acknowledged mineralogy either). Watson was an even less tolerant version of Charles

Rothschild's Cambridge professor who not only wondered at anyone's interest in sticking insects between two bits of glass but would quite happily get rid of the microscopes and collections altogether. Wilson would later recall that Watson characterized naturalists as "stamp collectors who lacked the wit to transform their subject into a modern science."[146]

The increasingly common comparison of taxonomy and systematics with stamp collecting contained even more insidious suggestions than that taxonomists just gathered facts or, even worse, "made" species. Thomas Mann had used the hobby of stamp collecting in the novel *The Magic Mountain* as symbolic of the decadent state of the prewar European middle class.[147] ("Everybody pasted, haggled, exchanged, took in philatelic magazines, carried on correspondence with special vendors, foreign and domestic, with societies and private owners; astonishing sums were spent for rare specimens."[148]) To someone unsympathetic to the naturalist tradition, Mann's description of stamp collecting could read just as easily if butterflies were substituted for stamps and naturalist history magazines for philatelic ones. The naturalist tradition's historical ties with wealthy hobbyists made it particularly suspect to those eager to cull certain activities from the honored roll of useful science.

To maintain a place for systematics in biology and defend it from characterizations of useless bourgeois pedantry, Ernst Mayr redrew its boundaries, narrowed its aims, and gave it some heroes. Charles Darwin, of course, was one champion. By 1955, Karl Jordan became another. But in many ways, Mayr's use of Jordan as a hero only emphasized how much the world had changed since the days when Jordan had first become fascinated by the endeavor to describe, order, and explain natural diversity.

The hero-creation began when Mayr received a letter from Harry Clench of the Carnegie Museum of Natural History in Pittsburgh requesting his help on a Festschrift in honor of Jordan's ninety-fourth birthday. "It has seemed to me that very inadequate recognition has been given publicly of Karl Jordan's worth," Clench wrote. He added that had Jordan been a nuclear physicist "it would have been a different matter!,"[149] a comment that tellingly reflected both systematists' conviction of their low status in the hierarchy of twentieth-century science and who, by the 1950s, stood at the top.

Reading Jordan's early papers in preparation for his contribution to the Festschrift, Mayr soon wrote to ornithologist David Lack that he found the "maturity of [Jordan's] viewpoint as put down in papers published between 1895 and 1910" astonishing. "There is no doubt," Mayr wrote, "that he was, in his thinking, far ahead of the contemporary geneticists."[150] Armed with Jordan's early papers, Mayr used his essay to correct what he called the "rather distorted" histories of biology in which geneticists claimed responsibility for everything biologists knew about speciation. "Anyone reading my quotations from Dr. Jordan's papers will learn that the truth is different," he crowed to Miriam Rothschild.[151] On the lookout for eloquent defenses of systematics for its own sake, he also took note of Jordan's warning that to press for quick results could lead to the equivalent of "quack" science, and he carefully copied the words for future use.[152]

Mayr believed that correctly identifying the respective historical roles of geneticists and naturalists in the "modern synthesis" of genetics and natural selection mattered for very specific reasons. He argued that a historical analysis of respective contributions to the synthesis was needed "not for purely historical reasons or to establish priorities for prestige reasons, but because the planning of future work will be helped by a clear recognition of the potential contributions that can be made by the various collaborating branches of biology."[153] The future was at stake in Mayr's understanding of history, as much as, if not more than, the past, as professions jockeyed for positions and square footage. In other words, Mayr knew that perspectives on the past influenced ideas about how the hierarchy of biological disciplines would be ordered in the years to come.

Indeed, Mayr portrayed Jordan's work as a model of what it meant to do good biology. He described Jordan as "one of the great biological thinkers of our time," while attributing Jordan's lack of renown to the fact that he worked "with a material (natural populations) and with methods (non-experimental) that were unpopular among laboratory biologists of his time." The essay contained a clear moral: those who had dispensed with such methods and material had been misguided, for it had been Jordan, quietly ordering beetles and butterflies, who had been right about species and speciation all along.[154]

Mayr's tribute to Jordan also served as a response to recent arguments by Edward O. Wilson and William L. Brown in the journal *Sytematic Zoology* that the practice of naming subspecies should be abandoned. This recent version of the antitrinomialist campaign obviously concerned Mayr. He wrote to Richard Meinertzhagen in 1954 that, while Wilson and Brown's paper took the criticisms of subspecies much too far, "it shows which way the wind blows."[155] Mayr used the timely opportunity of his essay for Jordan's Festschrift to direct such critics to Jordan's defenses of the focus on subspecies, composed half a century earlier. Yet the role Jordan played in this particular campaign highlights how difficult doing "Jordanian" systematics really was. For Wilson and Brown had agreed with many of the main tenets outlined in Jordan's early work, including his interbreeding criterion for determining species, and the basic concepts upon which Jordan had then based a defense of the use of trinomials. But they argued that, given the amount of data required to apply trinomials well and the inability of most naturalists to amass such data, using them in practice had proved highly problematic. They cited Mayr's own confessed frustration with systematists' inconsistent use of the trinomial and argued that workers needed to analyze trends of variation rather than "expending their energy on the describing and naming of trifling subspecies." Nature was too complex to be bound by nomenclatural rules based on the presumed nature of geographic variation, museums usually did not possess enough material to adequately assess such variation, and in any case the modern taxonomist had enough to do without worrying about giving subspecies trinomial names.[156] Taxonomists seemed to have forgotten, they warned, "the great complexities and disparities revealed in racial patterns by some really thorough analyses of geographical variation made in the past." They thus concluded that "the subspecies concept is the most critical and disorderly area of modern systematic theory."[157]

Those who joined Wilson and Brown did not argue that the research program outlined by Jordan, and championed by Mayr, was inherently useless or conceptually misguided. Rather, given the available information in collections and naturalists' varied ways, the practice of naming subspecies had, they said, only added to the chaos. William Gosline found Wilson and Brown's paper "a refreshing relief" after trying to live with the trinomial system for years. He, too, had no quarrel with the subspecies as a zoological concept; and he believed the adoption of trinomial nomenclature had "been instrumental in demonstrating the tremendous part that geography plays in speciation." But, he thought, as a rule-bound system that attempted to cover a range of heterogeneous phenomenon, trinomials had lowered the status of the systematist even further.[158]

Theodore H. Hubbell, of the Museum of Zoology at the University of Michigan, attributed entomologists' challenges in using the system in part to entomology's diversity. The field consisted of highly trained biologists, competent self-taught amateurs, workaday taxonomists whose main job was to get on with morphologically based description and naming, and dilettante collectors. "Obviously it is not likely," he pointed out, "that entomologists, so diverse in background, viewpoint, and objectives, will ever fully agree on how to treat subspecific entities." To make matters worse, many entomologists working to name the "estimated one to several million species of insects still undescribed" continued to work with small numbers of specimens and few localities. They thus could not deal properly with geographical variation. In other words, the diversity among entomologists combined with that in the insect world to create lists "filled with arrays of... nominal 'subspecies,' formed unintentionally, the actual status of which is unknown." Only revisions based on adequate collections could remedy this situation. But "such a revision," Hubbell noted, "may not be forthcoming for years or for generations."[159]

Jordan's organizational efforts, intended to ensure entomologists had both the time and infrastructure required to complete revisions in a timely manner, had constantly been hampered by the blows of twentieth-century economic and political turmoil. This organization work had in part been aimed at raising the status of systematic entomology within the sciences as a means of ensuring its future. Ultimately, Jordan's wish to study geographical variation, his concept of scientific systematics, and his organizational efforts were all intertwined. For, to those who ridiculed the taxonomic wing of the naturalist tradition as unscientific, Jordan had countered that there was a biological point to "species-making," a point he tied directly to the knowledge gained by careful attention to geographical variation. He once wrote that although the describer of a new form may be indifferent to the factors that had caused the modification leading to that form, "his very act of describing it as a member of a chain of allied forms inhabiting each a different district connects a new modification with the environment where it was found, and thus the describer adduces a causation."[160] Jordan and his colleagues at Tring had realized that in order for the connections between geography and forms to be visible within the collections, the facts of importance had to change. They may only rarely have commented on debates over the mechanism of evolution. But the museum's focus on retraining their collectors and colleagues to amass long series of specimens with detailed locality data in order to study geographical variation all combined to create what Mayr later praised as a more "biological" systematics.

In practice the ideal proved elusive, even under the best of circumstances. Mayr certainly knew the challenges of doing Jordanian systematics as well as anyone. He acknowledged that applying the criterion of interbreeding (a basic component of Mayr's "biological species concept") often proved extremely difficult in practice due to a lack of material. But given he was involved in a fight for the status of systematics as well as biological concepts, he could hardly emphasize that Jordan's work reflected an ideal research program that proved extraordinarily difficult in practice. Within the changing context of institutional support and availability of specimens for entomology, it could sometimes prove impossible, even for Jordan. As a result, and despite both the scientific intentions of its early proponents and Mayr's insistence that the description of subspecies represented a "new," more "biological" systematics, it often seemed that the practice of describing and naming geographical varieties had turned taxonomists into "subspecies-makers," rather than biologists.[161]

That given enough time and resources, Jordan's methods could produce extraordinary and robust results is evident from the praise of both his contemporaries and modern entomologists. Describing his work on fleas, a fellow specialist, Robert Traub, noted in 1955:

> It is axiomatic that the affinities of such species (as fleas) are enormously difficult to determine, particularly since the phenomena of adaptation and convergence have further obscured the relationships. Authors, basing their views upon small and unrepresentative collections, have frequently been led astray for these reasons. The outstanding exception throughout the history of the study of fleas has been Karl Jordan.[162]

More recently, the keeper of entomology at the Natural History Museum, Malcolm Scoble, noted that the *Revision of the Sphingidae* was colossal, standing over all later work in the group. There are a number of monographs like that, he explained, including Darwin's *Monograph on Cirripedia*, and those who produce such works can be contrasted to those who produce lots of smaller papers but never synthesize their work into a revision.[163]

Modern Sphingidae experts continue to see the Rothschild and Jordan revision as marking "a revolution ... in our understanding of the higher classification of Sphingidae." The revision's emphasis on the detailed study of the morphology and variation, Ian Kitching and Jean-Marie Cadiou explain, "developed a solid foundation for future work on the family" and placed many of the genera "onto a firm footing for the first time."[164] Gazing at the revision on his bookshelf at the Natural History Museum in London, Ian Kitching exclaimed, "What I want to know is how the hell they did so much work in so little time?"[165]

Jordan's position as curator to Walter Rothschild's museum made his position ideal in many ways. As an expert on Papilios exclaimed when discussing the amount of good systematics Jordan was able to do, "Well, he did have a Rothschild!"[166] During the time in which Tring's famous revisions appeared, Jordan also had access to a healthy network of fellow naturalists who did not have to spend much time defending their work before potential patrons. One is tempted to imagine how much more Jordan could have produced in subsequent decades had he not had to deal with the heady claims of geneticists, the rise of applied entomology, or the chaotic diversity of his fellow entomologists and their long legacy of "species-making." On the other hand, each of these challenges inspired creative organizational efforts, eloquent manifestos, and careful reassessments of the aims and methods of entomological taxonomy as a whole.

Jordan himself expressed a rather tempered view of his accomplishments. Miriam Rothschild recounted to Mayr how, as he finished reading Mayr's contribution to the Festschrift, Jordan exclaimed, "Well, Mayr has read my papers!!! But he is too kind—he only refers to those in which I was right!"[167] Jordan's friends often noted the aura of philosophical humility that pervaded his work and his dealings with others, drawing him into the background rather than the spotlight at both congresses and commissions, unless some important work needed to be done. "It was characteristic of the man," wrote Francis Hemming, his colleague at the ICZN, "that he should do good work by stealth leaving to others the credit due for advances achieved."[168] He allowed Walter Rothschild to subsume his best work under citations that read either "Rothschild" or "Rothschild and Jordan," and gave of his time, ideas, and knowledge so generously that both Riley and Miriam Rothschild felt his intellectual contributions were rather needlessly obscured. In contrast to Mayr's agenda-driven insistence that Jordan's lack of notoriety arose from his objects and methods of study, Riley attributed it to his being a "naturally rather shy man," never particularly interested in personal prestige. Moreover, "he was not the type," Riley wrote, "to stump the country in support of any theory or to engage in polemics."[169]

Jordan's preference for working "behind the blinds"[170] may have been at least partly rooted in his knowing all too well that systematists had long been ridiculed for "making species" for the sake of personal notoriety. In 1895, the entomologist Augustus Radcliffe Grote lamented, in noting the tendency to prefer one's own names or those of one's countrymen, that "not all systematists are gentlemen in the sense that Don Quixote was one, willing to forget themselves and to break a lance for that which is and which must seem to them everlastingly right."[171] In contrast to those who back-dated publications to ensure that their names received priority, Jordan often played down his role in describing new forms. Zimmerman once roundly contested Jordan's own estimate of the number of species he had described. "I do not feel that to say that you have described nearly 2,000 species of Anthribidae is stretching the point too far," Zimmerman insisted.[172] When pressed, Jordan argued that the entomologist's best memorial was his work.[173] Throughout his life, friends and colleagues were struck by his example as a humble and objective searcher for truth, an impression Jordan cultivated often. "My work and my hobby have been identical and I married the girl with whom I fell in love when still a schoolboy," he wrote at the age of eighty, "What more can one want?"[174]

Yet as Jordan lay on his deathbed in the hospital at the age of ninety-eight, his friend Miriam Rothschild was astonished to learn from one of Jordan's daughters that, when he had "not been himself" during his final illness, the philosophical Karl Jordan had become distressed that he was going to be late for his knighthood.[175] The revelation astonished because Jordan had seemed so uninterested in such trappings of social recognition, apart from his silent disdain when Hartert received an honorary PhD; but that, of course, was the one symbol he had of his ascent to a superior class defined by intellectual merit and educational achievement.

Jordan once noted that if you wanted to get at the fundamentals in a person's character and at his "true likes and dislikes," you must observe him "when reasoned control is low or absent, as it is, under the influence of narcotics or great excitement, or during sleep."[176] Did Jordan, despite his aura of humility, really wish for some kind of official recognition of his work? For a museum entomologist to dream of a knighthood was not absurd. He had watched directors of the British Museum, including E. Ray Lankester, William Flower, and Sidney Harmer, receive knighthoods for service as administrators of that scientific extension of the British Empire. E. B. Poulton, too, had been knighted in 1935. Like them, Jordan had directed a museum, one that, by the time he dreamt of a knighthood, had been the property of the British nation for more than two decades. But Jordan's occasional references to Don Quixote and his windmills hint that he may have been dreaming of the knights of legends, rather than his fellow civil servants.

When Jordan expressed a certain affinity, as he often did, with the knights of literature, it was with those who pursued what seemed to everyone else an aimless quest. His knights were always well-meaning, but rather misguided, as when he confessed that the energy spent on the book research necessary to disentangle entomological nomenclature reminded one "too much of the famous fight against windmills." They were also clumsy. "When looking at the White Knight of Alice in the Looking Glass," he once confessed to Riley, "I have to smile broadly, because a comparison with him is so very apt; I too fall about rather when setting traps and I am not looking where I put my feet."[177]

In his heart, though, and no matter how clumsy the results may have appeared to enthusiastic mineralogists or geneticists, Jordan did not believe that his quest had been aimless. More than ninety years after the field trip with his brother, Jordan urged the importance of knowledge about specific distinctions in a tape-recorded address to the tenth International Congress of Entomology in Montreal in 1956. He did not feel downhearted when remembering some species he had described, he said, for "taxonomy, if reliable, gives biology a sound basis."[178] It was another reply of the so-called species-maker to Günther, to the mineralogist, to the Cambridge professor, and to a whole army of new critics, yet again.

The difficulty of making the taxonomy reliable, due to both the complexity of nature and the often chaotic legacy of the naturalist tradition, had sometimes made the jibe "species-maker" seem all too accurate. Faced by those who would thus dispense with the whole endeavor, the American naturalist David Starr Jordan once admonished the geneticist Thomas Hunt Morgan: "No modern taxonomist regards species as 'arbitrary collections of individuals assembled for purposes of classification.' Many species are obscured for lack of material or lack of accuracy in published accounts. This is not Nature's fault. Charge it up to the weakness of humanity."[179] Karl Jordan believed the answer to this "weakness of humanity" could be found in a research program based on the examination of as many facts, individuals, and species as possible. To make taxonomy reliable and avoid falling clumsily into the traps set by either the poisonous snake of premature philosophy or human nature, this research must be slow, cautious, meticulous, and tentative. And given the diversity of those studying insects, it must proceed in concert with efforts to organize the "nation of Entomologists."

Jordan's methods and organizational efforts completely depended on a stable context in which large natural history collections could be amassed, taxonomists could communicate freely, and time could be spent processing specimens. But during the first half of the twentieth century, the science and society of the West had been anything but stable. This instability meant entomologists and taxonomists constantly had to shift their ground to find a place within a changing world, to the point that, by the end of Jordan's life, they had offered an at times bewildering variety of accounts of what precisely their quest entailed. By 1959 one might fairly wonder whether Jordan and the entire taxonomic division of the naturalist tradition had, like Don Quixote, pursued their quest—whatever it was—into an age that could no longer sustain the old ideals, much less properly equip its required army.

Conclusion

More than four decades would pass between Jordan's last scientific publication, "A Contribution to the Taxonomy of Stenoponia, a Genus of Palaearctic and Nearctic Fleas" (1958), and the call by the All Species Foundation (ASF) in 2000 to finally complete the catalog of life within twenty-five years. As a taxonomist who described, alone or with a coauthor, 3,426 species of insects, Karl Jordan would seem to have brought taxonomists precisely (and only) 3,426 steps closer to the ASF's goal. We can now address the astonishment expressed by the founders of the ASF that the catalog of life was not yet completed. By looking at the life and context of a taxonomist who for seven decades described thousands of species, we can ask what allowed him to describe those 3,426 species and what prevented him from describing more. In other words, taking as examples the various groups on which Jordan worked, how do we know what we know about biodiversity? Why, conversely, do we seem to know so little? And furthermore, what precisely are we counting as knowledge, either possessed or desired?

First, why could Jordan describe the species he did? At the most foundational level, a robust naturalist tradition valued by a certain strata of society supported his work. That tradition meant that a child interested in beetles might respectably pursue that interest along certain lines outlined by generations of naturalists. For a young German teenager in the 1870s, this could even mean earning a university degree in disciplines devoted to the study of the living world, in large part due to the role of natural history in the grand intellectual project of understanding the natural world. This particular young naturalist's contacts with professors of forestry hinted that this project could have an economic component as well, once "grand intellectual projects" were simply not good enough. Meanwhile, the presence of mentors committed to a certain way of doing natural history proved enormously important. Jordan lived in a world in which science was valued as a part of elite culture, and taxonomy—at least done a certain way—counted as part of science. Critics existed, to be sure, inspiring much of the "how to do entomology" debates that Jordan encountered upon arriving in Britain. But as a student, Jordan learned that the facts of biological diversity, and natural history collections, mattered. Museum curation and describing new species, so long as one did these tasks well and in the context of a broader quest, could be combined with Jordan's strong sense of identity as an academically trained scientist.

One of the defining factors in Jordan's experience in the naturalist tradition, and one that made his life seem ideal to other taxonomists, was his position as curator of insects at one of the largest private natural history museums in the world. Jordan could not only survive doing taxonomy, he could pursue a respectable, middle-class existence. This brings us to one of the most important determinants of Jordan's ability; namely, concrete financial support. Indeed, Jordan's position as a curator governed by a patron's wishes may in fact have led him to describe more species than he would have liked. By virtue of the task given him by Walter Rothschild, he found hundreds of previously undescribed species of beetles in Rothschild's cabinets. He could ultimately justify the time spent composing the resulting descriptions by appealing to the importance of taxonomy to biology. As he often pointed out, an accurate understanding of relationships, and therefore the origin of species, depended on a knowledge of those species, and the more species one knew, the better one could check generalizations. Concentrated wealth proved crucial to this project, for it ensured that Jordan could deliver, at least for a time, on the ideals he outlined for those species to be described well. Rothschild could pay Jordan's way as he visited collections on the Continent to compare type specimens, buy any book desired in order to amass the literature required to deal with the nomenclatural windmills, purchase specimens, and fund expeditions. Access to Rothschild's wealth ensured that when Jordan did name and describe species, those determinations could be reliable. Under such ideal circumstances, Jordan's famous caution in the name of good science need not become a liability, nor prevent the confidence required to coin each new name. It meant Jordan could describe new species within the context of unprecedented revisions and monographs rather than in isolation from the comparative work required to avoid mistakes and maintain his credentials as a scientist. "Well, he did have a Rothschild!" Indeed.

Although completing such robust systematics certainly depended greatly on Rothschild's financial resources and support, it also relied on much, much more. The Tring revisions, while completed at museum desks, relied on a network of collectors in the field, fellow private collection owners and museum curators, and natural history agents. Each individual in this diverse network pursued the naturalist tradition in their own way according to different means and goals, but all were united by the tiny objects of exchange upon which knowledge, income, and even status, hinged. Naturalists exchanged and discussed these tiny objects—the butterflies, moths, beetles, and fleas—by way of the infrastructure of European expansion and exploration and, ultimately, the seemingly robust framework of nineteenth-century liberal capitalism.

During Tring's heyday in the 1890s, the movement of specimens, letters, and entomologists proceeded with relatively few restrictions. And over time, specimens coming in from new sources such as the Imperial Bureau of Entomology seemed to provide yet another avenue for natural history knowledge, even as worrisome tensions arose regarding the goals of species work and how taxonomists' time would be spent. Meanwhile, institutions devoted to natural history, in the form of places such as the British Museum, conferred an organizational reality to the task, even as taxonomists increasingly wondered what that task was and what it was becoming. Finally, all of this work describing new species required paper, open postal routes, access to other natural history museums and far-off lands, and the ability to sit for hours, days, and years studying birds, moths, beetles, and fleas with a clear conscience. These final points may seem mundane until one recalls first, the number of new species described by any naturalist, not just Jordan, dropped to almost zero in both 1914 and 1939,[1] and second, Jordan's countless hours in his wartime garden, "digging for victory."

This brings us rather abruptly to the various factors that prevented Jordan from describing more species. First, there are those that Jordan himself often acknowledged as delaying his scientific work. He constantly lamented the time and effort required to clear up the nomenclatural chaos of his predecessors—and, infuriatingly, his contemporaries. Countering this legacy of the naturalist tradition, Jordan devoted a huge amount of time to clearing up synonymy and to the national and international commissions aimed at ensuring nomenclatural stability in the future. More mundanely, countless "other matters" took up his time as curator, and later director, of a natural history museum. His chores ranged from receiving packages of specimens and unpacking them himself (so the insects would not be damaged by someone else) to taking a group of schoolchildren or political dignitaries around the museum.

Undoubtedly, Walter Rothschild's wealth and interests made it possible for Jordan to describe thousands of new species, but other aspects of Rothschild's life, which resulted in the loss of much of that wealth, lessened those opportunities. As Miriam Rothschild wrote, Jordan was forced to "carry the can" for

Rothschild in many instances, whether that entailed diplomatically consoling an offended entomologist in London or making excuses to newspapermen for the sale of Rothschild's birds. Such things took time, even as Rothschild's scandals took much-needed resources away from the museum. In a different way, Jordan's attempts to help Charles Rothschild kept him from his museum desk and the collection for a number of years. Both forms of service were, in their own way, carried out in devotion to brothers who had allowed him to make his hobby his work. The details of these more personal battles are obscured by the official nature of the museum archives, but they at least hint at the obviously important yet nebulous role that personalities and human relationships play in influencing how scientists spend their days.

Building and maintaining the network of naturalists so important to the revisions also took up time. Jordan knew he had one of the few "nice berths" in natural history, and he became famous, like Hartert, for generously giving up afternoons to help visitors work through the collection or compose dozens of letters answering queries regarding specimens in Rothschild's cabinets. Many naturalists did not have access to the kind of resources available to the Tring triumvirate. And while Jordan tried, either through correspondence or visits, to remedy that lamentable fact in the interest of helping everyone produce good work, entomologists comprised a diverse, often isolated network, with divergent views of how to describe species, do taxonomy, or even be an entomologist. Thus, Jordan founded the international congresses of entomology. This in turn required time that might otherwise have been spent describing new species. Meanwhile, the incentives to continue devoting time to organizational efforts multiplied, even after those efforts were destroyed by world wars. For with the rise of both genetics and applied entomology, entomologists had to decide how (and whether) to maintain a role for the old within the new.

In the end, Jordan's definition of the primary task of systematics prevented him from describing more species because quite simply he believed that task—the quest—was most definitely not merely to find and name new species. Here, his academic training and his identity as a scientist combined to emphasize certain methods of working that slowed down the work of describing and naming species considerably in the name of "higher" goals. The naming, in other words, must be maintained as an ally of the quest to order, and ultimately explain, the diversity of living beings. To avoid the jibe of "species-maker," Jordan had pursued methods that would ensure the taxonomist based his distinctions between species and varieties on nonarbitrary methods. These methods, including the collection and comparison of long series of well-labeled specimens of the same species, considerably increased the amount of resources and time needed. It also resulted in the fact that Jordan, while he was most definitely not a "species-maker," was a proud "species *un*maker," as he worked to revise the conclusions of his predecessors.

Finally, and perhaps most importantly, Jordan outlined his vision of how and why to do scientific systematics—a vision completely dependent on the availability of large collections—during the Tring Museum's "golden age," when Rothschild had hundreds of collectors sending him material from all over the globe. Even given impressive financial resources, a network of correspondents, and a stable environment for specimen accumulation and exchange, Jordan had always found drawing conclusions from the incomparable material in Rothschild's museum difficult. As years passed, Walter Rothschild's allowance was reined in and the Great Depression and two world wars destroyed—temporarily for some nations, and permanently for others—much of the natural history trade. The war also focused the entomological endeavor on applied concerns, drawing resources and prestige away from systematics for its own sake. As the world changed around entomologists, their increasing workloads often prevented taxonomists from doing Jordanian systematics.

Jordan once described how he spoke on species "as a zoologist who has spent the larger part of his life battling in a museum with the species problem in its great variety."[2] He might also have added that he had battled with the isolation of his colleagues, the diversity of the naturalist tradition's methods and goals, the impact of two world wars, and the winds of social, economic, and political change. As biology in general changed, the naturalist tradition in which systematics had formed became increasingly sidelined from the most prestigious realms of the life sciences. The rise of genetics and molecular biology is often given as the main source of the decline in status and manpower of modern taxonomy, but all of the above changes profoundly influenced Jordan's ability to describe, and ultimately classify, species. They also help explain why Ernst Mayr was on the defense in the 1950s and why he needed Jordan as a champion.

In analyzing the history of the Rothschild Banking House, Niall Ferguson has asked whether, in continuing to carry out business in the anachronistic rooms of the banking house in London after the First World War, the Rothschild banks were "gently prolonging the nineteenth century?"[3] Given the profound changes that had taken place since Jordan first carried his microscope to the sunlit window in Tring, the same question may be asked regarding natural history museums. Given the specific context in which naturalists created these grand collections, does the loss of that context warrant their dismissal from the scientific halls of modern biology? In answering his question regarding the bank, Ferguson argues that in fact the impression of inactivity and redundancy can easily be overestimated and that "it would perhaps be more historically accurate to regard the memory of 'immobility' as a consequence of the two great economic traumas of the inter-war period," rather than problems peculiar to the Rothschilds.[4] For natural history museums, one might also add to these traumas that of enlistment into a new patronage system in which work must be defended before a tax-paying public. (Symbolic of this shift has been the recent demise of the 150-year-old tradition of separating the public from the study collections at the Natural History Museum, London. The workers may still be glorified office boys—now joined by a few girls—turning to information-sharing on the Internet to revitalize their field,[5] but they can no longer be kept in the back room.)

In the two centuries preceding the appearance of the grand ambitions of the All Species Foundation, Jordan and his fellow taxonomists named and described between 1.4 million and 1.75 million animals, more than half of which were insects. The production of this scientific knowledge about biodiversity has depended on distinct priorities, methods, and values in science, as well as a certain amount of stability in the social networks and economic infrastructure in which that work is done. Jordan's lifelong effort to learn more about beetles, butterflies, moths, and fleas demonstrates how learning about biodiversity depends upon building communities, adjusting practices and techniques to conceptual changes and vice versa, finding financial support, establishing organization, determining what counts as science and what it is for, and navigating complex social and political changes. Though quite familiar to any working scientist, the daily activities that result from these factors have often been lost in historical work focused on the "great ideas" and "great men" of science. Yet these activities are those on which naturalists, like scientists in general, must spend much of their time. They are also those aspects of scientific life that, first, can be profoundly influenced by events taking place outside the walls of museums and laboratories, and second, influence scientists' ability to deliver on the particular justifications given for their work.

In reflecting on the potential lessons of Jordan's experience with the naturalist tradition, one must begin by noting that, by the time some of the big names in modern taxonomy signed on to the All Species Foundation's quest (described colorfully by E. O. Wilson as a "biological moon shot"),[6] some of the central challenges faced by previous generations of taxonomists had been surmounted. As the ASF noted, advances in transportation made travel to remote sites less difficult, the Internet facilitated the flow of information around the world to an astonishing degree, and online databases ended the need for biologists to physically examine type specimens in remote museums. New tools such as rapid DNA sequencing and phylogenetic analysis seemed to provide solutions to many of the problems of comparing and analyzing species, and electronic publication ended the expense and delay of traditional publication.[7]

While some of these new tools are roundly contested, others are indeed already solving some of the fundamental problems faced by Jordan and his colleagues. In 1915, one entomologist lamented how "every new serial publication adds to the difficulty of the future student. At the present time it is well nigh impossible for any individual worker by himself to make a complete search in a question which for the time being he is dealing with."[8] Walther Horn and Karl Jordan would, quite simply, be astounded if they could see the wonders and potential of the Google Age for solving the organizational and informational issues tackled by the early congresses.

Given these technological resources, Wilson called his recent revision of *Pheidole* ants, in which, like Jordan, he meticulously drew the minute details of the ants himself through endless hours at the microscope, "the last of the great sailing ships." Careful drawings using the microscope, he noted, will now, for example, give way to digital photography.[9] Wilson's colleague at Harvard, the curator of entomology, Brian Farrell, has been digitizing the specimens in the Museum of Comparative Zoology. Furthermore, when collecting in the field in places like the Dominican Republic, he is using such methods to place photographs immediately on the Web, and then leaving the original specimens in the institutions of the nations in which they were found, in the hope that this will increase the ability of scientists in developing nations to study their own biodiversity (and, no doubt, improve relations sullied by the tradition's roots in the age of imperialism). As Farrell describes it, the digitization of collections "is a way to basically jumpstart the description and knowledge of the biota."[10]

This is just one example of how the practice of modern systematics is in many ways that of a different era. There are, it seems, good reasons for "thinking big," particularly when one adds the seemingly unassailable justification that, as Wilson explains, this "may be the last generation with the opportunity to inventory much of our planet's biodiversity before it is forever gone."[11] Indeed, the "biodiversity crisis" is perhaps the single biggest addition to taxonomists' armory since Jordan put away his last drawer of specimens. Ernst Mayr first linked his own calls for the importance of systematics to concerns with conservation soon after Jordan's death. In a circular entitled "The Systematics Program in the International Biological Program," Mayr described the disappointment felt by taxonomists "the world over" that the first plans of the International Biological Program made no provision for the needs of systematics. He pointed out that "this seemed the more regrettable since the current population explosion gives an enormous urgency to accelerate research in the developing countries. The native biota in the tropical zone is being destroyed before our eyes at an appalling rate. It is no one's obligation but that of our present generation to save as much knowledge about this biota as is humanly possible."[12]

Eventually this became the distinctly modern justification of both taxonomy and natural history museums on the grounds that one cannot protect and manage species' conservation if they have not been named and described.[13] Taxonomy is now defended, for example, as crucial to conservation biology and environmental management decisions.[14] Museum administrators have seen the possibility of returning taxonomy and natural history collections to the center of modern biology through their new-found use for biodiversity studies and protection.[15] Systematists have accordingly worked to incorporate this focus into their methodology and priorities, and workers have extensively evaluated their role in biological conservation.[16] In view of this new quest, spokesmen for taxonomy such as Quentin Wheeler, of Arizona State University, have insisted that "the time is at hand for taxonomy to take its rightful place among big sciences." Taxonomists must, Wheeler writes, "lead society in a meaningful response to the biodiversity crisis." Furthermore, he insists that taxonomists now have "the scientific perspective, theoretical framework, and technological tools to succeed."[17] Meanwhile, taxonomists have continued Walther Horn's manifestos under new banners, this time calling for an end to the "taxonomic impediment" that has resulted from the incomplete knowledge of taxa (and a lack of taxonomists).[18]

It was Mayr's younger colleague at Harvard, E. O. Wilson, who really brought the argument that systematics is a critical tool in the fight to save biodiversity to the fore. After decades campaigning for biodiversity conservation, Wilson enthusiastically joined ASF's call for a "world biota survey" that would catalog all life forms in the next few decades. Having lived through his own version of Jordan's and Poulton's tussle with early Mendelians during what he termed "The Molecular Wars" at Harvard University in the 1950s, Wilson insisted that "such a project would be far less expensive and far more important for humanity than the human genome project."[19] Others have joined the campaign, turning the old cataloging project of the naturalist tradition into a quest to save the world. "One of the greatest scientific goals of the twenty-first century," writes a former director of London's Kew Botanical Gardens, "should be to describe and classify all living organisms. This challenge must be taken up if conservation is to be effective, bioprospecting successful and environmental change effectively monitored."[20] The old quest, it seems, has a new uniform.

But some things, on the other hand, have not changed. Little time passed before journalists casting about for comments on the ASF's project turned up skeptics on grounds that, in view of the lessons conveyed by Jordan's life in taxonomy, certainly make historical sense. Some taxonomists described the proposal as impossibly ambitious, even quixotic. An "all species count" simply could not be done in twenty-five years, they argued. More importantly, others pointed out that some of the basic assumptions behind the project posed a threat to the scientific status of taxonomy. Fourteen well-known biologists sent an open letter to the All Species Foundation's Science Advisory Board urging them not to ignore the fact that biologists had a notoriously difficult time defining what a species is. Indeed, they pointed out that "uncertainty around species" formed a basic component of biological research.[21] (Another reader of the ASF's proposal was more blunt, asking in a letter to the editor of *Science*: "Did I miss out on something? Did we all finally reach agreement on what a 'species' is?")[22]

Biologist Cristián Samper, then deputy director of the Smithsonian Tropical Research Institute in Panama, pointed to the complex sociopolitical factors involved in biological inventory that the foundation ignored at its peril. He found the project "well intentioned but incredibly naïve." "Unlike counting genes or galaxies," he explained, "counting species involves complex political, educational, and legal issues—ranging from the shortage of trained researchers in developing countries to concerns about biopiracy."[23] The ASF must acknowledge, for example, that the United Nations Convention on Biological Diversity—the 1992 establishment of national sovereignty over biodiversity—had restricted the "free flow of biological specimens around the globe."[24] There is now, others note, a "regulatory maze associated with collecting biological samples."[25] This particular maze may be new, but it is rooted firmly in the past, a continued legacy of the naturalist tradition's historical ties to imperialism.

Not surprisingly, taxonomists' dependence on financial support also continues to stubbornly loom over all initiatives to study the natural world. The grand ambitions of the All Species Foundation, which came with a price tag of anywhere from $3 billion to $20 billion,[26] were abruptly halted after the recent economic downturn, and its presence on the Internet evaporated into a few sentences in Wikipedia. It seemed, at least until quite recently, that the Rothschilds and Carnegies of the world had other tasks on their hands; for example, malaria. (This may justify some Diptera taxonomists, as it has since the nineteenth century, but it most definitely does not entail a full inventory of life.)

In other words, and like Walther Horn's naïve mathematician, the All Species Foundation's ambitious, twenty-five-year goal for completion of species description had a few variables missing from its formula. New species needed to be arriving at a constant rate from desired regions, yet the world in which this could be done with relative ease had vanished. (When asked if he had taken into account things such as civil wars interfering with taxonomists' ability to gain access to certain species, one taxonomist responded with astonishment, "I'd never thought of that, but it sounds like a brilliant argument for digitization!"). Despite the optimistic claims of the All Species Foundation that renewed interest in and support for systematics was increasing the number of individuals available to carry out the inventory, the enormous gap between the army needed and the work to be done also continues. Botanist Les Landrum has pointed out that the reasons the army available to do descriptive systematics has remained small is rooted in the current reward system in which scientists live. Taxonomy is a time-consuming endeavor in which good work does not necessarily lead to hundreds of publications nor hundreds of thousands of dollars in grant money, yet these are often the criteria by which scientists are judged, hired, and promoted. Furthermore, when students are attracted to taxonomy, they are often absorbed into the one area of the field that has garnered a respectable amount of support; namely, phylogenetic systematics.[27] But as Jordan pointed out long ago, and as many modern taxonomists continue to insist, good descriptive work is not only hypothesis-driven, but provides the foundation upon which robust phylogeny must be based.

Finally, just as entomologists argued over their aims in the 1890s, taxonomists continue to debate what, precisely, their quest actually is. Wilson's call for the completion of the exploration of Earth's biodiversity "not as a destination eventually to be reached but as a concrete goal with a timeline"[28] has made many taxonomists nervous. Like all scientists, they know that the terms of the quest can profoundly influence the kind of work done. The profound tension between the knowledge that results when drawing up an inventory of species versus focusing on the analysis of relationships persists, for example. Mayr once argued that the catalog of nature that has been such a central part of the naturalist tradition's aim must not be the main thing. He confessed that "the fact that there are still ten thousands of undescribed nematodes, mites, and other lower invertebrates in various parts of the world does not disturb me." And he insisted that the interpretation of known forms must now be key, including evolutionary and ecological studies.[29]

Mayr and Jordan would no doubt have balked (as did many biologists) at the All Species Foundation's explicit dismissal of any discussion of species concepts as a basis for completing the catalog. As ichthyologist Anthony Gill notes, "an Inventory service implies you can complete it eventually, dismissing many of the questions that have qualified systematics as science, i.e. geographical variation, species concepts questions, etc."[30] Both Jordan and Mayr insisted that the description of species must not be the primary quest of systematics. Rather, to be included in the halls of science, systematists must make the analysis of relationships the central aim, analysis firmly based in robust revisionary taxonomy and careful engagement with the "species problem." And for this, what happens within species is far more important than amassing a catalog. Some critics have called the very idea that we can catalog all of life as based on a "static, atheoretical view of species,"[31] while others have pointed out, and as Jordan knew all too well, that a "strict focus on species taxon discovery and on the counting of taxa will yield numbers that are in large part a function of investigator effort.... Such numbers are particularly misleading if they are generated under strong limitations of time and resources."[32] Still, debates about species concepts and analyzing where subspecies meet does not sound quite as exciting as "completing the catalog." Nor do they allow one to avoid mentioning evolution in press releases for distribution to a public of which, at least in the United States, only 40 percent of those surveyed "believe in the theory of evolution."[33]

Since Linnaeus first justified his search for the order in nature on the grounds it would reveal the divine order of God's creation, taxonomists have been able to provide new justifications for an old, diverse project in response to changes in both science and society. Indeed, Jordan's life in taxonomy demonstrates how the naturalist tradition's diversity provided a useful reservoir of potential ways and reasons to study nature, from which naturalists could draw new methods and goals. But as Jordan had confessed in his flippant response to the female relative bemused by his interest in insects, most came to entomology through an inherent fascination with animal and plant diversity. The terms in which that quest is now placed, whether to one's relatives or potential patrons, is often driven, not by scientific questions, but by social concerns. As one trio of entomologists explain quite frankly, natural history museums today must focus on "biodiversity, faunistics, medical and veterinary studies, and molecular work, all in the pursuit of contemporary relevance."[34]

CONCLUSION 311

Recently, the new packaging seems to finally be paying off, thanks in part to a call by E. O. Wilson at the 2007 TED conference for an "Encyclopedia of Life." Wilson envisioned a Web page for every one of the known 1.8 million species of animals and plants, an ambitious solution to the long-standing complaints of "chaos! chaos! chaos!" voiced by taxonomists for decades. As the president of the John D. and Catherine T. MacArthur Foundation that pledged a $10 million seed grant to the endeavor, explains: "Technology is allowing science to grasp the immense complexity of life on this planet. Sharing what we know, we can protect Earth's biodiversity and better conserve our natural heritage."[35]

Justifying the study of biodiversity on the grounds of human or planetary welfare has not, of course, been merely an opportunistic move on the part of taxonomists fighting for resources and prestige. Jordan acknowledged that an emotional pressure to justify one's work and wishes existed in anyone with a conscience in modern society. Returning home from the Paris congress in the midst of the depression, he described to a friend how he and his daughters did not quite know how to accomplish all they needed to do, but he admitted that the complaint sounded "almost funny at a time where millions have no work."[36] As modernist visions of creating a better world doggedly persist, fighting tiny but dangerous carriers of deadly disease or destroyers of foodstuffs or documenting an imperiled biodiversity provide a set of tangibly constructive activities for which so many faced by the traumatic events and trends of the twentieth and twenty-first centuries have yearned.

Perhaps the most useful question inspired by Jordan's quest to order insects and naturalists is whether our most ambitious visions, the ones that inspire the public and justify time and resources, match both the complex reality of what it means to do science day-to-day and the somewhat nebulous drive to study the natural world apart from any practical value? Even Wilson, whose list of reasons to finish the catalog sometimes waxes apocalyptic, occasionally laments that "people are forgetting the value of understanding a group of organisms for its own sake."[37] The problem is not limited to taxonomy. Scientists today live in an age in which the dynamics of both patronage and competition with other professionals often drives what they say and do. It might not always be what they would have chosen, were they given the funds of a Rothschild and the isolation of an incredible museum or laboratory out in the country. (But then, as we have seen, Jordan did not necessarily spend his days doing exactly what he would have liked.)

In the final analysis, taxonomy, like all science, is an endeavor of an elite professional class, making its way in a society that purports to have left elitism in the nineteenth century. Solving the mysteries of nature must therefore be done along with decisions that determine which mysteries must be solved first, given a limited amount of time and money. Making this journey successful will depend on the presence of workers such as Jordan, who, while they may not make the most groundbreaking theoretical innovations or complete the catalog of life, carefully reflect on the path, deliberate regarding what might make it more successful, debate what counts as success, and adapt in the interest of ensuring the traditions they value persist amid constant change.

Acknowledgments

Since this book ultimately stems from dilemmas raised during travels in the tropics when I was an undergraduate, almost two decades ago, a number of people have helped, advised, and inspired me during my time studying the naturalist tradition. Placing them in some sort of chronological order, I thank Scott Barton and Paul Salaman for inspiring me to delve into the history of the study of biodiversity in the first place and Keith Benson and Bruce Hevly for helping me figure out where to start. Thank you to the faculty and staff of the history department at Oregon State University, especially Bill Husband, Mark Largent, Mary Jo Nye, and Robert Nye. I am very grateful to my fellow graduate students and to Ginny Domka and Vreneli Farber for their friendship and conversation.

I would like to acknowledge the help of the following individuals in carrying out the archival research for this study: at the Natural History Museum in London, Susan Snell and Polly Parry, of the Archives, Julie Harvey and Sharon Touzel, of the Entomology Library, and Christopher Mills, of Special Collections. My thanks for help with archival research also go to Linda Birch at the Alexander Library of the Edward Grey Institute of Field Ornithology; to Stella Brecknell at the Hope Entomological Archives at the Oxford University Museum of Natural History; Alison Harding and Effie Warr at the Tring Library of the Natural History Museum, London; Berit Pedersen, of the Royal Entomological Society of London; Bernadette G. Callery, of the Carnegie Museum of Natural History Archives; Julia Lindkvist, Maria Asp, and the staff of the Center for the History of Science of the Royal Swedish Academy of Science in Stockholm; Melanie Aspey, of the Rothschild Archive; Barbara Mathe, of the American Museum of Natural History; Pamela Henson, of the Smithsonian Institution; Andrea Goldstein, of the Harvard University Archives; Roberto Poggi, of the Museo Civico di Storia Naturale, Genoa; Lee R. Hiltzik, of the Rockefeller Archive Center; and the archivists at the University of Göttingen.

For conversations and encouragement during my research, I am grateful to Phil Ackery, Daniel Alexandrov, David Allen, Jean-François Auger, Andrew Brower, Robert C. Dagleish, Fritz Davis, Channah Farber, Elihu M. Gerson, Jonathan Harwood, Pamela Henson, Graham Howarth, Connie Johnson, William Kimler, Ian Kitching, Jim Mallet, Don Opitz, Robert Prys-Jones, Malcolm Scoble, Michael Walters, and Effie Warr. I would like to particularly acknowledge the help of the late Ernst Mayr, Miriam Rothschild, and Elwood C. Zimmerman, who shared their memories of Jordan, and Jürgen Haffer, who encouraged me to tell the entomologists' story.

The fellows and staff of the Dibner Institute for the History of Science and Technology and the faculty, staff, and students of Arizona State University's Center for Biology and Society provided enormous support. At the latter, I am especially grateful to Christofer Bang, Matt Chew, Antony Gill, Marie Glitz, Andrew Hamilton, Manfred Laublicher, Jane Maienschein, Kathleen Pigg, David Pearson, and Mary Sunderland. The members of the 2010 MBL-ASU Seminar in History of Biology provided helpful insight during the final stages of the project. I am grateful to all my colleagues at the University of Puget Sound for their encouragement and friendship and would particularly like to thank James Evans, Mott Greene, Suzanne Holland, and Peter Wimberger for helping to ensure I could find time to finish this book. Most recently, I would like to thank Bob Brugger, Kara Reiter, and Ashleigh McKown at the Johns Hopkins University Press.

To those who do not fit any particular place on a timeline because they are everywhere, thank you Donna and Larry Johnson, Christa and Gary Ellis, Shannon Dixon, Tamra Erickson, and (because correct grammar fails to provide me with a strong enough superlative) *most especially* Erik Ellis. Finally, I am very grateful to Paul Farber for his patient and encouraging mentorship. Thank you, Paul, for all your guidance during my journey into the history of the naturalist tradition.

This project was supported through National Science Foundation Grant SES-0218289, the Dibner Institute for the History of Science and Technology, NSF Post-Doctoral Fellowship # SES-0324033, and a University of Puget Sound Martin Nelson Junior Sabbatical Grant. All material from the archives of the Natural History Museum, London, is used by permission of the Trustees of the Natural History Museum, for which I am very grateful. I would also like to acknowledge the staffs of the archives (listed in the notes) for permission to use and quote material from their collections.

Notes

ABBREVIATIONS

AMNH	American Museum of Natural History
APS	American Philosophical Society
AWNM	Archiv für Wissenschafschichte, Naturhistorisches Museum, Vienna
CfHS, RSAS	Center for History of Science, Royal Swedish Academy of Sciences
CMNH	Carnegie Museum of Natural History, CMNH 2010-1, Administrative Records of the Director's Office, Part Two, 1920–2004.
EMM	*Entomologist's Monthly Magazine*
ERJV	*The Entomologist's Record and Journal of Variation*
ESL	Entomological Society of London
GA	Göttingen Archives
HLA	Hope Library Archives
HUA	Harvard University Archives
ICE	*International Congress of Entomology* (various years)
JICA	John Innes Centre Archives
MSNG	Museo Civico di Storia Naturale "G. Doria" Archives, Genoa
NARA	National Archives and Records Administration
NHM	Natural History Museum, London
DF 306	Department of Entomology, Keeper of Entomology's Subject Files, Zoological Museum, Tring
KJC	Karl Jordan Correspondence (Box #)
TM	Tring Museum Correspondence (Series # / Box #)
TMLB	Tring Museum Letter Book (Vol. #)
NS	*Natural Science: A Monthly Review of Scientific Progress*
NZ	*Novitates Zoologicae*
PESL	*Proceedings of the Entomological Society of London* (after 1933, PRESL, *Proceedings of the Royal Entomological Society of London*)
PRO	Public Records Office, Kew
RAC	Rockefeller Archive Center
SIA	Smithsonian Institution Archives
TESL	*Transactions of the Entomological Society of London* (after 1933, TRESL, *Transactions of the Royal Entomological Society of London*)

Introduction

1. A. Lawler, "Up for the Count," *Science* 294 (2001): 769–770.
2. Graham Howarth, personal communication, Sept. 2002.
3. Elwood Zimmerman, "Karl Jordan's Contribution to Our Knowledge of the Anthribid Beetles," *TRESL* 107 (1955): 67–68, at 67.
4. Richard Goldschmidt, "Short Appreciations," *TRESL* 107 (1955): 10–14, at 13, and Ernst Mayr, "Jordan's Contributions to Current Concepts in Systematics and Evolution," *TRESL* 107 (1955): 45–66, at 45, respectively.
5. Jordan, "The President's Address, 1930," *PESL* 5 (1931): 128–142, at 134.
6. W. F. Kirby, "An Entomologist's Jubilee," *Entomologist* 26 (1893): 233–235, at 234.
7. Charles Darwin, *The Origin of Species by Means of Natural Selection* (New York: Penguin, 1985; originally published by John Murray, 1859), 456.
8. A. R. Wallace, *My Life: A Record of Events and Opinions* (London, 1905), 60.
9. P. B. Mason, "Variation in the Shells of the Mollusca," *Journal of Conchology* 7 (1894): 328–346, at 345.
10. N. D. Riley, "Heinrich Ernst Karl Jordan," *Memorials of the Fellows of the Royal Society* 6 (1960): 107–133, at 110.
11. Jordan, "The Systematics of Some Lepidoptera which Resemble Each Other, and Their Bearing on General Questions of Evolution," *1er Congrès International d'Entomologie, Bruxelles, 1–6 août 1910*, ed. G. Severin (Brussels: Hayez, 1911–12), 385–404, at 385.
12. Ernst Hartert, "Miscellanea Ornithologica: Critical, Nomenclatural, and Other Notes, Mostly on Palearctic Birds and Their Allies," *NZ* (1906): 386–405, at 388.
13. Garland E. Allen, *Thomas Hunt Morgan: The Man and His Science* (Princeton, NJ: Princeton University Press, 1978).
14. Jordan, "The President's Address, 1929," *PESL* 4 (1930): 128–141, at 133.
15. Stephen D. Hopper, "New Life for Systematics," *Science* 316 (2007): 1097.
16. Transcript, "A Little Known Planet, Part 1," National Public Radio, Apr. 30, 2004.

Chapter 1: Joining the Naturalist Tradition

1. E. R. Bankes to Lord Walsingham, May 22, 1898, Walsingham Correspondence. NHM.
2. Miriam Rothschild, "Jordan: A Biography," *TRESL* 107 (1955): 1–9, at 1–2.
3. Walter Rothschild and Jordan, "A Revision of the American Papilios," *NZ* 13 (1906): 411–754, at 431.
4. M. S. Schenkling, "Junk's Coleopterum Catalogus," *1er Congrès International d'Entomologie* (Brussels: Académies Royales, 1911), 177–178.
5. Alfred Kelly, *The Descent of Darwin: The Popularization of Darwinism in Germany, 1860–1914* (Chapel Hill: University of North Carolina Press, 1981), 65.
6. Jordan, "The President's Address, 1929," *PESL* 4 (1930): 128–141, at 134.
7. Stewart A. Stehlin, *Bismarck and the Guelph Problem, 1866–1890* (The Hague: Martinus Nijhoff, 1973), 183 and 211.
8. Jordan's childhood memories can be found in untitled TS. Karl Jordan Correspondence (KJC) #2, NHM.

9. Margarita Bowen, *Empiricism and Geographical Thought: From Francis Bacon to Alexander von Humboldt* (Cambridge: Cambridge University Press, 1981), 68 and 106.

10. Lister to Ray, Dec. 22, 1669, quoted in John Ray, *The Correspondence of John Ray*, ed. Edwin Lankester (London: C. & J. Adlard, 1848), 49.

11. Gordon Craig, *Germany, 1866-1945* (Oxford: Clarendon Press, 1978), 190-191.

12. Quoted in Konrad Jarausch, *Students, Society, and Politics in Imperial Germany* (Princeton, NJ: Princeton University Press, 1982), 103.

13. James Albisetti, *Secondary School Reform in Imperial Germany* (Princeton, NJ: Princeton University Press, 1983), 30-31, 77.

14. Jordan, "Reminiscences of an Entomologist," *Proceedings of the Tenth ICE: Montreal, August 17-25, 1953*, ed. Edward C. Becker, 59-60 (Canada: s.n., 1958), 59.

15. Albisetti, *Secondary School Reform*, 310.

16. Jarausch, *Students, Society, and Politics*, 28, 36, 105-106, and 136.

17. Lewis Pyenson, *Neohumanism and the Persistence of Pure Mathematics in Wilhelmian Germany* (American Philosophical Society, 1983), 43.

18. Stehlin, *Bismarck and the Guelph Problem*, 189.

19. Lynn K. Nyhart, *Biology Takes Form: Animal Morphology and the German Universities, 1800-1900* (Chicago: University of Chicago Press, 1995).

20. Hugh Hawkins, "Transatlantic Discipleship: Two American Biologists and Their German Mentor," *Isis* 71 (1980): 196-210, at 199.

21. Nyhart, *Biology Takes Form*, 314-315 and 321.

22. Hawkins, "Transatlantic Discipleship," 199.

23. Jordan, "Anatomie und Biologie der Physapoda," *Zeitschrift für wissenschaftliche Zoologie* 47 (1888): 541-620.

24. Jordan, *Die Schmetterlingsfauna Göttingens*, vi, 52 (1); Alfeld. (inaug. dissert. Erlangung der phil. Doctorwürde Univ. Göttingen, 1885), and "Die Schmetterlingsfauna. Nordwest-Deutschlands," *Zoologische Jahrbücher* Suppl. 1, i-xviii, 1-164 (1886).

25. D. E. Showalter, "Army, State, and Society in Germany, 1871-1914: An Interpretation," in *Another Germany: A Reconsideration of the Imperial Era*, ed. J. R. Dukes and Joachim Remak, 1-18 (Boulder, CO: Westview Press, 1988), 9.

26. Jordan, "In Memory of Lord Rothschild," *NZ* 41 (1938): 1-16, at 1.

27. R. Hertwig, "Zoologie und vergleichende Anatomie" (1893), noted in Nyhart, *Biology Takes Form*, 170.

28. Jarausch, *Students, Society, and Politics*, 149.

29. Richard Goldschmidt, *Portraits from Memory* (Seattle: University of Washington Press, 1956), 20.

30. Jordan, "Nachtrag zum Verzeichniss der Käfer Hildescheims," *Societas entomologica*, Zürich (1886), 1, 121, 130, 139, 145, 154, 161, 172, 169; "Nachtrag zum Verzeichniss der Käfer Hildescheims," *Societas entomologica*, Zürich (1887), 2, 4, 20, 29, 60, 73, 83, 178, and 185; "Nachtrag zum Käfer-Verzeichniss Hildesheims," *Societas entomologica*, Zürich, 3 (1888):1-2, 11, 21, 58, and 65.

31. Nyhart, "Civic and Economic Zoology in Nineteenth-Century Germany: The Living Communities of Karl Mobius," *Isis* 89 (1998): 605-630, at 629.

32. Jordan, "Zur Käferfauna der nordwestdeutschen Tiefebene," *Entomologische Zeitschrift*, 1 Apr. 1891, 2.

33. John Croumbie Brown, *Schools of Forestry in Germany, with Addenda Relative to a Desiderated British National School of Forestry* (Edinburgh: Oliver & Boyd, 1887), 75.

34. Bernhard Eduard Fernow, *A Brief History of Forestry in Europe, the United States, and Other Countries* (Toronto, ON: University Press of Toronto, 1911), 114, 119, and 132.

35. Brown, *Schools of Forestry in Germany*, 64 and 70.

36. Sw. "Forestry," *Nature* 36 (1887): 193–194.

37. Jordan, "The President's Address, 1929," 133.

38. Fernow, *A Brief History of Forestry in Europe*, 121.

39. P. L. Sclater, "On the General Geographical Distribution of the Members of the Class Aves," *Journal of the Proceedings of the Linnean Society of London: Zoology* 2 (1858): 130–145, at 132.

40. Jane Camerini, "Evolution, Biogeography, and Maps: An Early History of Wallace's Line," *Isis* 84 (1993): 700–727.

41. T. A. Chapman, "Sidelights on the Lepidopterological Work of the Nineteenth Century," *ERJV* 13 (1901): 31–36, at 33.

42. Ludwig von Graff, "Zoology since Darwin, Part III," *NS* 9 (1896): 364–368, at 365.

43. Edgar Feuchtwanger, *Imperial Germany, 1850–1918* (London: Routledge, 2001), 91.

44. Brigitte Hope, "Naturwissenschaftliche und zoologische Forschungen in Afrika während der deutschen Kolonialbewegung bis 1914," *Berichte zur Wissenschaftsgeschichte* 13 (1990): 193–206.

45. Max Lenz, *Geschichte der Königlichen Friedrich-Wilhelms-Universität zu Berlin* (Halle a.d.S.: Buchhandlung des Waisenhaus, 1910), 387.

46. "The Association of German Naturalists and Physicians," *Nature* 30 (1884): 625–627, at 626.

47. Jörg Adelberger, "Eduard Vogel and Eduard Robert Flegel: The Experiences of Two Nineteenth-Century German Explorers in Africa," *History in Africa* 27 (2000): 1–29, at 18, 20, and 22.

48. Janet Browne, "Biogeography and Empire," *Cultures of Natural History*, ed. N. Jardine, J. A. Secord, and E. C. Spary (Cambridge: Cambridge University Press, 1996), 305–321.

49. Alfred Newton, "Opening Address to the British Association, Section D. Biology," *Nature* 36 (1887): 462–465, at 465.

50. Albert Gunther, *A Century of Zoology at the British Museum through the Lives of Two Keepers, 1815–1914* (London: Dawsons, 1975), 420.

51. Gunther, *A Century of Zoology*, 365.

52. Colin Holmes, *Anti-Semitism in British Society, 1876–1939* (New York: Holmes & Meier, 1979), 10.

53. Hartert to Mrs. E. J. Hutchinson, Oct. 28, 1893, TMLB #1, NHM.

54. Niall Ferguson, *The House of Rothschild*, vol. 2: *The World's Banker, 1849–1999* (New York: Penguin, 2000), 479.

55. Jonathan Schneer, *London, 1900: The Imperial Metropolis* (New Haven, CT: Yale University Press, 1999), 67, 71, 89.

56. Pauline Anderson, *The Background of Anti-English Feeling in Germany, 1890–1902* (Washington DC: American University Press, 1939), 264–265.

57. Ferguson, *The House of Rothschild*, 361.

58. Caroline Shaw, "Rothschilds and Brazil: An Introduction to Sources in the Rothschild Archive," *Latin American Research Review* 40 (2005): 165–185, at 173.

59. Miriam Rothschild, *Dear Lord Rothschild: Birds, Butterflies, and History* (London: Hutchinson, 1983), 329 n. 11.

60. Jordan to Ehlers, Dec. 31, 1892, Ernst Ehlers Papers, Manuscript Division, University Library, GA.

61. Schneer, *London, 1900*, 7.

62. Fritz K. Ringer, *The Decline of the German Mandarins: The German Academic Community, 1890–1933* (Middletown, CT: Wesleyan University Press, 1990), 5.

63. K. H. Jarausch, "The Universities: An American View," in J. R. Dukes and J. Remak, eds., *Another Germany: A Reconsideration of the Imperial Era* (Boulder, CO: Westview Press, 1988), 187–190.

64. Jordan, "The President's Address, 1930," *PESL* 5 (1931): 128–142, at 136.

65. Jordan, "The President's Address, 1930," 133.

66. Nyhart, *Biology Takes Form*, 318.

67. Gunther, *A Century of Zoology*, 460–463.

68. Gunther, *A Century of Zoology*, 138.

69. Miriam Rothschild, "Jordan: A Biography," *TRESL* 107 (1955): 1–9 at 2–3.

70. Gunther, *A Century of Zoology*, 417.

71. See Rothschild, *Dear Lord Rothschild*, 72.

72. Jordan, "In Memory," 4.

73. Rothschild, *Dear Lord Rothschild*, 104.

74. Hartert to Jordan, Mar. 17, 1908 and May 9, 1908, TM1/120, NHM.

75. C. R. Riedermann to Jordan, Aug. 12, 1910, TM2/12, NHM.

76. Rothschild, *Dear Lord Rothschild*, 107.

77. Gunther, *A Century of Zoology*, 422–423, and *Science Gossip*, 1892, 88.

78. Rothschild, *Dear Lord Rothschild*, 22.

79. Rothschild, *Dear Lord Rothschild*, 130 and 124.

80. Rothschild, *Dear Lord Rothschild*, 23.

81. Jordan, "In Memory," 3.

82. Rosamund W. Purcell and Stephen Jay Gould, *Finders Keepers: Eight Collectors* (New York: W. W. Norton, 1992), 80.

83. Rothschild, *Dear Lord Rothschild*, 127.

84. Hartert to Alfred Everett, July 31, 1894, TMLB #2, NHM.

85. Charles McClelland, "Republics within the Empire: The Universities," in Dukes and Remak, *Another Germany*, 169–180, at 172.

86. Rothschild, *Dear Lord Rothschild*, 130.

87. Rothschild, *Dear Lord Rothschild*, 130.

88. Jordan to Otto Kleinschmidt, Feb. 1, 1952, KJC#1, NHM.

89. Jordan, "In Memory," 10.

90. Rothschild, *Dear Lord Rothschild*, 145.

91. Rothschild, *Dear Lord Rothschild*, 132.

92. Hartert to Richard Dawes, May 15, 1895, TMLB #3, NHM.

93. J. F. Roxburgh to Jordan, Oct. 26, 1926, TM2/58, NHM.

94. Hartert to S. M. Klages, Oct. 3, 1910, TM2/11, NHM.
95. Jordan to C. E. F. Manson, Dec. 2, 1912, TM2/21, NHM.
96. Jordan, "In Memory," 9.
97. Jordan to W. J. Holland, July 22, 1899, Director's Correspondence, pt. 2, 1897–1961, CMNH.
98. Hans von Berlepsch and Hartert, "On the Birds of the Orinoco Region," *NZ* 9 (1902): 1–135, at 2.
99. Jordan, "Some New Butterflies and Moths from Eastern New Guinea," *NZ* 35 (1930): 277–287, at 277.
100. Walter Rothschild and Hartert, "List of the Collections of Birds Made by Albert S. Meek in the Lower Ranges of the Snow Mountains, on the Eilanden River, and on Mount Goliath during the Years 1910 and 1911," *NZ* 20 (1913): 473–527.
101. Walter Rothschild, "Notes on *Saturnidae*: With a Preliminary Revision of the Family Down to the Genus *Automeris*, and Descriptions of Some New Species," *NZ* 2 (1895): 35–51, at 35.
102. Martin Jacoby to Jordan, May 1, 1893, TM1/3, NHM.
103. Ferguson, *The House of Rothschild*, 369–410.
104. W. R. Ogilvie-Grant, "On the Birds of Southern Arabia," *NZ* 7 (1900): 243–273, at 243.
105. Hartert, "On Some Birds from the Congo Basin," *NZ* 2 (1895): 55–56.
106. Rothschild, "Jordan: A Biography," 3.
107. Quoted in Gunther, *A Century of Zoology*, 260.
108. Jordan, "In Memory," 11.
109. Jordan to Hans Zerny, June 13, 1929, AWNM.
110. Jordan to Walter Rothschild, Nov. 27, 1933, DF306/3, NHM.
111. Harold Perkin, *The Rise of Professional Society: England since 1880* (London: Routledge, 1989), xxiii.

CHAPTER 2: REFORMING ENTOMOLOGY

1. J. W. Tutt, "Retrospect of a Lepidopterist for 1894," *ERJV* 6 (1895): 8–14, at 8.
2. "News of Universities, Museums, and Societies," *NS* May, 1893, 393.
3. Tutt, "Collectors," *ERJV* 1 (1890/1891): 99–100, at 99.
4. Tutt, "Retrospect of a Lepidopterist for 1894," 8.
5. Jordan to unknown correspondent, Aug. 9, 1950, KJC#1, NHM.
6. Harry Eltringham to Jordan, Aug. 27, 1911, TM2/15, NHM.
7. Anon, "A Philosopher of the Scissors," *Punch*, Apr. 8, 1907, 250.
8. Ludwig von Graff, "Zoology since Darwin, Part III" *NS* 9 (1896): 364–368, at 366.
9. Phil Ackery, Kim Goodger, and David Lees, "The Bürgermeister's butterfly," *Journal of the History of Collections* 14 (2002): 225–230.
10. From a letter to Hugh Birrell, quoted in Miriam Rothschild, *Dear Lord Rothschild: Birds, Butterflies, and History* (London: Hutchinson, 1983), 123.
11. E. R. Bankes to Lord Walsingham, May 7, 1897, Walsingham Correspondence, Entomology Library, NHM.
12. Jordan to N. D. Riley, Feb. 10, 1955, DF306/9, NHM.

13. See Albert Gunther, *A Century of Zoology at the British Museum through the Lives of Two Keepers, 1815–1914* (London: Dawsons, 1975), 419–429.
14. Hartert to Herbert Goss, Nov. 5, 1892, TMLB #1, NHM.
15. Hartert to Otto Staudinger, Nov. 5, 1892, TMLB #1, NHM.
16. Hartert to Staudinger, Nov. 7 and Nov. 10, 1892, TMLB #1, NHM.
17. Rothschild, *Dear Lord Rothschild*, 75–79.
18. A Country Cousin, "Lead Us Not into Temptation," *ERJV* 4 (1893): 325–329, at 328.
19. Jordan to John Hartley Durrant, Oct. 3, 1896, Walsingham Correspondence, Entomology Library, NHM.
20. "Rothschild and Cassowary," *New York Times*, Apr. 21, 1901.
21. F. Primrose Stevenson to the Museum, Nov. 16, 1904, TM1/93, NHM.
22. Hartert to the Editors, *Popular Science Siftings*, May 14, 1910, TM2/11, NHM.
23. Hartert to John E. Sherlock, July 7, 1909, TM2/5, NHM.
24. Tutt, "Variation," *ERJV* 2 (1891): 11–15, at 15.
25. Tutt, "Notes and Comments: The Modern Scientific (?) Explorer," *NS* 9 (1897): 217–218, at 217.
26. Jordan to Miles Moss, Jan. 11, 1936, MSS MIL C 7:7, Entomology Library, NHM.
27. F. G. Foetterle 12-11-1906, TM1/96, NHM.
28. Jordan, "On African *Longicorns*," *NZ* 1 (1894):139–266, at 145.
29. David Sharp to Hartert, Apr. 6, 1895, TM1/16. NHM.
30. Entomological Society of London to Jordan, May 5, 1895, TM1/16, NHM.
31. Richard South to Moss, Nov. 6, 1900, MSS MIL C 7:7, Entomology Library, NHM.
32. P. P. Dodd to John Hartley Durrant, Nov. 22, 1901, Walsingham Correspondence, Entomology Library, NHM.
33. Hartert to A. Noaker, Apr. 28, 1893, TMLB #1, NHM.
34. D. E. Allen, "On Parallel Lines: Natural History and Biology from the Late Victorian Period," *Archives of Natural History* 25 (1998): 361–371, at 361 and 365.
35. H. Rowland Brown, "Dealers and Stealers," *ERJV* 5 (1894): 92–95.
36. Günther to Hartert, May 7, 1907, TM1/120.
37. Nicolaas Rupke, *Richard Owen: Victorian Naturalist* (New Haven: Yale University Press, 1994), 43.
38. Günther to Haldane, Apr. 3, 1902, Günther Collection, Box 23, NHM. Rothschild would be elected in 1911.
39. Gunther, *A Century of Zoology*, 422.
40. Walsingham to August Busck, Mar. 29, 1908, August Busck Papers, 1902–1933, Series I. Correspondence, SIA 7129.
41. Tutt, "Collectors," *ERJV* 1 (1890/1891): 99–100, at 99.
42. See Frederick H. Burkhardt, "England and Scotland: The Learned Societies," in T. F. Glick, ed., *The Comparative Reception of Darwinism* (Austin: University of Texas Press, 1974).
43. "Papers &c., Read," *PESL* (1890): iii–vii, at vi–vii.
44. Tutt, "The President's Address to the British Association Considered in Its Relation to Entomology," *ERJV* 9 (1893): 261–265, at 265.
45. Circular "To the Fellows of the Entomological Society of London," Jan. 7, 1893, signed by E. B. Poulton and Charles Swinhoe, ESL.

46. H. J. Elwes, "Presidential Address," *PESL* (1893): xlvi–lviii, at xlvii.

47. Frederick DuCane Godman, "President's Address," *PESL* (1891): xliv–xlviii.

48. R. Meldola, "President's Address: The Speculative Method in Entomology," *PESL* (1895): lii–lxvii, at lx–lxi.

49. Tutt, "Address by the Vice-President to the City of London Entomological and Natural History Society," *ERJV* 6 (1895): 59–69, at 67.

50. Tutt, "Reviews," *ERJV* 4 (1893): 67–68, at 68.

51. Tutt, "Current Notes," *ERJV* 5 (1894): 178.

52. Tutt, "Classification by Neuration," *ERJV* 3 (1892): 50–51.

53. G. H. Carpenter, "Some Recent Researches on Insects and Arachnids," *NS* 1 (1892): 53–56, at 53.

54. "Correspondence: The Retort of the Systematist," *NS* 9 (1896): 70–71.

55. Meldola, "President's Address," lvi and lx.

56. G. H. Carpenter, "Some New Books," *NS* 8 (1896): 123–126, at 123.

57. A Country Cousin, "Lead Us Not into Temptation," *ERJV* 4 (1893): 325–329, at 325.

58. "Newspaper Science," *Natural History* 1 (1892): 5.

59. "Some New Books," *NS* 8 (1896): 274.

60. "Notes and Comments: Species-Making and Species-Taking," *NS* 5 (1894): 175–176, at 175.

61. Quoted in Poulton, "The Influence of Darwin upon Entomology," *ERJV* 13 (1901): 72–76, at 72.

62. Jordan, "Anthribidae from the Islands of Engano, Mentawei and Sumatra, Collected by Dr. E. Modigliani," *Annali del Museo Civico di Storia Naturale di Genova* 38 (1897): 623–643, at 624.

63. "Notes and Comments: 'Mutato Nomine ⸺' Date," *NS* 12 (1898): 218.

64. Jordan, "The President's Address, 1929," *PESL* 4 (1930): 128–141, at 141.

65. Jordan, "On Mechanical Selection and Other Problems," *NZ* 3 (1896): 426–525, at 426.

66. C. Claus, *Elementary Text-book of Zoology*, trans. Adam Sedgwick and F. G. Sinclair (London: Swan Sonnenschein, 1884–1885), 141.

67. Charles Darwin, *The Origin of Species by Means of Natural Selection* (New York: Penguin, 1985; originally published by John Murray, 1859), 103–104.

68. Darwin, *On The Origin of Species*, 455.

69. Darwin, *On The Origin of Species*, 47.

70. Addison Gulick, ed. *Evolutionist and Missionary: John Thomas Gulick, Portrayed Through Documents and Discussions* (Chicago: University of Chicago Press, 1932), 410.

71. Tutt, "Editor's Note: Societies," *ERJV* 1 (1891): 215.

72. William Bateson, *Materials for the Study of Variation, Treated with Especial Regard to Discontinuity in the Origin of Species* (London: MacMillan, 1894), vi.

73. Otto Kleinschmidt, *The Formenkreis Theory and the Progress of the Organic World* (London: H. F. and G. Witherby, 1930), 97.

74. "Zoological Nomenclature," *Nature* 30 (1884): 256–259 and 277–279, at 279.

75. Erwin Stresemann, *Ornithology: From Aristotle to the Present* (Cambridge, MA: Harvard University Press, 1975), p. 248.

76. Jordan, "Presidential Address," *Lepidopterists' News* 7 (1953): 3–4, at 3.

77. Note from the Editors, *NZ* 1 (1894) i.
78. R. B. Sharpe, *A Hand-list of the Genera and Species of Birds*, vol. 5 (London, BMNH, 1909), v.
79. "Zoological Nomenclature," *Nature* 30 (1884): 256–259 and 277–279, at 258.
80. P. L. Sclater to Hartert, Nov. 30, 1903, TM1/77, NHM.
81. Alfred Newton to J. A. Harvie-Brown, July 20, 1904, quoted in A. F. R. Wollaston, *Life of Alfred Newton: Late Professor of Comparative Anatomy, Cambridge University, 1866–1907* (New York: Dutton, 1921), 218.
82. Arthur Erwin Brown, "Notes on Anthropoid Apes," *Science* 22 (1905): 12–14, at 14.
83. "Zoological Nomenclature," *Nature* 30 (1884): 256–259 and 277–279, at 278.
84. "Current Notes," *ERJV* 23 (1911): 217.
85. "Notes and Comments: Trinomial Nomenclature," *NS* 6 (1895): 158–159.
86. "Zoological Nomenclature," *Nature* 30 (1884): 256–259 and 277–279, at 257.
87. "Some New Books," *NS* 4 (1894): 218–219.
88. Hartert to C. Davies, Mar. 3, 1894, TMLB #2. NHM.
89. Rothschild, *Dear Lord Rothschild*, 132.
90. Gunther, *A Century of Zoology*, 215.
91. A. R. Wallace, *Darwinism: An Exposition of the Theory of Natural Selection with Some of its Applications* (London: MacMillan, 1889), 13, 45, and 127.
92. Winsor, *Reading the Shape of Nature*, 78.
93. Frederick Du Cane Godman, "The President's Address," *PESL* (1891): xliv–lii, at xlvii.
94. N. D. Riley, "Obituary of Jordan," Typescript draft of memoriam for the Royal Society for London, page 3. DF ENT/332/1/81, Department of Entomology, Norman D. Riley Correspondence and Papers, 1922–1976, NHM.
95. Gulick, *Evolutionist and Missionary*, 413 (from a letter to Romanes, Dec. 29, 1888).
96. Louis Agassiz, *Essay on Classification*, ed. Edward Lurie (Mineola, NY: Dover Publications, 2004; originally published in 1857), 178.
97. Mary P. Winsor, *Reading the Shape of Nature: Comparative Zoology at the Agassiz Museum* (Chicago: Chicago University Press, 1990), 40.
98. See Kleinschmidt, *The Formenkreis Theory*, 40.
99. Jordan, "On Mechanical Selection," 446.
100. Alfred Newton to Rothschild, Dec. 16, 1891, quoted in Rothschild, *Dear Lord Rothschild*, 77.
101. Newton to T. Southwell, Mar. 14, 1856, quoted in Wollaston, *Life of Alfred Newton*, 55.
102. Quoted in Rothschild, *Dear Lord Rothschild*, 366.
103. Dodd to Walsingham, Sept. 30, 1898, Walsingham Correspondence, Entomology Library, NHM.
104. Wallace to Jordan, Apr. 3, 1905, TM1/94, NHM.
105. Jordan, "The Species Problem as Seen by the Systematist," *Proceedings of the Linnean Society of London* 150 (1938): 241–247, at 243.
106. Riley, "Obituary of Jordan," 77.
107. W. Rothschild, "Notes on 'Sphingidae,' with Descriptions of New Species," *NZ* 1 (1894): 65–98, at 66.
108. Burkhardt, "England and Scotland," 64.

109. Hartert to G. Cherrie, Mar. 30, 1897, TMLB #4, NHM.

110. Jordan, in the introduction to Walter Rothschild, "A Revision of the *Papilios* of the Eastern Hemisphere, exclusive of Africa," *NZ* 2 (1895): 167–463, at 169.

111. Hartert, "On Ornithological Collections made by Alfred Everett in Celebes and on the Islands South of It. Part II. The Birds of Saleyer, Djampea, and Kalao," *NZ* 3 (1896): 165–183, at 181.

112. Tutt, "Address by the Vice-President to the City of London Entomological and Natural History Society," *ERJV* 6 (1895): 59–69, at 61.

113. Hartert to Messrs. Burns, Philip, undated. TMLB #2, NHM.

114. Walter Rothschild and Jordan, "Lepidoptera Collected by Oscar Neumann in North-East Africa," *NZ* 10 (1903): 491–542, at 492.

115. Jordan to J. Schröder, May 22, 1893, TMLB #1, NHM.

116. Hartert to Alan Ouston, Mar. 28, 1894, TMLB #2, NHM.

117. Hartert to A. Noakes, Dec. 12, 1893, TMLB #1, NHM.

118. Rothschild and Hartert, "Notes on Papuan Birds," *NZ* 10 (1903): 65–116, at 98.

119. Hartert to Charles Hose, Mar. 17, 1893, TMLB #1, NHM.

120. Hartert to Alan Ouston, Mar. 28, 1894, TMLB #2, NHM.

121. Ernst Mayr to Miriam Rothschild, Nov. 21, 1983, Papers of Ernst Mayr, Correspondence, 1931–1982, Folder 1371, HUGFP 74.7, HUA.

122. Rothschild and Jordan, "Lepidoptera Collected by Oscar Neumann in North-East Africa," 501.

123. Dodd to J. H. Durrant, May 14, 1896, Walsingham Correspondence, Entomology Library, NHM.

124. Rothschild and Jordan, "A revision of the American Papilios," *NZ* 13 (1906): 411–752, at 416.

125. Jordan to Frederick William Frohawk, Feb. 16, 1937, TM2/73, NHM.

126. Mark Barrow, *A Passion for Birds: American Ornithology after Audubon* (Princeton, NJ: Princeton University Press, 1998), 502.

127. *Entomologist* 26 (1893): 22.

128. The Kny-Scheerer Co., Dept. of Nat. Science, New York, to Hartert, May 3, 1905, TM1/90, NHM.

129. Jordan, "In Memory of Lord Rothschild," *NZ* 41 (1938): 1–16, at 6.

130. Janson & Sons, Natural History Agents to Rothschild, Feb. 12, 1918, KJC #2, NHM.

131. Hartert to Walter L. Butler, Apr. 2, 1895, TMLB #2, NHM.

132. Hartert to Captain Webster, July 6, 1896, TMLB #3, NHM.

133. Hartert to G. K. Cherrie, Feb. 3, 1897, TMLB #4, NHM.

134. Hartert to Edward Gerrard, Jan. 24, 1894, TMLB #1, NHM.

135. A Looker-On, "The Sale of the Late Mr. Machin's Macro-Lepidoptera," *ERJV* 6 (1895):134–137, at 136.

136. Elwes to W. Rothschild, Feb. 26, 1897, TM1/26, NHM.

137. Elwes to W. Rothschild, Feb. 4, 1897, TM1/26, NHM.

138. Anon., "The Congress of Zoology: Visit to Tring Museum," *Standard* (London), Tuesday, Aug. 30, 1898.

139. Jordan, "In Memory," 13.

140. Rothschild and Jordan, "Lepidoptera Collected by Oscar Neumann," 501.
141. Darwin, *On the Origin of Species*, 101.

CHAPTER 3: ORDERING BEETLES, BUTTERFLIES, AND MOTHS

1. Jordan, "On *Xenopsylla* and Allied Genera of Siphonaptera," in *Verhandlungen des III Internationalen Entomologen-Kongresses, Zürich, 19.–25. Juli 1925*, ed. Walther Horn and Karl Jordan, 593–627, vol. 2 (Zürich: Druck von G. Uschmann, 1926), 593.
2. Leonhard Stejneger, "Review of *Das Tierreich. Eine Zusammenstellung und Kennzeichnung der rezenten Tierformen*," *Science* 5 (1897): 846–848.
3. Harry Eltringham, "A Monograph of the African Species of the Genus *Acraea*, Fab., with a Supplement on Those of the Oriental Region," *TESL* (1912): 1–374, at 1.
4. W. E. Sharp, "The New Entomology," *Entomologist* 27 (1894): 81–86 and 110–116, at 83, 85, and 114–116.
5. Alfred Russel Wallace, *My Life: A Record of Events and Opinions* (New York: Dodd, Mean, 1905), 404.
6. Albert Gunther, *A Century of Zoology at the British Museum through the Lives of Two Keepers, 1815–1914* (London: Dawsons, 1975), 460.
7. Poulton, "The Influence of Darwin upon Entomology," *ERJV* 13 (1901): 72–76, at 73.
8. Quoted in J. F. M. Clark, *Bugs and the Victorians* (Yale University Press, 2009), 127.
9. W. B. Grove in the *Midland Naturalist*, quoted in Allen, *A Naturalist in Britain*, 174.
10. Mary P. Winsor, *Reading the Shape of Nature: Comparative Zoology at the Agassiz Museum* (Chicago: University of Chicago Press, 1991), 67 and 90.
11. "The Nomenclature of Insects," *NS* 1 (1892): 572–573.
12. Thomas R. R. Stebbing, "On Random Publishing and Rules of Priority," *NS* 5 (1894): 337–444, at 337.
13. J. W. Tutt, Review of "Die Palearktischen Gross-Schmetterlinge und ihre Natur-Geschichte," by Fritz Rühl, *ERJV* 4 (1893): 68.
14. "The Speciesmonger in High Places," *NS* 8 (1896): 4–5, at 5.
15. "The Preliminary Notice," *NS* 8 (1896): 73–75.
16. Jordan, "In Memory of Lord Rothschild," *NZ* 41 (1938): 1–16, at 7.
17. Miriam Rothschild, "Jordan: A Biography," *TRESL* 107 (1955): 1–9, at 4.
18. Jordan to H. Kühn, June 3, 1897, TMLB #4, NHM.
19. TS, Jordan, with handwritten notes on the Coleoptera collection, KJC #3, NHM. Although the collection had been valued at £3,935, Jordan noted that "some buyers did not pay at all." Charles Oberthür took all families not sold.
20. Jordan to N. D. Riley, Sept. 30, 1937, DF306/3, NHM.
21. Jordan to Ernst Mayr, Dec. 22, 1954, Ernst Mayr Papers, General Correspondence, 1931–1982, Folder 591, HUGFP 74.7, HUA.
22. Jordan to E. B. Poulton, Nov. 5, 1906, Poulton Correspondence, HLA.
23. Walter Rothschild, "A Revision of the *Papilios* of the Eastern Hemisphere, Exclusive of Africa," *NZ* 2 (1895):167–463, at 167.
24. Jordan to Harry Clench, Dec. 23, 1954, Correspondence of Harry A. Clench, Department of Invertebrate Zoology, CMNH.

25. Jordan to Kleinschmidt, Feb. 1, 1952, KJC#1, NHM.
26. Jordan to Kleinschmidt, Feb. 1, 1952, KJC#1, NHM.
27. Rothschild, "A Revision of the *Papilios* of the Eastern Hemisphere," 167.
28. See Ernst Mayr, "Jordan's Contribution to Current Concepts in Systematics and Evolution," *TRESL* 107 (1955): 45–66. The introductory notes include, for example, the words "When I began to study the Papilios more closely under Mr. Walter Rothschild's guidance..." (p. 168).
29. Draft obituary of Durrant. August Busck Papers, 1902–1933, Series 1, Correspondence, SIA 7129.
30. W. J. Holland to Jordan, Mar. 16, 1903, TM1/74, NHM.
31. Jordan's introduction to Rothschild, "Revision of the *Papilios* of the Eastern Hemisphere," 168, 180–181.
32. Hartert to Grose Smith, Feb. 14, 1894, TMLB #2, NHM.
33. E. D. Cope, "The Formulation of the NSs," *American Naturalist* 30 (1896):101–112, at 111.
34. Tutt, "The Sub-genera Viminia, Cuspidia, and Bisulcia, Chapman," *ERJV* 2 (1891): 82.
35. W. Rothschild and Jordan, "A Revision of the Lepidopterous Family Sphingidae," *NZ* 9 (1903): Suppl., cxxv+972, xi.
36. W. Rothschild and Jordan, "A Revision of the Lepidopterous Family Sphingidae," xxii.
37. Jordan's introduction to Rothschild, "Revision of the *Papilios* of the Eastern Hemisphere," 168.
38. Gunther, *A Century of Zoology*, 422–423, and *Science Gossip*, 1892, 88.
39. Jordan, "In Memory," 9.
40. Rothschild, "Revision of the *Papilios* of the Eastern Hemisphere," 221. Note signed KJ.
41. Jordan's introduction to Rothschild, "Revision of the *Papilios* of the Eastern Hemisphere," 179.
42. W. Rothschild, "A Revision of the *Papilios* of the Eastern Hemisphere," 327.
43. Jordan's introduction to Rothschild, "Revision of the *Papilios* of the Eastern Hemisphere," 169.
44. R. Lydekker, *Sir William Flower* (London: J. M. Dent, 1906), 97–98.
45. Jordan, "Contributions to the Morphology of Lepidoptera," *NZ* 5 (1898): 374–415.
46. Walter Rothschild and Jordan, "Lepidoptera Collected by Oscar Neumann in North-East Africa," *NZ* 10 (1903): 491–542, at 493–496.
47. Jordan's introduction to Rothschild, "Revision of the *Papilios* of the Eastern Hemisphere," 177.
48. "Notes and Comments: Reminiscences of Huxley," *NS* 7 (1895): 298–299.
49. Jordan to Kleinschmidt, Feb. 1, 1952, KJC#1, NHM.
50. "Notes and Comments: Habits as Diagnostic of Species," *NS* 9 (1896): 2–4.
51. F. Buchanan White, "The Value of the Genitalia in Determining Species," *ERJV* 2 (1891): 82–83, at 83.
52. W. H. Edwards, "The Value of the Genitalia in Determining Species," *ERJV* 2 (1891): 13–14, at 14.
53. W. Bloomfield, "Current Notes: Coremia Ferrugata and Unidentaria," *ERJV* 3 (1892): 242.

54. Edwards, "The Value of the Genitalia in Determining Species," 14.

55. Otto Blüh, "Ernst Mach: His Life as a Teacher and Thinker," in *Ernst Mach: Physicist and Thinker*, ed. R. S. Cohen and Raymond J. Seeger, Boston Studies in the Philosophy of Science, 1–22.

56. Jordan's introduction to Rothschild, "A Revision of the *Papilios* of the Eastern Hemisphere," 182.

57. W. F. de Vismes Kane, "The 'Melanism' Controversy," *Entomologist* 26 (1893): 307–311, at 309.

58. J. T. Cunningham, "Origin of Species among Flat-Fishes," *NS* 6 (1895): 169–177, at 170.

59. David Sharp, "Mr. W. Bateson on Variation," *Entomologist* 27 (1894): 162–164, at 163–164.

60. Alfred Russel Wallace, "The Method of Organic Evolution," *Fortnightly Review* 57 (1895): 211–224 and 435–445, at 218 and 223.

61. William Bateson, *Materials for the Study of Variation* (London: Macmillan, 1894), 574–575.

62. William Bateson, "Heredity and Evolution," *Popular Science Monthly* 65 (1904): 522–531, at 530.

63. Jordan, "On Mechanical Selection and Other Problems," *NZ* 3 (1896): 426–525, at 426–428.

64. Cunningham, "Origin of Species among Flat-Fishes," 169–170.

65. Jordan, "On Mechanical Selection," 441.

66. Jordan, "On Mechanical Selection," 441.

67. Jordan, "On Mechanical Selection," 431–437.

68. Jordan, "On Mechanical Selection," 436–442.

69. Quoted in Phillip R. Sloan, "Buffon, German Biology, and the Historical Interpretation of Biological Species," *British Journal for the History of Sciences* 12 (1979): 109–153, at 139.

70. Louis B. Prout, "Scientific Notes," *ERJV* 3 (1892): 149–154, at 151.

71. "Recent Literature," *Entomologist* 26 (1893): 171–172, at 172.

72. Jordan to Mr. Scheben, June 7, 1946, KJC#5, NHM.

73. Jordan, "On Mechanical Selection," 442.

74. Jordan and N. C. Rothschild, "Revision of the Non-combed Eyed Siphonaptera," *Parasitology* 1 (1908): 1–100, at 3.

75. Jordan, "On Mechanical Selection," 450–451.

76. Robert S. Cohen, "Ernst Mach: Physics, Perception, and the Philosophy of Science," in *Ernst Mach: Physicist and Philosopher*, 126–164, at 129.

77. Rothschild and Jordan, "A Revision of the Lepidopterous Family Sphingidae," xlii.

78. Jordan, "On Mechanical Selection," 446.

79. Rothschild and Jordan, "Lepidoptera Collected by Oscar Neumann," 492.

80. Geo. H. Carpenter, "Some New Books: A Study in Variation," *NS* 10 (1897): 335–337, at 337.

81. Tutt, "Reviews and Notices of Books," *ERJV* 9 (1897): 130.

82. Tutt, "Our Century Number," *ERJV* 13 (1901): 1–3, at 2.

83. Louis B. Prout, "The Lepidopterological Books of the Nineteenth Century," *ERJV* 13 (1901): 20–26, at 25–26.

84. G. C. Champion to Jordan, Dec. 9, 1905, TM1/88, NHM.
85. Poulton to Jordan, Feb. 29, 1904, TM1/84, NHM.
86. Jordan, "The President's Address, 1930," *PESL* (1931): 128–142, at 133.
87. Gunther, *A Century of Zoology*, 461–463.
88. Jordan, "On Mechanical Selection," 428.
89. Theodor Eimer, *On Orthogenesis and the Importance of Natural Selection in Species Formation* (Leyden: Congress of Zoologists, 1895; Chicago: Open Court, 1898, trans. Thomas J. McCormack), 37.
90. J. W. Spengel to Jordan, Feb. 14, 1896, TM1/23.
91. Jordan, "An Examination of the Classificatory and Some Other Results of Eimer's Researches on Eastern Papilios: A Review and Reply," *NZ* 5 (1898): 435–455, at 435–437, 447, and 454.
92. Jordan, "On Mechanical Selection," 432 and 451.
93. Rothschild and Jordan, "A Revision of the Lepidopterous Family Sphingidae," xci.
94. Jordan to Harry Clench, Dec. 20, 1955, Correspondence of Harry A. Clench, Department of Invertebrate Zoology, CMNH.
95. Jordan, "On Mechanical Selection," 451.
96. Rothschild and Jordan, "A Revision of the Lepidopterous Family Sphingidae," xxx.
97. E. Munroe, "Jordan's the Lepidopterist," *TRESL* 107 (1955): 69–75, at 72.
98. Rothschild and Jordan, "A Revision of the Lepidopterous Family Sphingidae," xxix.
99. Rothschild and Jordan, "A Revision of the Lepidopterous Family Sphingidae," vii.
100. Ian J. Kitching and Jean-Marie Cadiou, *Hawkmoths of the World: An Annotated and Illustrated Revisionary Checklist* (*Lepidoptera: Sphingidae*) (London: Comstock Publishing, a division of Cornell University Press, 2000), 27.
101. Rothschild and Jordan, "A Revision of the Lepidopterous Family Sphingidae," i.
102. Rothschild and Jordan, "A Revision of the Lepidopterous Family Sphingidae," ix.
103. C. M. Stuart to Jordan, Apr. 27, 1903, TM1/77, NHM.
104. Walsingham to August Busck, Mar. 29, 1908, August Busck Papers, 1902–1933, Series I, Correspondence, SIA 7129.
105. Holland to Jordan, Apr. 16, 1924, TM2/50, NHM.
106. Riley, "Heinrich Ernst Karl Jordan," 110.
107. W. J. Kaye to Jordan, Mar. 13, 1904, TM1/82, NHM.
108. Rothschild and Jordan, "A Revision of the American Papilios," *NZ* 13 (1906): 411–752, at 427.
109. Quoted in Jordan, "An Examination of the Classificatory and Some Other Results of Eimer's Researches," 445.
110. Charles Oberthür, *Études de Lépidoptérologie Comparée*, vol. 17 (Rennes, 1920), vi and viii.
111. Rothschild and Jordan, "Lepidoptera Collected by Oscar Neumann," 500.
112. Hampson to W. Rothschild, Oct. 8, 1894, TM1/7, NHM.
113. Hampson to W. Rothschild, Oct. 8, 1894, TM1/7, NHM.
114. W. Rothschild and Jordan, "A Revision of the Lepidopterous Family Sphingidae," xii.
115. "Memorandum on the Tring Museum," Nov. 19, 1937, DF306, NHM.
116. Rothschild and Jordan, "A Revision of the American Papilios," 421.

117. Rothschild and Jordan, "Lepidoptera Collected by Oscar Neumann," 491–492.

118. Foetterly to Jordan, Nov. 12, 1906, TM1/96, NHM.

119. Moss to Jordan, Oct. 18, 1910, TM2/11, NHM.

120. Jordan, "Der Gegensatz zwischen geographischer und nichtgeographischer Variation," *Zeitschrift für wissenschaftliche Zoologie* 83 (1905): 151–210, at 182.

121. Rothschild and Jordan, "A Revision of the American Papilios," 422 and 430.

CHAPTER 4: ORDERING NATURALISTS

1. Charles Maxwell Stuart to Jordan, Apr. 28, 1903, TM1/77, NHM.

2. Ehlers to Jordan, May 10, 1903, TM1/73, NHM.

3. Holland to Jordan, Mar. 24, 1903, TM1/74, NHM.

4. Chapman, "A Revision of the Sphingides," *ERJV* 16 (1903): 309–312; 17 (1904): 5–10, 44–47, and 75–78, at 311.

5. William James Kaye to W. Rothschild, Apr. 24, 1903, TM1/75, NHM.

6. Severin to Jordan, May 11, 1903, TM1/77.

7. Holland, "Scientific Books," *Science* 18 (1903): 15–16.

8. Distant to Jordan, Sept. 3, 1902, TM1/62, NHM.

9. Butterfield to "Gentlemen," Apr. 26, 1903, TM1/72, NHM.

10. Harry Eltringham, "A Monograph of the African Species of the Genus *Acraea*, Fab., with a Supplement on Those of the Oriental Region," *TESL* (1912): 1–374, at 2.

11. F. G. Foetterle to Hartert, Oct. 28, 1906, TM1/96, NHM.

12. A. R. Wallace to Jordan, Dec. 9, 1907, TM1/109, NHM.

13. Wallace, "The Social Quagmire and the Way Out of It," *Arena* 7 (1893): 525–542, at 539.

14. Wallace, *Bad Times: An Essay on the Present Depression of Trade, Tracing It to Its Sources in Enormous Foreign Loans, Excessive War Expenditure, the Increase of Speculation and of Millionaires, and the Depopulation of the Rural Districts; with Suggested Remedies* (London: MacMillan, 1886), 26.

15. Niall Ferguson, *The House of Rothschild: The World's Banker, 1849–1999* (New York: Penguin, 2000; first published in Britain, 1998), 295.

16. Miriam Rothschild, *Dear Lord Rothschild: Birds, Butterflies, and History* (London: Hutchinson, 1983), 224.

17. Wallace, "If There Were a Socialist Government—How Should It Begin?" *Clarion* (1905): 5.

18. Wallace, *Bad Times*, 67–68.

19. Wallace, "Is It Peace or War? A Reply," *Public Opinion* 94 (1908): 202–203.

20. Wallace, "The World of Life: As Visualised and Interpreted by Darwinism," 85 (1909): 411–434, at 420.

21. Wallace, *Bad Times*, and "The Social Quagmire," 400.

22. Wallace, "Darwinism *versus* Wallaceism," *Contemporary Review* 94 (1908): 716–717, at 717.

23. Wallace, "The Present Position of Darwinism," *Contemporary Review* 94 (1908): 129–141, at 139.

24. Bateson, *Mendel's Principle of Heredity* (Cambridge: Cambridge University Press, 1902), x.

25. William Bateson, *Materials for the Study of Variation, Treated with Especial Regard to Discontinuity in the Origin of Species* (London: MacMillan, 1894), 574–575.

26. Poulton, *Essays on Evolution, 1889–1907* (Oxford: Clarendon Press, 1908), 579 and 575.

27. Wallace, "The Present Position of Darwinism," 140.

28. W. Rothschild and Jordan, "A Revision of the Lepidopterous Family Sphingidae," *NZ* 9 (1903): Suppl., cxxv + 972, xlii.

29. Poulton to Jordan, Dec. 12, 1905, TM1/91, NHM.

30. Poulton. "The President's Address," *Proceedings of the Entomological Society of London* (1903): lxxiii–cxvi.

31. Poulton to Jordan, Mar. 13, 1908, TM1/115, NHM.

32. Poulton, "A Hundred Years of Evolution," *Science* 74 (1919): 345–360, at 349.

33. Poulton, *Essays on Evolution*, xvi.

34. Wallace, "By Cable to the Editors of *The World*," *World* (1907): 2.

35. Wallace, *The Revolt of Democracy* (London: Cassell, 1913), 4.

36. Caroline Shaw, "Rothschilds and Brazil: An Introduction to Sources in the Rothschild Archive," *Latin American Research Review* 40 (2005): 165–185, at 175.

37. H. H. Corbet, "Entomology, Evolution, and Romance: A Plea for a New Departure. A Criticism of Mr. Frost's Article," *ERJV* 8 (1896): 29–31, at 30.

38. W. E. Sharp, "The New Entomology," *Entomologist* 27 (1894): 81–86 and 110–116, at 15.

39. Tutt, "Societies," *ERJV* 8 (1895): 163–168, at 165.

40. Tutt, "Current Notes," *ERJV* 17 (1905): 80–82, at 81.

41. Rothschild and Jordan, "A Revision of the Lepidopterous Family Sphingidae," xv.

42. David Sharp, "The Cost and Value of Insect Collections," *TESL* (1893): 419–424, at 419 and 424.

43. Quoted in James J. Sheehan, *German Liberalism in the Nineteenth Century* (Chicago: University of Chicago Press, 1978), 231.

44. Jordan to Riley, Feb. 10, 1955, DF306/9, NHM.

45. Elisabeth Crawford, *Nationalism and Internatonalism in Science, 1880–1939: Four Studies of the Nobel Population* (Cambridge: Cambridge University Press, 2002), 61.

46. "News of Universities, Museums, and Societies," *Zoologist* 9 (1896): 137–141, at 139.

47. "Premièr Séance Plénière," in *Compte-Rendu des Séances du Troisième Congrés International de Zoologie, Leyden, 16–21 Septembre, 1895*, ed. P. P. C. Hoek (Leyden: E. J. Brill, 1896), 29–31.

48. *Congrès International de Zoologie. Deuxième Session, à Moscou 1892* (Moscow: Laschkevitsch Znamensky, 1892–1893), 59.

49. "Erste allgemeine Sitzung," *Verhandlungen des V. Internationalen Zoologen-Congresses zu Berlin, 12–16 August 1901*, ed. Paul Matschie (Jena: Gustav Fischer, 1901), 87–115, at 115.

50. See L. Stejneger, *A Chapter in the History of Zoological Nomenclature* (Washington DC: Smithsonian Institution, 1924), 5, and chapter 3 of Debra Adrienne Everett-Lane, "International Scientific Congresses, 1878–1913: Comunity and Conflict in the Pursuit of Knowledge" (PhD diss., Columbia University, 2004).

51. Henryk Arctowski, "The Genealogy of the Sciences as the Basis of Their Bibliography," *NS* 10 (1897): 395–405, at 395.

52. "Premièr Séance Plénière," in *Troisième Congrés International de Zoologie, Leyden*, 29–31.

53. "Postage of Natural History Specimens," *NS* 7 (1895): 311.

54. "Mr. Harvie-Brown on a Colour Code," *Proceedings of the (Fourth) International Congress of Zoology, Cambridge, 22 August, 1898*, ed. Adam Sedgwick (London: C. J. Clay), 155–156.

55. O. C. Marsh, "The Value of Type Specimens and Importance of Their Preservation," *Proceedings of the (Fourth) International Congress of Zoology, Cambridge*, 158–163.

56. Clark to Theodor Mortensen, Sept. 27, 1916, Austin Clark Papers, Box 10, RU 7183, SIA.

57. Rothschild and Jordan, "A Revision of the Lepidopterous Family Sphingidae," xv.

58. "Nomenclature of Lepidoptera: Correspondence Relating to Questions Circulated by Sir George F. Hampson," 273–342 in *Proceedings of the (Fourth) International Congress of Zoology*, at 338.

59. Hartert, "The Principle Aims of Modern Ornithology," *Proceedings of the Fourth International Ornithological Congress, London, June, 1905*, ed. R. Bowdler Sharpe, Hartert, and J. Lewis Bonhote, 265–270 (London: Dulau, 1907), 268.

60. Jordan to James Fletcher, Mar. 9, 1907, reprinted in E. C. Becker, ed., *Proceedings of the Tenth International Congress of Entomology, Montreal, August 17–25, 1956* 1 (1958), 48.

61. Jordan to R. Gestro, Nov. 24, 1906, ACSN.

62. Jordan to Riley, Aug. 28, 1946, DF306/8, NHM.

63. "International Entomological Congress in 1908," *ERJV* 19 (1907): 166–167.

64. These included Chr. Aurivillius, E. L. Bouvier, I. Bolivar, L. Bedel, T. Becker, M. Bezzi, P. Bachmetjew, S. Bengtssen, J. C. Bradley, W. Beutenmueller, C. J. S. Bethune, T. A. Chapman, G. H. Carpenter, T. D. Cockerell, Ph. P. Calvert, K. Daniel, F. A. Dixey, W. L. Distant, E. C. Van Dyke, H. Druce, Ed. Everts, A. Forel, J. Fletcher, H. C. Fall, L. Ganglbaür, A. Giard, R. Gestro, F. Du Cane Godman, W. Horn, A. Handlirsch, K. M. Heller, G. Horvàth, H. J. Kolbe, G. Kraatz, F. Klapalek, P. Mabille, J. C. U. de Meijere, A. L. Montadon, P. Magretti, F. Merrifield, L. W. Mengel, Chas. Oberthür, R. Oberthür, H. Osborn, E. B. Poulton, H. Rebel, W. Ris, W. Rothschild, H. Schouteden, A. v. Schulthess-Rechberg, G. Severin, F. Silvestri, Y. Sjöstedt, H. Skinner, J. B. Smith, M. Standfuss, S. Schenkling, J. W. Tutt, G. H. Verrall, E. Wassmann, Chas. O. Waterhouse, "and others." See "International Entomological Congress in 1908," *ERJV* 19 (1907): 166–167.

65. Durrant to Jordan, Aug. 28, 1907, TM1/103; Percy Lathy to Jordan, Dec. 18, 1907, TM1/105; Edward Meyrick to Jordan, July 13, 1907, TM1/105; Malcolm Burr to Jordan, Oct. 3, 1907, and Dec. 2, 1907, TM1/102; and Holland to Jordan, Sept. 2, 1907, TM1/104, NHM.

66. C. Grist to Jordan, Mar. 26, 1908, TM1/112, NHM.

67. Jordan, "In Memory," 11.

68. Rothschild, *Dear Lord Rothschild*, 220. Miriam Rothschild provides an account of what had happened in chapter 25 (entitled "The Great Row").

69. Rothschild, *Dear Lord Rothschild*, 221.

70. Ferguson, *The House of Rothschild*, 443. His debts were rumored to be at £750,000 (p. 524, fn. 1).

71. Rothschild, *Dear Lord Rothschild*, 92 and 222.

72. Rothschild, *Dear Lord Rothschild*, 93.

73. Jordan to F. Birch, Jan. 25, 1908, TM1/110, NHM.

74. Jordan to Standfuss, Oct. 8, 1908, TM2/1, NHM.

75. Wallace, "If There Were a Socialist Government," 5.

76. Jordan to Standfuss, Oct. 8, 1908, TM2/1, NHM.
77. Rothschild, *Dear Lord Rothschild*, 222.
78. Walter Rothschild and Hartert, "Ornithological Explorations in Algeria," *NZ* 18 (1911): 456–550, at 456.
79. Jordan to G. Severin, Sept. 21, 1908, TM2/1, NHM.
80. Guy K. Marshall to R. Trimen, Feb. 16, 1908, Trimen Correspondence, HLA.
81. Poulton to Jordan, Apr. 7, 1908, TM1/115, NHM.
82. Poulton to Jordan, Jan. 27, 1910, TM2/11, NHM.
83. Jordan to Leigh, Dec. 2, 1910, TM2/12, NHM.
84. Hartert to Jordan, June 30, 1900, TM1/50, NHM.
85. A. F. R. Wollaston, *Life of Alfred Newton: Late Professor of Comparative Anatomy, Cambridge University, 1866–1907* (New York: Dutton, 1921), 60.
86. "Premièr Séance Plénière," in *Troisième Congrés International de Zoologie, Leyden*, 29–31.
87. James Buckland, "The Destruction of Wild Birds in General Throughout the World," in *Verhandlungen des V. Internationalen Ornithologen-Kongresses in Berlin 1910*, ed. Herman Schalow (Berlin: Deutsche ornithologische gesellschaft, 1911), 847–857, at 855.
88. "Meetings of Section IV," *Proceedings of the Fourth International Ornithological Congress*, 56.
89. Frank M. Chapman, "What Constitutes a Museum Collection of Birds?" *Proceedings of the Fourth International Ornithological Congress*, 144–156, at 145.
90. Hartert, "The Principle Aims of Modern Ornithology," *Proceedings of the Fourth International Congress of Ornitholgy*, 265–266.
91. Walter Rothschild, "*Mirounga angustirostris* (Gill)," *NZ* 15 (1908): 393–394.
92. J. E. Sherlock, courtesy Colonial Office of Jamaica, to Hartert, Oct. 27, 1907, TM1/108, NHM.
93. Sherlock to Hartert, Oct. 20, 1907, TM1/108, NHM.
94. "Notes and Comments," *NS* 3 (1893): 401–422, at 419.
95. Harold Perkin, *The Rise of Professional Society: England since 1880* (London: Routledge, 1989), xiii.
96. Frank M. Turner, "Public Science in Britain, 1880–1919," *Isis* 71 (1980): 589–608, at 593.
97. Fred Birch to Jordan, Nov. 21, 1907, TM1/102, NHM.
98. Ernst Hartert, "On the Birds Collected on the Tukang-Besi Islands and Buton, Southeast of Celebes," *NZ* 10 (1903): 18–38, at 20.
99. Jordan to Riley, Aug. 28, 1946, DF306/8, NHM.
100. Jordan to Eugene Boullet, Feb. 8, 1910, TM2/11, NHM.
101. Longstaff to Dixey, May 29, 1909, Dixey Letters, Box 4, HLA. Longstaff was at the 1912 meeting as secretary.
102. Dr. Buckell, "Specific Nomenclature: Present, Past, and Future," *ERJV* 4 (1893): 127–140, at 128.
103. "Notes and Observations: High-flat Setting," *Entomologist* 30 (1897): 45–47, at 46.
104. W. H. Harwood, "High-flat Setting," *Entomologist* 30 (1897): 142–143.
105. Tutt, "Preface," *ERJV* 5 (1894): i–ii, at ii.
106. Augustus R. Grote, "Generic Names in the Noctuidae," *ERJV* 6 (1895): 77–81, at 81.

107. James E. Russell, *German Higher Schools* (New York: Longmans, Green, 1913), 351.

108. Peter Alter, *The Reluctant Patron: Science and the State in Britain, 1850–1920* (Oxford: Berg, 1987), 93.

109. Wallace, "Letter to Keir Hardie," *Labour Leader* (1896): 251.

110. Tutt, "Current Notes," *ERJV* 17 (1905): 245–246, at 246.

111. Blanchard's "Deuxième rapport sur la nomenclature des être organizes," *Congrès International de Zoologie. Deuxième Session, à Moscou 1892*, 305.

112. Première Séance Plénière," in *Troisième Congrés International de Zoologie, Leyden*, 28.

113. "First General Meeting," *Proceedings of the (Fourth) International Congress of Zoology, Cambridge*, 50.

114. "First General Meeting," *Proceedings of the (Fourth) International Congress of Zoology, Cambridge*, 49.

115. "Zweite allgemeine Sitzung," *Verhandlungen des V. Internationalen Zoologen-Congresses zu Berlin, 12–16 August 1901*, ed. Paul Matschie (Jena: G. Fischer, 1902), 120.

116. Robert Fox, *The Culture of Science in France, 1700–1900* (Aldershot, UK: Variorum, 1992), 7–10.

117. *Congrès International de Zoologie. Deuxième Session, à Moscou 1892*, 65.

118. Jordan to Riley, Jan. 29, 1946, DF306/8, NHM.

119. William Bateson, "Facts Limiting the Theory of Heredity (Address)," in *Proceedings of the Seventh International Zoological Congress, Boston, 19–24 August 1907* (Cambridge: University Press, 1912), 306–319, at 306, 307, and 319.

120. Richard Goldschmidt, "Fifty Years of Zoology," *Scientific Monthly* 71 (1950): 359–369, at 361.

121. William Bateson to Beatrice Bateson, Aug. 24, 1907, Bateson Letters, vol. 1. nos 1–96, C52, JICA.

122. T. A. Cockerell, "Aspects of Modern Biology," *Popular Science Monthly* 72 (1908): 540–548.

123. Wm. E. Kellicott, "The Ninth International Congress at Monaco," *Science* 37 (1913): 593–595, at 594.

124. Cockerell, "Aspects of Modern Biology," *Popular Science Monthly* 72 (1908): 540–548.

125. Garland E. Allen, *Thomas Hunt Morgan: The Man and His Science* (Princeton, NJ: Princeton University Press, 1978), 104.

126. Poulton, *Essays on Evolution*, xiv and xlii.

127. Auguste Lameere, "Discours d'ouverture par M. le Prof ^{r} A. Lameere, president du Congrès," *1er Congrès International d'Entomologie, Bruxelles, 1–6 août 1910*, ed. G. Severin (Brussels: Hayez, 1911–12), 69–84, at 83.

128. "The Banquet," in *Proceedings of the Second International Congress of Entomology, Oxford, August 1912*, ed. Karl Jordan and Henry Eltringham (Oxford: Hazell, Watson & Viney, 1914), 129–143, at 138.

129. A. R. Wallace, "Imperial Might and Human Right," *Clarion* (1900): 230.

130. Jordan, "Historique de la création du 1er Congrès International d'Entomologie," *1er Congrès International d'Entomologie, Bruxelles*, 5–7, at 5.

131. Jordan to N. D. Riley, Aug. 28, 1946. DF 306, Folder 8, NHM.

132. Riley, "Jordan and the International Congresses of Entomology," *TRESL* 107 (1955): 15–24, at 16.

133. Jordan, "The Systematics of Some Lepidoptera Which Resemble Each Other, and Their Bearing on General Questions of Evolution," *1ᵉʳ Congrès International d'Entomologie, Bruxelles*, 385-404, at 385.

134. Lameere, "Discours d'ouverture," 81-83.

135. E. Ernest Green, "A Plea for the Centralization of Diagnostic Descriptions," *Proceedings of the Second International Congress of Entomology, Oxford*, 216-219, at 216.

136. "Section de Nomenclature," *1ᵉʳ Congrès International d'Entomologie, Bruxelles*, 137-148.

137. Jordan, "The Systematics of Some Lepidoptera which Resemble Each Other, and Their Bearing on General Questions of Evolution," *1ᵉʳ Congrès International d'Entomologie, Bruxelles*, 385-404, at 385.

138. "Sektion de Bibliographie," *1ᵉʳ Congrès International d'Entomologie, Bruxelles*, 104-107.

CHAPTER 5: A DESCENT INTO DISORDER

1. H. Turner, "Nomenclature," *ERJV* 28 (1916): 146-148, at 147.

2. G. W. Kirkaldy, "Evolution of Our Present Knowledge of the British Rynchota," *ERJV* 13 (1901): 59-62, at 60.

3. *PESL* (1910): lxxx.

4. Auguste Lameere, "Banquet," *1ᵉʳ Congrès International d'Entomologie, Bruxelles, 1-6 août 1910*, ed. G. Severin (Brussels: Hayez, 1911-12), 268-271, at 269.

5. G. W., "Current Notes," *ERJV* 23 (1911): 155-158, at 156 and 158.

6. "Resolutions," *PESL* (1911) xviii-xix.

7. F. D. Morice, "The President's Address," *PESL* 59 (1912): clix-cc, at cxviii and cxx.

8. "Current Notes," *ERJV* 10 (1910): 242-244, at 243.

9. "Our Century Number," *ERJV* 13 (1901): 1-3, at 3.

10. A. Shipley to Jordan, Jan. 25, 1905, TM2/93, NHM.

11. Jordan and N. C. Rothschild, "Revision of the Non-combed Eyed Siphonaptera," *Parasitology* 1 (1908): 1-100, at 2.

12. "Historique de la création du 1ᵉʳ Congrès International d'Entomologie," *1ᵉʳ Congrès International d'Entomologie, Bruxelles*, 5-7, at 6.

13. First International Congress of Entomology. Circular. Office of the Secretary, 1890-1929, Series 1. Box 15, Folder 10, RU 45, SIA.

14. F. Merrifield, "Comments," *1ᵉʳ Congrès International d'Entomologie, Bruxelles*, 94.

15. "Séance Générale," *1ᵉʳ Congrès International d'Entomologie, Bruxelles*, 245.

16. F. A. Dixey, "African Entomological Research Committee," *Nature* 81 (1909): 278.

17. Lameere, "Discours d'ouverture par M. le Profᵗ A. Lameere," *1ᵉʳ Congrès International d'Entomologie, Bruxelles*, 69-84, at 77-78.

18. "The Banquet," *Proceedings of the Second International Congress of Entomology, Oxford*, August, 1912, ed. K. Jordan and H. Eltringham (London: Hazell, Watson, & Viney, 1914), 129-143, at 129-133.

19. Jordan, "The Systematics of Some Lepidoptera which Resemble Each Other, and Their Bearing on General Questions of Evolution," *1ᵉʳ Congrès International d'Entomologie, Bruxelles*, 385-399.

20. Poulton, "A Hundred Years of Evolution. Address to BAAS Section D—Zoology, for 1931," *Science* 74 (1931): 345–360, at 355.

21. Summary of Poulton's "Mr. C. A. Wiggins's and Dr. G. H. Carpenter's Researches in Mimicry in the Forest Butterflies of Uganda," *Proceedings of the Second International Congress of Entomology, Oxford,* 51–52, at 52.

22. Jordan, "The Systematics of Some Lepidoptera," 386.

23. Bateson, *Problems of Genetics* (New Haven, CT: Yale University Press, 1913), 248.

24. Bateson, *Problems of Genetics,* 9–11.

25. Bateson to N. Heribert-Nilsson, Jan. 29, 1919, Bateson Letters, G78-26-27, JICA.

26. Jordan, "Presidential Address," *Lepidopterists' News* 7 (1953): 3–4, at 3.

27. R. C. Punnett, "Mendelisme," *Proceedings of the Second International Congress of Entomology, Oxford,* 151–152, at 152.

28. Walter Rothschild, "On the Term 'Subspecies' as Used in Systematic Zoology," *NZ* 19 (1911): 135–136.

29. Benjamin Preston-Clark to Bror Yngve Sjöstedt, July 4, 1921, Sjöstedt Correspondence, CfHS, RSAS.

30. Jordan to Moss, July 3, 1917, MSS MIL C 7:7, Entomology Library, NHM.

31. Jordan to Moss, Oct 12, 1917, MSS MIL C 7:7, Entomology Library, NHM.

32. Preston Clark to Sjöstedt, June 5, 1920, Sjöstedt Correspondence, CfHS, RSAS.

33. Jordan, "Contributions to the Morphology of Lepidoptera," *NZ* 5 (1898): 374–415, at 377.

34. Jordan, "Species Problem as Seen by the Systematist," 243.

35. Letter from Walsingham to Shipley, *Proceedings of the Second International Congress of Entomology, Oxford,* 58–60, at 60.

36. Jordan, "The Systematics of Some Lepidoptera," 386.

37. C. W. Stiles, "Report on the International Commission on Zoological Nomenclature," *Science* 38 (1913): 6–19, at 8.

38. Francis Hemming, "Karl Jordan and Zoological Nomenclature," *TRESL* 107 (1955): 25–32, at 32.

39. Michael A. Salmon, *The Aurelian Legacy: British Butterflies and their Collectors* (Berkeley: University of California Press, 2000), 198.

40. Stejneger, diary entries, Mar. 26–28, 1913, Leonhard Stejneger Papers, 1867–1943, Series 6, Box 26, Folder 13, RU 7074, SIA.

41. Stiles, "Report on the International Commission on Zoological Nomenclature," 9–12.

42. Hemming, "Karl Jordan and Zoological Nomenclature," 26.

43. Hemming, "Karl Jordan and Zoological Nomenclature," 27.

44. Stiles, "Report on the International Commission on Zoological Nomenclature," 8.

45. "The Banquet," *Proceedings of the Second International Congress of Entomology, Oxford,* 129–143, at 140–142.

46. Form letters sent Feb. 12, 1914, Jordan, H. E. K. Correspondence, 1914–1927, regarding ICZN, MSS JOR 1:1, Entomology Library, NHM.

47. Jordan to Handlirsch, Apr. 18, 1913, AWNM.

48. Jordan to C. Schrottky, Aug. 8, 1914, TM2/27, NHM.

49. Jordan to Sjöstedt, Sept. 8, 1914, Sjöstedt Correspondence, CfHS, RSAS.

50. Miriam Rothschild, "Jordan: A Biography," *TRESL* 107 (1955): 1–9, at 6.
51. Niall Ferguson, *The House of Rothschild*, vol. 2: *The World's Banker, 1849–1999* (New York: Penguin: 2000), 447.
52. Jordan to Moss, July 3, 1917, MSS MIL C 7:7, Entomology Library, NHM.
53. Jordan to Moss, Sept. 2, 1914, MSS MIL C 7:7, Entomology Library, NHM.
54. Bethune-Baker, "The President's Address," *PESL* (1914): cxix–clxviii, at clxiv.
55. K. G. Blair, "Notes from the Trenches," *ERJV* 27 (1915): 199–200.
56. Noel S. Sennett, "Ants from the Front," *PESL* (1916): iii–iv and D. A. J. Buxton, "Lepidoptera from Gallipoli," *PESL* (1916): xiii.
57. Poulton, "Notes on the Migration of Lepidoptera, with a Suggestion as to the Cause of the Backward and Forward Flight Occasionally Observed," *PESL* (1921): xii–xxvii, at xvi.
58. "Letter from a Fellow Interned in Germany," *PESL* (1917): xlv.
59. Colonel Manders, "A Day in the ___," *ERJV* 27 (1915): 149–150, at 149.
60. "Lepidopterology," *ERJV* 27 (1915): 203–205, at 203.
61. "Librarian's Report," *PESL* (1915): cxxvii.
62. Jordan to Sjöstedt, Sept. 8, 1914, Sjöstedt Correspondence, CfHS, RSAS.
63. Hartert to W. E. Clyde Todd, Oct. 7, 1915, TM2/30, NHM.
64. "Annual Meeting," *PESL* (1916): cxxxv–cxxxviii, at cxxxvii.
65. "Resolution on the Closing of the Natural History Museum," *PESL* (1916): ii.
66. "The Annual Meeting," *PESL* (1917): ciii–cviii, at cv.
67. C. G. Gahan, "The President's Address," *PESL* (1917): cix–cxxii, at cxxii.
68. Anonymous letter, *Times*, Feb. 4, 1916.
69. Jordan to Sjöstedt, Oct. 13, 1914, Sjöstedt Correspondence, CfHS, RSAS.
70. Jordan to Gestro, Aug. 27, 1914, MSNG.
71. Jordan to Sjöstedt, Dec. 29, 1915, Sjöstedt Correspondence, CfHS, RSAS.
72. T. A. Chapman, "Lepidopterology," *ERJV* (1916): 186–188, at 187.
73. Gilbert J. Arrow, "Walther Horn," *EMM* 75 (1939): 204–205, at 205.
74. George F. Hampson, "The Determination of Generic Types in the Lepidoptera," *Entomological News* 28 (1917): 463–467.
75. Walter Rothschild, "On the Naming of Local Races, Subspecies, Aberrations, Seasonal Forms, etc." *TESL* (1918): 115–116.
76. W. J. Holland, "Shall Writers upon the Biological Sciences Agree to Ignore Systematic Papers Published in the German Language since 1914?" *Science* 48 (1918): 469–471, at 469.
77. Hemming, "Jordan and Zoological Nomenclature," 27.
78. Jordan to Sjöstedt, Sept. 8, 1914, Sjöstedt Correspondence, CfHS, RSAS.
79. *Verhandlungen des V. Internationalen Zoologen-Congresses zu Berlin, 12–16 August 1901*, ed. Paul Matschie (Jena: G. Fischer, 1902), 76–77.
80. Ferguson, *The House of Rothschild*, 436–437 and 454.
81. Jordan to Moss, Sept. 2, 1914, MSS MIL C 7:7, Entomology Library, NHM.
82. Walter Rothschild, "Supplemental Notes to Mr. Charles Oberthur's *Fauna des Lépidoptères de la Barbarie*, with Lists of the Specimens Contained in the Tring Museum," *NZ* 24 (1917): 61–120, at 66.
83. Jordan to Raphael Gestro, Aug. 27, 1914, MSNG.
84. Jordan to Sjöstedt, Dec. 29, 1915, Sjöstedt Correspondence, CfHS, RSAS.

85. Jordan to Sjöstedt, June 6, 1916, Sjöstedt Correspondence, CfHS, RSAS.
86. Walsingham to Busck, Aug. 20, 1916, United States National Museum, Division of Insects, Correspondence, 1909–1963, Box 48, RU 140, SIA.
87. Jordan, "On the Species of *Somabrachys* in the Tring Museum," *NZ* 23 (1916): 350–358, at 350.
88. N. C. Rothschild to Sjöstedt, Jan. 18, 1915, Sjöstedt Correspondence, CfHS, RSAS.
89. C. S. Baker to Sjöstedt, Feb. 8, 1919, Sjöstedt Correspondence, CfHS, RSAS.
90. N. D. Riley to Clark, Dec. 6, 1946, Austin Clark Papers, Box 10, RU 7183, SIA.
91. C. S. Baker to Sjöstedt, Feb. 8, 1919, Sjöstedt Correspondence, CfHS, RSAS.
92. Quoted in Giuliano Pancaldi, "Scientific Internationalism and the British Association," in *The Parliament of Science: The British Association for the Advancement of Science, 1831–1981*, ed. R. MacLeod and P. Collins (London: Science Reviews, 1981), 145–169, 162.
93. "Current Notes and Short Notices," *ERJV* 27 (1915): 211–214, at 211–212.
94. Jordan to Moss, Dec. 29, 1915, MSS MIL C 7:7, Entomology Library, NHM.
95. Ferguson, *The House of Rothschild*, 455.
96. Jordan to Moss, Aug. 31, 1916, MSS MIL C 7:7, Entomology Library, NHM.
97. Schaus to C. H. Lankester, Nov. 5, 1923, William Schaus Papers, 1917–1939, Box 7, RU 7100, SIA.
98. Skinner to Schaus, May 26, 1920, William Schaus Papers, 1917–1939, Box 8, RU 7100, SIA.
99. Schaus to George A. Ehrman, May 16, 1925, William Schaus Papers, 1917–1939, Box 3, RU 7100, SIA.
100. Schaus to J. T. Mason, Oct. 21, 1924, William Schaus Papers, 1917–1939, Box 6, RU 7100, SIA.
101. Miriam Rothschild, *Dear Lord Rothschild: Birds, Butterflies, and History* (London: Hutchinson, 1983), 93.
102. Jordan to Moss, July 3, 1917, MSS MIL C 7:7, Entomology Library, NHM.
103. W. Rothschild to Hartert, July 21, 1917, TM2/33, NHM.
104. Rothschild, "Supplemental Notes to Mr. Charles Oberthur's *Fauna des Lépidoptéres de la Barbarie*," 86.
105. W. Rothschild to Hartert, June 30, 1918, TM2/34, NHM.
106. Jordan to Moss, July 3 and Oct. 12, 1917, MSS MIL C 7:7, Entomology Library, NHM.
107. Melou's lawyer to Tring, Sept. 12, 1918, TM1/123, NHM.
108. Jordan to Hartert, Mar. 15, 1917, TM2/33, NHM.
109. Jordan to Hartert, Aug. 27, 1917, TM2/33, NHM.
110. Speiser to unnamed addressee, probably Sjöstedt, July 24, 1920, Sjöstedt Correspondence, CfHS, RSAS.
111. Herbert Gold, "The Art of Fiction XL: Vladimir Nabokov," *Paris Review* 41 (1967): 92–111.
112. M. Rothschild, personal communication, 2002.
113. "Regarding the Collections and Property of Peters of Göttingen, Collected in German East Africa, 1919." KJC #2, NHM.
114. E. Study to Jordan, Mar. 5, 1923, TM2/47, NHM.
115. Horn to Jordan, Jan. 31, 1924, TM2/50, NHM.

116. Horn to Jordan, Feb. 26, 1924, TM2/50, NHM.

117. G. J. Arrow, "Walther Horn," *EMM* 74 (1939): 204–205, at 205.

118. Jordan to Louwerens, July 8, 1946, KJC #4, NHM.

119. James A. G. Rehn to Sjöstedt, Sept. 14, 1920, Sjöstedt Correspondence, CfHS, RSAS.

120. E. Study to Jordan, Mar. 5, 1923, TM2/47, NHM.

121. Jordan to Holland, Apr. 27, 1921, Director's Correspondence, pt. 2, 1897–1961, CMNH.

122. Max Draudt to Schaus, Feb. 18, 1922, William Schaus Papers, Box 3, RU 7100, SIA.

123. Jordan to Ehlers, Aug. 29, 1919, Ernst Ehlers Papers, Manuscript Division, University Library, GA.

124. Jordan to Zerny, Dec. 13, 1920, AWNM.

125. Hampson to Schaus, Feb. 19, 1919, William Schaus Papers, Box 4, RU 7100, SIA.

126. A. G. Cock, "Chauvinism and Internationalism in Science: The International Research Council, 1919–1926," *Notes and Records of the Royal Society of London* 37 (1983): 249–288, at 249–251 and 257.

127. *Verhandlungen des III Internationalen Entomologen-Kongresses, Zürich, 19.–25. Juli 1925*, ed. Walther and Karl Jordan, 593–627, vol. 2 (Zurich: Druck von G. Uschmann, 1926), 11–12.

128. Hemming, "Jordan and Zoological Nomenclature," 27.

129. John Merton Aldrich to Major W. S. Patton, Jan. 17, 1923, John Merton Aldrich Papers, Box 1, RU 7305, SIA.

130. Rothschild, "Jordan—A Biography," 4.

131. Rothschild, *Dear Lord Rothschild*, 147.

132. Miriam Rothschild, personal communication, 2003.

133. Rothschild, *Dear Lord Rothschild*, 147.

134. Jordan to Handlirsch, Apr. 9, 1925, AWNM.

135. Jordan to Gestro, Dec. 15, 1924, MSNG.

136. Jordan to N. D. Riley, undated, DF306, NHM.

137. Jordan to Handlirsch, Apr. 9, 1925, AWNM.

138. Jordan to Holland, June 4, 1925, Director's Correspondence, pt. 2, 1897–1961, CMNH.

139. N. D. Riley, "H. E. K. Jordan," in E. T. Williams and H. M. Palmer, eds., *Dictionary of National Biography 1951–1960* (Oxford: Oxford University Press, 1971), 560–562, at 561.

140. Jordan to N. D. Riley, undated, DF306, NHM.

141. *Verhandlungen des III. Internationalen Entomologen-Kongresses, Zürich*, 29.

142. *Verhandlungen des III. Internationalen Entomologen-Kongresses, Zürich*. 26–27.

143. "Section II: Morphology and Anatomy," *Proceedings of the Second International Congress of Entomology, Oxford*, 95–96, at 96.

144. H. J. Turner, "Reviews and Notices of Books: *Sex-linked Inheritance in Drosophila*, by T. H. Morgan and C. B. Bridges," *ERJV* 29 (1917): 196–200.

145. Poulton, "Mendelian Heredity in Relation to Selection," *PESL* (1917): lxxxv–lxxxix.

146. Poulton, "The 'Fruit-fly' Drosophila and the Inheritance of Small Variations," *PESL* (1918): xxii.

147. F. A. E. Crew, "Recollections of the Early Days of the Genetical Society," in *The Genetical Society: The First Fifty Years*, ed. John Jinks, 9–15 (Edinburgh: Oliver & Boyd, 1969), 13–14.

148. Poulton, "The Conception of Species as Interbreeding Communities," *Proceedings of the Linnean Society of London* 150 (1938): 225–226, at 226.

149. Richard Goldschmit, *In and Out of the Ivory Tower* (Seattle: University of Washington Press, 1960), 189.

150. R. K. Webb, *Modern England: From the Eighteenth Century to the Present*, 2nd ed. (New York: Harper & Row, 1980), 386.

151. T. A. Cockerell to William Schaus, Dec. 31, 1922, William Schaus Papers, 1917–1939, Box 3, RU 7100, SIA.

152. Miles Moss, "Sphingidae of Para, Brazil: Early Stages, Food-plants, Habits, etc.," *NZ* 27 (1920): 333–415, at 354.

153. T. A. Cockerell to William Schaus, Dec. 31, 1922, William Schaus Papers, 1917–1939, Box 3, RU 7100, SIA.

CHAPTER 6: TAXONOMY IN A CHANGED WORLD

1. S. A. Neave, "The Relations between Mankind and the Insect World," *PRESL* 10 (1936): 112–121, at 119.

2. E. Ray Lankester, "Inaugural Address before the BAAS," *Science* 24 (1906): 225–238, at 237.

3. T. A. Cockerell, "Recollections of English Naturalists," *Natural History* 19 (1919): 325–329, at 325.

4. Neave, "The Relations between Mankind and the Insect World," 117.

5. Frank M. Turner, "Public Science in Britain, 1880–1919," *Isis* 71 (1980): 589–608, at 593.

6. Turner, "Public Science in Britain," 596.

7. Austin H. Clark, "Selling Entomology," *Scientific Monthly* 32 (1931): 527–536, at 527.

8. "Centenary Meeting," *PRESL* 8 (1933): 25–94, at 42.

9. L. O. Howard, "The Recent Progress and Present Conditions of Economic Entomology," *Proceedings of the Seventh International Zoological Congress, Boston, 19–24 August 1907* (Cambridge: Cambridge University Press, 1912), 574–600, at 577.

10. "Societies," *Entomologist* 27 (1894): 199–204, at 202.

11. W. Conner Sorensen, *Brethren of the Net: American Entomology, 1840–1880* (Tuscaloosa: University of Alabama Press, 1995), 86.

12. S. J. Capper, "Societies: Lancashire and Cheshire Entomological Society President's Address," *ERJV* 4 (1893): 56–65, at 61.

13. W. E. Sharp, "The New Entomology: Part II," *Entomologist* 27 (1894): 110–116, at 114.

14. "The Banquet," *Proceedings of the Second International Congress of Entomology, Oxford, August, 1912*, ed. K. Jordan and H. Eltringham (London: Hazell, Watson & Viney, 1914), 129–143, at 129.

15. J. F. Clark, "Bugs in the System: Insects, Agricultural Science, and Professional Aspirations in Britain, 1890–1920," *Agricultural History* 75 (2001): 83–114, at 95–96.

16. Peter Alter, *The Reluctant Patron: Science and the State in Britain, 1850–1920* (Oxford: Berg, 1987), 64.

17. John J. McKelvey, *Man against Tsetse: Struggle for Africa* (Ithaca, NY: Cornell University Press, 1973), 107.

18. Clark, "Bugs in the System," 97, 100 and 109.

19. James J. Walker, "The President's Address," *PESL* (1920): cviii–cxxxviii, at cviii.

20. Busck to Walsingham, Feb. 4, 1916, United States National Museum, Division of Insects, Correspondence, 1909–1963, Box 48, RU 140, SIA.

21. Sent to Aldrich by Rennie Wilbur Doane, undated, John Merton Aldrich Papers, 1916–1934, Box 1, RU 7305, SIA.

22. Tutt, "Address by the Vice-President to the City of London Entomological and Natural History Society," *ERJV* 6 (1895): 59–69 at 64.

23. Schaus to Holland, Oct. 4, 1923, William Schaus Papers, 1917–1939, Box 5, RU 7100, SIA.

24. Schaus to Howard, Feb. 7, 1924, William Schaus Papers, 1917–1939, Box 5, RU 7100, SIA.

25. Schaus to M. Draudt, July 18, 1928, William Schaus Papers, 1917–1939, Box 3, RU 7100, SIA.

26. Neave, "The Relations between Mankind and the Insect World," 117.

27. A. J. T. Janse to Schaus, Feb. 19, 1923, William Schaus Papers, 1917–1939, Box 6, RU 7100, SIA.

28. W. T. Calman, "The Taxonomic Outlook in Zoology," *Science* 72 (1930): 279–284, at 280.

29. Fred Muir to E. D. Ball, Apr. 30, 1921, Elmer Darwin Ball Papers, 1915–1938, Box 3, RU 7121, SIA.

30. Jordan and Horn, "Der Kongress," *Verhandlungen des III Internationalen Entomologen-Kongresses, Zürich, 19.–25. Juli 1925*, ed. Walther Horn and Karl Jordan (Zürich: Druck von G. Uschmann, 1926), 25–54, at 25, 26, and 30.

31. Jordan, "On *Xenopsylla* and Allied Genera of Siphonaptera," *Verhandlungen des III. Internationalen Entomologen-Kongresses, Zürich*, 593–633, at 594.

32. "Zur Vorgeschichte," *Verhandlungen des III. Internationalen Entomologen-Kongresses, Zürich*, 12.

33. *Verhandlungen des III. Internationalen Entomologen-Kongresses, Zürich*, 29.

34. H. S. Fremlin, "The Necessity for More General Education in Entomology," *Verhandlungen des III. Internationalen Entomologen-Kongresses, Zürich*, 344–351, at 351.

35. Howard, contributing to the General Session, in *Verhandlungen des III. Internationalen Entomologen-Kongresses, Zürich*, 37.

36. Jordan, "On *Xenopsylla* and Allied Genera of Siphonaptera," 594, 596.

37. Dampf to Schaus, Nov. 2, 1928, William Schaus Papers, 1917–1939, Box 3, RU 7100, SIA.

38. R. T. Leiper, "Some Outstanding Questions in Medical Entomology," and discussion, *Verhandlungen des III. Internationalen Entomologen-Kongresses, Zürich*, 631–632.

39. Jordan "On *Xenopsylla* and Allied Genera of Siphonaptera," 596.

40. Ball to E. P. Van Duzee, Sept. 1, 1927, Elmer Darwin Ball Papers, 1915–1938, Box 4, RU 7121, SIA.

41. Van Duzee to Ball, Aug. 17, 1935, Elmer Darwin Ball Papers, 1915–1938, Box 4, RU 7121, SIA.

42. Walther Horn, "Über die Notlage der systematischen Entomologie, mit besonderer Berücksichtigung der Verhältnisse in Deutschland, und Reformvorschläge," *Verhandlungen des III. Internationalen Entomologen-Kongresses, Zürich*, 53–69.

43. Circular for the Zurich Congress, CfHS, RSAS.

44. *Verhandlungen des III: Internationalen Entomologen-Kongresses, Zürich*, 45–49.

45. Alison Kraft, "Pragmatism, Patronage, and Politics in English Biology: The Rise and Fall of Economic Biology, 1904-1920," *Journal of the History of Biology* 37 (2004): 213-258.

46. "Discussion on the 'Place of the Systematist in Applied Biological Work,'" *Annals of Applied Biology* 13 (1926): 466-485, at 467 and 483.

47. "Discussion on the 'Place of the Systematist in Applied Biological Work,'" 475-476.

48. "Discussion on the 'Place of the Systematist in Applied Biological Work,'" 479-480.

49. Walter Rothschild and Jordan, "A Revision of the American Papilios," NZ 13 (1906): 411-754, at 429.

50. G. W. Herrick, "The Fourth International Congress of Entomology," *Science* 68 (1928): 237-244, at 244.

51. Holland to Jordan, Mar. 6, 1925, Director's Correspondence, pt. 2, 1897-1961, CMNH.

52. Holland to Jordan, Jan. 12, 1928, Director's Correspondence, pt. 2, 1897-1961, CMNH.

53. Jordan to Holland, Jan. 30, 1928, Director's Correspondence, pt. 2, 1897-1961, CMNH.

54. J. Jablonowski, "The Black Locust-Tree-Scale, Lecanium robiniarum Dougl., and the European Corn Borer, Pyrausta nubilalis Hubn., a Biological Parallel," *Transactions of the Fourth International Congress of Entomology, Ithaca, August 1928*, ed. Jordan and Walther Horn (Naumburg A/Saale: G. Pätz, 1929-30), 455-462, at 462.

55. Jablonowski, "The Black Locust-Tree Scale," at 462.

56. Royal N. Chapman, "The Measurement of the Effects of Ecological Factors," *Transactions of the Fourth International Congress of Entomology, Ithaca*, 408-411, at 408-409.

57. "Program," *Proceedings of the Fourth International Congress of Entomology, Ithaca, August 1928*, ed. Jordan and Walther Horn (Naumburg A/Saale: G. Pätz, 1929-30), 51-78, at 73.

58. "Program," *Proceedings of the Fourth International Congress of Entomology, Ithaca*, 71-72.

59. J. P. Kryger, "Some Remarks on the Keys of the European Chalcids," *Transactions of the Fourth International Congress of Entomology, Ithaca*, 1020-1023, at 1020.

60. Horn, "The Future of Insect Taxonomy," *Transactions of the Fourth International Congress of Entomology, Ithaca*, 34-51.

61. "Forum on Problems of Taxonomy: Collections," *Transactions of the Fourth International Congress of Entomology, Ithaca*, 797-800, at 797.

62. Holland, "The Mutual Relations of Museums and Expert Specialists," *Transactions of the Fourth International Congress of Entomology, Ithaca*, 278-285.

63. J. B. Corporaal, "Forum on Problems of Taxonomy: Determinations," *Transactions of the Fourth International Congress of Entomology, Ithaca*, 795-796 at 795.

64. "Cornell Is Host to Entomologists," *New York Times*, Aug. 12, 1928.

65. Jordan, "On Some Problems of Distribution, Variability, and Variation in North American Siphonaptera," *Transactions of the Fourth International Congress of Entomology, Ithaca*, 489-499, at 491.

66. Horn to Sjöstedt, Oct. 27, 1922, Sjöstedt Correspondence, CfHS, RSAS.

67. Horn to Jordan, Feb. 26, 1924, TM2/50, NHM.

68. Horn, "Appendix: Entomological Institute for International Service," *Proceedings of the Fourth International Congress of Entomology, Ithaca, August 1928*, 82-83, at 82.

69. Memorandum related to preliminary program, Fourth ICE, Ithaca, NY, Aug 12-18, 1928, CfHS, RSAS.

70. Untitled, minutes of executive committee, annotated by Jordan, DF306/3, NHM.

71. Jordan to Frederick Laing, Dec. 31, 1928, DF304, Department of Entomology, Sectional Correspondence, NHM.

72. Jordan to Holland, Mar. 28, 1929, Director's Correspondence, pt. 2, 1897–1961, CMNH.

73. Jordan to Holland, July 1, 1929, Director's Correspondence, pt. 2, 1897–1961, CMNH.

74. Malloch, Illinois State Natural History Survey Division, to Aldrich, Oct. 6, 1920 (the chief he refers to is Stephen A. Forbes), John Merton Aldrich Papers, 1916–1934, Box 1, RU 7305, SIA.

75. James Edward Collin, "The President's Address," *PESL* 3 (1929): 102–111, at 110.

76. Jordan to Holland, July 1, 1929, Director's Correspondence, pt. 2, 1897–1961, CMNH.

77. Aldrich to Townsend, Feb. 24, 1933, John Merton Aldrich Papers, Box 1, RU 7305, SIA.

78. Aldrich to Townsend, May 9, 1933, John Merton Aldrich Papers, Box 1, RU 7305, SIA.

79. Jordan to Moss, Jan. 3, 1932, MSS MIL C 7:7, Entomology Library, NHM.

80. Fred Muir to Jordan, Apr. 28, 1930, TM2/68, NHM.

81. Neave, "The Relations between Mankind and the Insect World," 119.

82. Roger Verity, "On the Necessity of a Revision of the Rules of Entomological Nomenclature Concerning Groups of Lower Rank than the Specific One," *Transactions of the Fourth International Congress of Entomology, Ithaca*, 479–480, at 479.

83. Clarence Hamilton Kennedy, "The Theory of Nomenclature," *Transactions of the Fourth International Congress of Entomology, Ithaca*, 665–670.

84. Francis Hemming, "Karl Jordan and Zoological Nomenclature," *TRESL* 107 (1955): 25–32, at 28.

85. Muir to Jordan, Apr. 28, 1930, TM2/68, NHM.

86. Jordan, "A Survey of the Classification of the American Species of *Ceratophyllus s. lat.*" *NZ*, 39 (1933): 70–79, at 70.

87. Jordan to Holland, Dec. 19, 1930, Director's Correspondence, pt. 2, 1897–1961, CMNH.

88. C. W. Stiles, "Is an International Zoological Nomenclature Practicable?" *Science* 73 (1931): 349–354.

89. C. W. Stiles, "The Future of Zoological Nomenclature, with an Appendix: History of Rules re Designation of Genotypes," *Transactions of the Fourth International Congress of Entomology, Ithaca*, 622–645, at 622.

90. J. Harold Matteson, "International Organization Needed," *Ve Congrès International d'Entomologie, Paris, 18–24 juillet 1932*, ed. Lucien Berland and René Gabriel Jeannel (Paris: Secrétariat du Congrès, 1933), 199–201.

91. R. K. Webb, *Modern England: From the Eighteenth Century to the Present*, 2nd ed. (New York: Harper & Row, 1980), 542.

92. Jordan to Dr. Wickliffe Rose, Jan. 11, 1927, International Education Board, Series !, Box 29, Folder 412, RAC.

93. Jordan, "The President's Address, 1930," *PESL* 5 (1931): 128–142, at 128.

94. Jordan, "Special Meeting, 7th May 1930: Opening of New Meeting Room," *PESL* 5 (1930): 31–37, at 35.

95. R. J. Fromols-Rakowski to Hartert, Mar. 9, 1925, TM2/53, NHM.

96. Wm. Forbes to Jordan, Aug. 19, 1931 and Jordan to Forbes, Sept. 10, 1931, TM2/69, NHM Archives.

97. F. C. R. Jourdain to Alexander Wetmore, Mar. 19, 1932, Alexander Wetmore Papers, Box 31, Folder B, RU 7006, SIA.

98. Walter Rothschild to L. C. Sanford, Sept. 18, 1931, 1209 Rothschild, Folder Jan.–Feb. 1932, AMNH.

99. Sanford to H. Fairfield Osborn, Nov. 25, 1931, 1209 Rothschild, Folder Jan.–Feb. 1932, AMNH.

100. Sanford to George H. Sherwood, Feb. 8, 1932, 1209 Rothschild, Folder Jan.–Feb. 1932, AMNH.

101. A. R. Wallace, *The Revolt of Democracy* (London: Cassell, 1913), 19.

102. Osborn to I. Van Meter, Nov. 28, 1932, *Time Magazine*, 1209 Rothschild, Folder June–Dec. 1932, AMNH.

103. Jordan to Holland, Dec. 23, 1931, Director's Correspondence, pt. 2, 1897–1961, CMNH.

104. Jordan to N. D. Riley, Jan. 14, 1946, KJC #5, NHM.

105. Miriam Rothschild, *Dear Lord Rothschild: Birds, Butterflies, and History* (London: Hutchinson, 1983), 92.

106. Walter Rothschild to Robert Cushman Murphy, Feb. 5, 1932, 1209 Rothschild, Folder Jan.–Feb. 1932, AMNH.

107. Jordan to Moss, Jan. 3, 1932, MSS MIL C 7:7, Entomology Library, NHM.

108. Jordan to Holland, Apr. 27, 1932, Director's Correspondence, pt. 2, 1897–1961, CMNH.

109. Jordan to Moss, Jan. 3, 1932, MSS MIL C 7:7, Entomology Library, NHM.

110. Jordan to Holland, Apr. 27, 1932, Director's Correspondence, pt. 2, 1897–1961, CMNH.

111. Rothschild, *Dear Lord Rothschild*, 304.

112. Murphy to Sherwood, Feb. 8, 1932, 1209 Rothschild, Folder Jan.–Feb. 1932, AMNH.

113. Murphy to Sherwood, Feb. 18, 1932, 1209 Rothschild, Folder Jan.–Feb. 1932, AMNH.

114. Murphy to Sherwood, May 2, 1932 and Sanford to Sherwood, Mar. 4, 1932, AMNH.

115. Sherwood to Osborn, Mar. 7, 1932, 1209 Rothschild, Folder Mar. 1932, AMNH.

116. D. W. Snow, "Robert Cushman Murphy and His 'Journal of the Tring Trip,'" *Ibis* 115 (1973): 607–611, at 609.

117. "Hard Times Force Rothschilds to Sell Rare Bird Collection," *New York Times*, Mar. 11, 1932.

118. Percy R. Lowe to Murphy, Mar. 16, 1932, 1209 Rothschild, Folder Mar. 1932, AMNH.

119. F. M. Chapman to Murphy, Apr. 20, 1932, 1209 Rothschild, Folder Mar. 17–May 31, 1932, AMNH.

120. *Museums Journal* 32 (1932): 3.

121. Murphy to Sherwood, Mar. 24, 1932, 1209 Rothschild, Folder Mar. 17–May 31, 1932, AMNH.

122. "Lord Rothschild's Birds: Ornithologists and Sale to America: British Museum Loss: Tring Curator's Statement on the Transaction," *London Morning Post*, Mar. 16, 1932.

123. "Museum No One Can Look After: Ld. Rothschild's Sale to U.S." *Daily Mail*, Mar. 12, 1932.

124. "Hard Times Force Rothschilds to Sell Rare Bird Collection," *New York Times*, Mar. 11, 1932.

125. "Lord Rothschild's Birds," Mar. 16, 1932.

126. G. F. Herbert Smith to Jordan, Mar. 16, 1932, TM2/69, NHM.
127. Jordan to Smith, Mar. 18, 1932, TM2/69, NHM.
128. Rothschild, *Dear Lord Rothschild*, 125.
129. *Field—the Country Newspaper*, Mar. 26, 1932.
130. Phyllis Barclay-Smith, "The British Contribution to Bird Protection," *Ibis* 101 (1959): 115–122, at 115.
131. Murphy to Sherwood, Jan. 23, 1932, 1209 Rothschild, Folder Jan.–Feb. 1932, AMNH.
132. Sanford to Osborn, Nov. 25, 1931, 1209 Rothschild, Folder Jan.–Feb. 1932, AMNH.
133. Jordan, "The President's Address, 1929," *PESL* 4 (1930): 128–141, 129.
134. Jordan, "The President's Address, 1930," 133.
135. Austin H. Clark, "Selling Entomology," *Scientific Monthly* 32 (1931): 527–536, at 527.
136. Erwin Stresemann, *Ornithology: From Aristotle to the Present* (Cambridge, MA: Harvard University Press, 1975), 268.
137. Murphy to Sherwood, Mar. 10, 1932, AMNH.
138. Hartert to Murphy, Mar. 27, 1932, AMNH.
139. "Discussion on the 'Place of the Systematist in Applied Biological Work,'" 484.

CHAPTER 7: THE RUIN OF WAR AND THE SYNTHESIS OF BIOLOGY

1. J. C. F. Fryer, "The President's Remarks," *PRESL* 3 (1938): 58–61, at 60.
2. Jordan, "Rapport du Secrétaire du Comité Exécutif," in *Ve Congrès International d'Entomologie, Paris, 18–24 juillet 1932*, ed. Lucien Berland, and René Gabriel Jeannel (Paris: Secrétariat du Congrès, 1933), 56–59.
3. Horn, "Gedanken über Entomologische Systematik, Mathematik, Genetik, Phylogenie und Metaphysik," in *Ve Congrès International d'Entomologie, Paris*, 134.
4. Jordan to Moss, Apr. 4, 1932, MSS MIL C 7:7, Entomology Library, NHM.
5. Jordan to Orazio Querci, Apr. 18, 1933, TM2/70, NHM.
6. A. M. Mistikawy, "The Locust Problem in Egypt," *Ve Congrès International d'Entomologie, Paris*, 617–625.
7. C. F. M. Swynnerton, "The Tsetse Flies of East Africa: A First Study of Their Ecology, with a View to Their Control," *TRESL* 84 (1936): 1–579, at 429 and 512.
8. Angus Buchanan, *Exploration of Air: Out of the World North of Nigeria* (New York: E. P. Dutton, 1922), xxi, xxiii, 1–2, 5, 38 and 39.
9. A. R. Wallace, "Imperial Might and Human Right," *Clarion* (1900): 230.
10. Janet Browne, "Biogeography and Empire," in N. Jardine, J. A. Secord, and E. C. Spary, *Cultures of Natural History* (Cambridge: Cambridge University Press, 1996), 305–321, at 305.
11. Robert S. Jansen, "Two Paths to Populism: Explaining Peru's First Episode of Populist Mobilization" (UC Irvine: Center for the Study of Democracy, 2008), 18.
12. C. H. T. Townsend to Aldrich, Feb. 17, 1931, John Merton Aldrich Papers, 1916–1934, Box 1, RU 7305, SIA.
13. Alf. Dampf to Aldrich, Apr. 15, 1925, John Merton Aldrich Papers, 1916–1934, Box 2, RU 7305, SIA.
14. Jordan to Moss, Mar. 31, 1933, MSS MIL C 7:7, Entomology Library, NHM.

15. Jordan to Registrar, Entomological Society of London, Nov. 3, 1937, ESL.
16. Jordan, "Dr. Karl Jordan's Expedition to South-West Africa and Angola: Narrative," *NZ* 40 (1936): 17–62, at 29.
17. Jordan, "Dr. Karl Jordan's Expedition," 50.
18. Miriam Rothschild, "Jordan: A Biography," *TRESL* 107 (1955): 1–9, at 7.
19. Ethelwynn Trewavas, "Dr. Jordan's Expedition to South-West Africa and Angola: The Freshwater Fishes," *NZ* 40 (1936–1937): 63–74.
20. Jordan, "Dr. Jordan's Expedition," 62.
21. Jordan, "On Mechanical Selection and Other Problems," *NZ* 3 (1896): 426–525, at 432.
22. Jordan, "The President's Address, 1930," *PESL* 5 (1931): 128–142, at 138.
23. Jordan, "Where Subspecies Meet," *VI Congreso Internacional de Entomología, Madrid 6–12 de septiembre de 1935* (Madrid: Laboratorio de Entomología del Museo Nacional de Ciencias Naturales, 1940), 145–151, at 151.
24. Jordan, "Where Subspecies Meet," *NZ* 41 (1938): 103–111, at 105.
25. Theodosius Dobzhansky, "The Origin of Geographical Varieties in Coccinellidae," *Transactions of the Fourth International Congress of Entomology, Ithaca, August 1928*, ed. Jordan and Walther Horn (Naumburg A/Saale: G. Pätz, 1929–30), 536.
26. William B. Provine, *Sewall Wright and Evolutionary Biology* (Chicago: University of Chicago Press, 1986), 330–332.
27. H. W. Bates, *The Naturalist on the Amazons* (London: John Murray, 1892, reprint; originally published 1863), 351.
28. E. O. Essig, "A Sketch History of Entomology," *Osiris* 2 (1936): 80–123, at 106.
29. Mark B. Adams, "The Founding of Population Genetics: Contributions of the Chetverikov School, 1924–34," *Journal for the History of Biology* 1 (1968): 23–39, at 23.
30. N. Krementsov, "Th. Dobzhansky and Russian Entomology: The Origin of His Ideas on Species and Speciation," in *The Evolution of Theodosius Dobzhansky: His Life and Thought in Russia and America*, ed. Mark B. Adams (Princeton, NJ: Princeton University Press, 1994), 31–48, at 33–34.
31. Provine, *Sewall Wright*, 333.
32. Theodosius Dobzhansky, "Geographical Variation in Lady-Beetles," *American Naturalist* 67 (1933): 97–126, at 124.
33. Dobzhansky, "Geographical Variation in Lady-Beetles," 100.
34. Jordan, "The Species Problem as Seen by a Systematist," 244.
35. Jordan, "Where Subspecies Meet," *NZ* 41 (1938): 103–111, at 103.
36. See Robert Kohler, *Landscapes and Labscapes: Exploring the Lab-Field Border in Biology* (Chicago: University of Chicago Press, 2002).
37. Paul Marchal, "Discours Inaugural," *Ve Congrès International d'Entomologie, Paris*, 37–47.
38. Zopp to Jordan, undated, but c.1947, KJC#5, NHM.
39. William Bateson, *Materials for the Study of Variation, Treated with Especial Regard to Discontinuity in the Origin of Species* (London: MacMillan, 1894), 574.
40. Imms to Snodgrass, Sept. 16, 1935, Robert E. Snodgrass Papers, Box 1, RU 7132, SIA.
41. J. Chester Bradley, "At the ICE and Zoology, of 1935," *Entomological News* 47 (1936): 28–34.

42. Richard Goldschmit, *In and Out of the Ivory Tower* (Seattle: University of Washington Press, 1960), 294.
43. Jordan, "Spolia Mentawiensia: Papilionidae," *NZ* 38 (1937): 315–330, at 319.
44. Jordan to Horn, Oct. 12, 1936, TM2/72, NHM.
45. Jordan to Avinoff, June 15, 1937, TM2/73, NHM.
46. Avinoff to Jordan, Oct. 1, 1937, TM2/73, NHM.
47. Jordan to Farrell, Aug. 30, 1944, KJC #4, NHM.
48. Jordan to M. Aigner, Nov. 3, 1951, KJC#1, NHM.
49. Jordan to Moss, Jan. 11, 1936, MSS MIL C 7:7, Entomology Library, NHM.
50. Jordan to Moss, Oct. 8, 1936, MSS MIL C 7:7, Entomology Library, NHM.
51. Jordan to M. Aigner, Nov. 3, 1951, KJC#1, NHM.
52. Jordan to Hans Eggers, May 18, 1946, KJC #5, NHM.
53. Jordan to C. Tate Regan, July 30, 1937, TM2/73, NHM.
54. Mayr to Jordan, Oct. 9, 1937, TM2/73, NHM.
55. A. J. T. Janse (Pretoria) to Schaus, Jan. 18, 1928, and Sept. 29, 1925, William Schaus Papers, Series I, General Correspondence, Box 6, RU 7100, SIA.
56. Neave, "The Relations between Mankind and the Insect World," 118.
57. Stanley Gardiner to W. Rothschild, Nov. 4, 1931, TM2/69, NHM.
58. Perkin, *The Rise of Professional Society*, 25.
59. "Centenary Meeting," *PRESL* 8 (1933): 25–94, at 63.
60. Meinertzhagen to Wetmore, Jan. 22, 1944, Wetmore Papers, Box 40, Folder 5, RU 7006, SIA.
61. Jordan, "In Memory of Lord Rothschild," *NZ* 41 (1938): 1–16, at 10.
62. Jordan to Zerny, June 13, 1929, AWNM.
63. Jordan, "In Memory," 9–10.
64. Miriam Rothschild, *Dear Lord Rothschild: Birds, Butterflies, and History* (London: Hutchinson, 1983), 146.
65. Jordan to J. Drysdale, Aug. 20, 1936, TM2/72, NHM.
66. Goldschmidt, *In and Out of the Ivory Tower*, 39.
67. Jordan to Eugenio Giacomelli, Dec. 24, 1939, KJC #3, NHM.
68. Jordan to Charles Grist, Nov. 11, 1944, KJC #5, NHM.
69. John M. Geddes to Jordan, Apr. 23, 1949, KJC #2, NHM.
70. Jordan, "The President's Address, 1929," *PESL* 4 (1930): 128–141, at 134.
71. Jordan to Moss, Feb. 13, 1938, MSS MIL C 7:7, Entomology Library, NHM.
72. "Memorandum on the Tring Museum," Nov. 19, 1937, DF306, NHM.
73. N. D. Riley memo, Mar. 29, 1939, DF 306/5, NHM.
74. Hinston's Report on the Tring Museum, Nov. 22, 1937, DF306, NHM.
75. Jordan to Moss, Feb. 13, 1938, MSS MIL C 7:7, Entomology Library, NHM.
76. Jordan to John Levick, Esq., Oct. 11, 1938, TM2/74, NHM.
77. Jordan, "The Species Problem as Seen by a Systematist," *Proceedings of the Linnean Society of London* (1938): 241–247, at 245 and 247.
78. Jordan to Moss, Feb. 26, 1936, MSS MIL C 7:7, Entomology Library, NHM.
79. J. T. Bonner, *Lives of a Biologist: Adventures in a Century of Extraordinary Science* (Cambridge, MA: Harvard University Press, 2002), 122.

80. Victor Rothschild, *Meditations of a Broomstick* (London: Collins, 1977), 17.

81. Miriam Rothschild, *Nathaniel Charles Rothschild, 1877–1923* (Cambridge: Cambridge University Press, 1979), 9.

82. Niall Ferguson, *The House of Rothschild*, vol. 2: *The World's Banker, 1849–1999* (New York: Penguin, 2000), 468.

83. Suzanne Reeve, "Nathaniel Mayer Victor Rothschild," *The Biographical Memoirs of Fellows of the Royal Society* 39 (1994): 365–380.

84. Nicolai Krementsov, *International Science between the World Wars: The Case of Genetics* (London: Routledge, 2005).

85. Jordan to Moss, Feb. 13, 1938, MSS MIL C 7:7, Entomology Library, NHM.

86. Jordan to F. W. Wooddisse, July 19, 1939, KJC #3, NHM.

87. Jordan to Lucien J. Vinson, Apr. 14, 1939, KJC #4, NHM.

88. J. B. Poncelet to Jordan, Mar. 12, 1939, KJC #3, NHM.

89. Jordan to Zimmerman, July 20, 1939, KJC #4, NHM.

90. M. Rothschild to Jordan, Dec. 3, 1938, KJC #2, NHM.

91. M. Rothschild to Jordan, Jan. 13, 1939, KJC #2, NHM.

92. Horn to Jordan, Mar. 10, 1932, TM2/69, NHM.

93. Horn to Jordan, Dec. 24, 1931, TM2/69, NHM.

94. Horn to Jordan, Oct. 23, 1935, TM2/71, NHM.

95. Jordan to Dampf, May 26, 1936, TM2/72, NHM.

96. Jordan to Moss, Jan. 11, 1936, MSS MIL C 7:7, Entomology Library, NHM.

97. Report by J. C. F. Fryer to Ministry of Agriculture, MAF33/666, PRO.

98. Hering to Carl Heinrich, Mar. 12, 1948, Department of Entomology Records, 1909–1963, Box 23, RU 140, SIA.

99. Hering to Jordan, Sept. 29, 1946, KJC #5, NHM.

100. Jordan to P. Barclay-Smith, Aug. 6, 1937, TM2/73, NHM.

101. Jordan to Turner, May 18, 1937, TM2/73, NHM.

102. Turner to Jordan, May 19, 1937, TM2/73, NHM.

103. Horn to Jordan, July 30, 1935, TM2/71, and Jordan to Horn, Aug. 2, 1935, TM2/71, NHM.

104. Jordan to Horn, Feb. 23, 1937, and Horn to Jordan, Mar. 2, 1937, TM2/73, NHM.

105. Hering to Jordan, May 28, 1946, KJC #5, NHM.

106. Theodore W. Hower to Paul Engelhardt, Mar. 29, 1938, Paul Engelhardt Papers, Box 1, RU 7102, SIA.

107. Erwin Stresemann to Alexander Wetmore, Aug. 25, 1938, Wetmore Papers, Box 66, RU 7006, SIA.

108. Avinoff to Schaus, Oct. 14, 1938, Schaus Papers, Box 1, RU 7100, SIA.

109. Riley, "Jordan and the International Congresses of Entomology," *TRESL* 107 (1955): 15–24, at 22.

110. Report by J. C. F. Fryer to Ministry of Agriculture, MAF33/666, PRO.

111. Fryer, "The President's Remarks," 60.

112. Program of the 1938 IEC, Berlin, CO323/1620/1, PRO.

113. Ernst Mayr, personal communication, Nov. 25, 2003.

114. Rudolf Hans Braun to Jordan, Sept. 30, 1938, TM2/75, NHM.

115. Jordan to Braun, Nov. 26, 1938, TM2/75, NHM.
116. Miriam Rothschild, *Dear Lord Rothschild: Birds, Butterflies, and History*, 144, and Rothschild, "Jordan," 6.
117. Horn to Jordan, Oct. 1, 1938, TM2/74, NHM.
118. Adolf Hoffman to Jordan, Aug. 16, 1939, KJC #3, NHM.
119. Jordan to Mrs. Schweitzer-Junk, Aug. 17, 1941, KJC #3, NHM.
120. Jordan to Riley, Oct. 18, 1946, KJC #5, NHM.
121. Martin Schwartz to Jordan, Aug. 3, 1939, KJC #3, NHM.
122. Jordan to V. G. L. van Someren, Aug. 16, 1939, KJC #3, NHM.
123. Jordan to C. D. H. Carpenter, May 20, 1939, KJC #3, NHM.
124. Jordan to Mrs. M. E. Walsh, undated, KJC #3, NHM.
125. Jordan to Sjöstedt, Sept. 28, 1939, Sjöstedt Correspondence, CfHS, RSAS.
126. Jordan to Oscar Neumann, Apr. 7, 1942, KJC #3, NHM.
127. Riley to Jordan, Sept. 28, 1939, KJC #3, NHM.
128. Riley to Jordan, Oct. 17, 1940, KJC #3, NHM.
129. Jordan to A. G. Norris, June 28, 1941, KJC #3, NHM.
130. Jordan to Marshall, July 14, 1941, KJC #4, NHM.
131. Jordan to Riley, May 5, 1940, KJC #3, NHM.
132. Jordan to J. D. Sherman, July 12, 1940, KJC #4, NHM.
133. Jordan to Mayr, Apr. 29, 1942, KJC #3, NHM.
134. Jordan to W. H. T. Tams, Apr. 30, 1941, ESL.
135. Jordan to C. B. Williams, Oct. 7, 1940, KJC #3, NHM.
136. Jordan to Moss, May 4, 1942, MSS MIL C 7:7, Entomology Library, NHM.
137. Jordan to Mrs. M. E. Walsh, undated, KJC #3, NHM.
138. Jordan to Lanauer, Aug. 17, 1941, KJC #4, NHM.
139. Jordan to secretary, the British Council, Aug. 17, 1944, KJC #5, NHM.
140. Jordan to the president of the Entomological Society of London, Oct. 14, 1941, ESL.
141. W. Parkinson Curtis to B. S. Doubleday, Jan. 20, 1941, HLA.
142. John Smart, *A Handbook of the Identification of Insects of Medical Importance* (London: British Museum, 1943), v.
143. Jordan to Louwerens, Oct. 11, 1946, KJC #5, NHM.
144. Jordan to J. B. Corporaal, Sept. 25, 1946, KJC #5, NHM.
145. Jordan to Moss, Aug. 12, 1940, KJC #3, NHM.
146. Moss to Jordan, July 23, 1942, KJC #5, NHM.
147. Jordan to F. Zacher, Aug. 11, 1947, KJC #5, NHM.
148. Jordan to Moss, Aug. 12, 1940, KJC #3, NHM.
149. J. Stanley Gardiner to Jordan, Oct. 23, 1943, KJC #5, NHM.
150. Jordan, "On Some Problems of Distribution, Variability, and Variation in North American Siphonaptera," *Transactions of the Fourth International Congress of Entomology, Ithaca*, 489–499, at 489 and 491.
151. Jordan to A. G. Rehn, Dec. 12, 1943, KJC #4, NHM.
152. Jordan to W. G. Sheldon, Aug. 14, 1940, KJC #3, NHM.
153. Jordan to Riley concerning the Levick Collection, Dec. 21, 1937, DF306, NHM.
154. Jordan to Charles D. Radford, Dec. 15, 1948, KJC #5, NHM.

155. Jordan to Radford, Dec. 17, 1947, KJC #5, NHM.
156. Jordan to Moss, May 4, 1942, MSS MIL C 7:7, Entomology Library, NHM.
157. "Memorandum on the Tring Museum."
158. Jordan, "Notes on Arctiidae," *NZ* 23 (1916): 124–150, at 150.
159. Poulton to Jordan, Mar. 13, 1908, TM1/115, NHM.
160. Poulton, "A Hundred Years of Evolution," *Science* 74 (1919): 345–360, at 349.
161. Poulton, "The Conception of Species as Interbreeding Communities," *Proceedings of the Linnean Society of London* (1938): 225–226, at 226.
162. Jordan to Horn, Sept. 22, 1936, TM2/72, NHM.
163. Ernst Mayr, personal communication, Nov. 25, 2003.
164. Ernst Mayr, "My Dutch New Guinea Expedition, 1928," *NZ* 36 (1930–31): 20–26, at 22 and 25.
165. Ernst Mayr, "Reminiscences from the first Curator of the Whitney-Rothschild Collection," *BioEssays* 19 (1997): 175–179, at 177–178.
166. Walter Bock, "Ernst Mayr at 100: A Life Inside and Outside of Ornithology," *Auk* 121 (2004): 637–651, at 643.
167. Mayr to Jordan, Mar. 11, 1942, KJC #5, NHM.
168. K. Mather, "Genetics and the Russian Controversy," *Nature* 149 (1942): 427–430.
169. Jordan to Mayr, Apr. 4, 1942, KJC #3, NHM.
170. Mayr to Jordan, Mar. 25, 1942, KJC #3, NHM.
171. Mayr, *Systematics and the Origin of Species* (Cambridge, MA: Harvard University Press, 1999; originally published 1942), 1.
172. Mayr to Richard Meinertzhagen, Oct. 28, 1943, Ernst Mayr Papers, General Correspondence, 1931–1952, HUGFP 14.7, Box 2, HUA.
173. E. O. Wilson, *The Naturalist* (Washington DC: Island Press, 1994), 44.
174. Mayr, *Systematics and the Origin of Species*, 7.
175. Jordan, "Presidential Address," *Lepidopterists' News* 7 (1953): 3–4, at 4.
176. Mayr, *Systematics and the Origin of Species*, 4.
177. Mayr, *Systematics and the Origin of Species*, 6–7.
178. Julian Huxley, "Introductory: Towards the New Systematics," in *The New Systematics*, ed. Julian Huxley, 1–46 (Oxford: Clarendon Press, 1940), 38.
179. John Smart, "Entomological systematics examined as a practical problem," in Huxley, *The New Systematics*, 475–492, at 475, 478, 487.
180. Huxley, "Introductory: Towards a New Systematics," in *The New Systematics*, 20.
181. Zimmerman to Mayr, Sept. 25, 1942, Ernst Mayr Papers, General Correspondence, 1931–1952, HUGFP 14.7, Box 2, Folder 91, HUA.
182. Mayr, *Systematics and the Origin of Species*, 8.

CHAPTER 8: NATURALISTS IN A NEW LANDSCAPE

1. Jordan, "The President's Address, 1930," *PESL* 5 (1931): 128–142, at 129.
2. Miriam Rothschild to Mayr, Oct. 13, 1943, Ernst Mayr Papers, General Correspondence, 1931–1952, HUGFP 14.7, Box 2, Folder 86, HUA.
3. Jordan to secretary of the British Council, Aug. 17, 1944, KJC #5, NHM.

4. Jordan to Riley, Nov. 25, 1942, DF306/6, NHM.
5. Jordan to Gunder, May 25, 1946, KJC #5, NHM.
6. Hering to Jordan, Sept. 29, 1946, KJC #5, NHM.
7. René Jeannel to Jordan, Jan. 21, 1946, KJC #1, NHM.
8. Hering to Jordan, Nov. 10, 1945, KJC #5, NHM.
9. Jordan to Hans Eggers, Nov. 22, 1946, KJC #5, NHM.
10. Jordan to A. J. T. Janse, July 30, 1945, KJC #5, NHM.
11. Hering to Jordan, Nov. 10, 1945, KJC #5, NHM.
12. Wolfdietrich Eichler to Jordan, May 28, 1946, KJC #5, NHM.
13. Orazio Querci to Jordan, Feb. 2, 1945, and Roepke to Jordan, Mar. 29, 1946, KJC #5, NHM.
14. Hering to Jordan, Nov. 19, 1945, KJC #5, NHM.
15. Hering to Jordan, Nov. 10, 1945, KJC #5, NHM.
16. Jordan to Hering, May 5, 1946, KJC #5, NHM.
17. Jordan to Hering, Sept. 12, 1946, KJC #5, NHM.
18. Willi Nolte to Jordan, June 29, 1947, KJC #5, NHM.
19. Jordan to Hugh Scott, Feb. 3, 1947, KJC #4, NHM.
20. Report No. 102, "Japanese Ornithology and Mammalogy during World War II: An Annotated Bibliography," Jan. 30, 1948, by O. L. Austin, Masauji Hachisuka, Haruo Takashima, and Nagahisa Kuroda, NARA.
21. Stresemann to Wetmore, Oct. 24, 1947, Wetmore Papers, Box 66, Folder 7, RU 7006, SIA.
22. Jordan to Riley, Nov. 27, 1946, and Riley to Jordan, Nov. 14, 1946, KJC #5, NHM.
23. Jordan to Hering, May 5, 1946, KJC #5, NHM.
24. Jordan to Frau Horn, Mar. 16, 1947, KJC #5, NHM.
25. Jordan to H. Turner, Apr. 25, 1946, KJC #5, NHM.
26. Jordan to Riley, Nov. 6, 1946, KJC#5, NHM.
27. Jordan to Josef Bijok, Mar. 11, 1954, KJC #1, NHM.
28. Jordan to Frau Horn, Oct. 20, 1946, KJC #5, NHM.
29. Riley to Clark, Dec. 6, 1946, Austin Clark Papers, Box 10, RU 7183, SIA.
30. Jordan to Gunder, May 25, 1946, KJC #5, NHM.
31. O. Lundblad to Jordan, Nov. 17, 1945, KJC #4, NHM.
32. A. J. T. Janse to Jordan, June 22, 1945, KJC #5, NHM.
33. Jordan to Frau Horn, Oct. 20, 1946, KJC #5, NHM.
34. Jordan to Riley, Jan. 29, 1946, DF306/8, NHM.
35. Jordan to Riley, Aug. 9, 1946, DF306/7, NHM.
36. Jordan to Hering, May 5, 1946, KJC #5, NHM.
37. Hering to Jordan, May 28, 1946, KJC #5, NHM.
38. Jordan to Riley, Jan. 29, 1946, DF306, NHM.
39. N. D. Riley, "Karl Jordan and the International Congresses of Entomology," *TRESL* 107 (1955): 15–24, at 24.
40. Riley to Jordan, Feb. 4, 1946, DF 306, NHM.
41. Cyril F. dos Passos to W. F. Reinig, Oct. 20, 1947, Special Collections, D67, Box 43, AMNH.
42. Jordan to Riley, Mar. 10, 1948, DF306/8, NHM.
43. Jordan to R. N. Mathur, Aug. 9, 1950, KJC #1, NHM.
44. Keizo Yasumatsu to Weld, Jan. 10, 1959, Lewis Hart Weld Papers, Box 2, RU 7127, SIA.

45. Jordan to Eugenio Giacomelli, undated, KJC #3, NHM.
46. Jordan to Corporaal, Oct. 18, 1945, KJC #5, NHM.
47. Jordan to J. Vinson, Oct. 14, 1940, KJC #4, NHM.
48. Jordan to G. V. Hudson, July 6, 1941, KJC #4, NHM.
49. Jordan to Hudson, Jan. 27, 1940, and Feb. 19, 1940, KJC #4, NHM.
50. Jordan, "The President's Address, 1929," *PESL* 4 (1930): 128–141, at 139.
51. Jordan to Riley, Nov. 25, 1942, DF306/6, NHM.
52. Elwood Zimmerman to Jordan, July 3, 1945, KJC #4, NHM.
53. Zimmerman to Jordan, Apr. 19, 1937, KJC #4, NHM.
54. Jordan, "The President's Address, 1929," 138.
55. Jordan to Hudson, Jan. 27, 1940, and Feb. 19, 1940, KJC #4, NHM.
56. Jordan to Hudson, Dec. 23, 1943, KJC #4, NHM.
57. Jordan and N. Charles Rothschild, "On *Ceratophyllus fasciatus* and Some Allied Indian Species of Fleas," *Ectoparasites* 1 (1921): 178–198, at 182.
58. Jordan to Zimmerman, Mar. 28, 1949, KJC #4, NHM.
59. Jordan to Riley, Nov. 13, 1949, KJC #2, NHM.
60. Jordan to Zimmerman, Mar. 28, 1949, KJC #4, NHM.
61. Jordan to Peter C. Ting, Mar. 14, 1947, KJC #4, NHM.
62. M. A. Stewart to Jordan, Jan. 19, 1937, TM2/73, NHM.
63. Riley to Jordan, Nov. 26, 1947, DF306/8, NHM.
64. "Rothschild Collection of Siphonaptera," Trustees meeting: Committee Oct. 1931, DF306/3, NHM.
65. Riley to Jordan, Nov. 26, 1947, DF306/8, NHM.
66. Jordan to Hering, July 9, 1946, KJC #5, NHM.
67. George P. Holland to Jordan, Sept. 5, 1946, KJC #5, NHM.
68. Jordan to C. D. H. Carpenter, Sept. 16, 1946, KJC #5, NHM.
69. Jordan to Domingo de Las Barcenas (Spanish ambassador), Feb. 29, 1946, KJC #5, NHM.
70. Jordan to H. Turner, Apr. 25, 1946, KJC #5, NHM.
71. Jordan to Kleinschmidt, Feb. 1, 1952, KJC #1, NHM.
72. Jordan to Carl Holdhaus in Vienna, May 30, 1951, KJC #1, NHM.
73. Zimmerman to Jordan, Mar. 13, 1955, KJC #2, NHM.
74. Jordan to R. N. Mathur, May 11, 1950, KJC #1, NHM.
75. Jordan to E. J. Hall, Dec. 11, 1952, KJC #1, NHM.
76. Jordan to R. Frieser, Aug. 16, 1954, KJC #1, NHM.
77. Jordan to Riley, Apr. 3, 1955, DF306/9, NHM.
78. Zimmerman to Jordan, Mar. 31, 1955, KJC #2, NHM.
79. Rutimeyer Fry to Jordan, May 23, 1950, and Aug. 12, 1950, KJC #2, NHM.
80. Hering to Jordan, Dec. 1, 1949, KJC #2, NHM.
81. Hering to Jordan, Jan. 4, 1949, KJC #2, NHM.
82. Jordan to Rutimeyer Fry, Aug. 17, 1950, KJC #2, NHM.
83. Jordan to W. Kousnetzoff, of Belgium, Jan. 29, 1947, KJC #5, NHM.
84. Holland to Jordan, June 3, 1931, Holland Correspondence: Director's Correspondence, pt. 2, 1897–1961, CMNH.
85. Jordan to Lichy, Feb. 16, 1949, KJC #2, NHM.

86. Meinertzhagen to Mayr, Jan. 29, 1951, Ernst Mayr Papers, General Correspondence, 1931–1952, HUGFP 14.7, Box 10, Folder 433, HUA.

87. Rutimeyer Fry to Jordan, Dec. 19, 1951, KJC #1, NHM.

88. Harry Hoogstraal to G. H. E. Hopkins, Dec. 30, 1957, DF306/9, NHM.

89. W. E. China to Hopkins, Jan. 24, 1958, and Hopkins to W. E. China, Jan. 23, 1958, DF306/9, NHM.

90. Hopkins to China, Jan. 23, 1958, DF306/9, NHM.

91. Ralph E. Crabill to William J. Baerg, May 18, 1959, Department of Entomology Records, 1909–1963, Box 2, RU 140, SIA.

92. H. J. Elwes, "Presidential Address," *PESL* (1893): xlvi–lviii, at xlvii.

93. D. E. Allen, "On Parallel Lines: Natural History and Biology from the Late Victorian Period," *ANH* 25 (1998): 361–371, at 363.

94. Moreau to Mayr, Aug. 27, 1948, Society for the Study of Evolution Papers. Ms. Coll. No. 81, Series V. APS.

95. Mayr to Moreau, Sept. 17, 1948, Society for the Study of Evolution Papers. Ms. Coll. No. 81, Series V. APS.

96. Unsigned to Jordan, Apr. 24, 1929, DF304, Department of Entomology, Sectional Correspondence, 1880–1965, NHM.

97. Smart, "Entomological Systematics," 478.

98. Jordan, "The President's Address, 1929," 134.

99. Jordan, "The President's Address, 1930," 135–136.

100. Rothschild, *Dear Lord Rothschild*, 143.

101. Miriam Rothschild, personal communication, July 2002.

102. Miriam Rothschild, *Nathaniel Charles Rothschild, 1877–1923* (Cambridge: Cambridge University Press, 1979), 11.

103. Traub, "Jordan's Studies on Siphonaptera," 38.

104. Traub to Bob Lewis, Nov. 25, 1964, Hoogstraal Papers, Box 61, RU 7454, SIA.

105. Jordan to Miriam Rothschild, July 12, 1954, quoted in M. Rothschild, "The Distribution of *Ceratophyllus borealis* Rothschild, 1906 and *C. garei* Rothschild, 1902, with Records of Specimens Intermediate between the Two," *TRESL* 107 (1955): 295–317, at 311.

106. M. Rothschild, "The Distribution of *Ceratophyllus borealis*," 311.

107. Jordan, "The President's Address, 1930," 134–135.

108. F. G. A. M. Smit, "Siphonaptera from Bariloche, Argentina, Collected by Dr. J. M. de la Barrera in 1952–1954," *TRESL* 107 (1955): 319–339, at 319.

109. Jordan, "On *Xenopsylla* and Allied Genera of Siphonaptera," 601.

110. N. D. Riley, "Heinrich Ernst Jordan," *Biographical Memoirs of Fellows of the Royal Society* (1960): 107–133, at 110.

111. N. D. Riley, "Jordan, (Heinrich Ernst) Karl," *The Dictionary of National Biography*, ed. E. T. Williams and Helen M. Palmer (Oxford: Oxford University Press, 1971), 560–562, at 561.

112. "The Banquet," *Proceedings of the Second International Congress of Entomology, Oxford, August, 1912*, ed. K. Jordan and H. Eltringham (London: Hazell, Watson & Viney, 1914), 129–143, at 141.

113. Ivar Trägårdh, "Presidential Address," *Proceedings of the Eighth International Congress of Entomology* (Stockholm: Alef R. Eström, 1950), 35.

114. Trägårdh, "Final Speech by the President of the Congress," 50–53, *Proceedings of the Eighth International Congress of Entomology*, 51.

115. Trägårdh, "Final Speech," *Proceedings of the Eighth International Congress of Entomology*, 51.

116. Jordan, "The President's Address, 1930," 128.

117. J. R. de la Torre Bueno to John D. Sherman, July 21, 1947, John D. Sherman Jr. Papers, 1886–1960, Box 4, RU 7296, SIA.

118. Jordan, "Presidential Address, 1953," 4.

119. Frank M. Turner, "Public Science in Britain, 1880–1919," *Isis* 71 (1980): 589–608, at 592–593.

120. Jordan to F. Primrose Stevenson, Aug. 29, 1947, KJC #5, NHM.

121. Riley, "Heinrich Ernst Jordan," 119.

122. Jordan to Riley, May 31, 1940, KJC #4, NHM.

123. Jordan to Moss, Feb. 26, 1936, MSS MIL C 7:7, Entomology Library, NHM.

124. Riley to William Schaus, June 4, 1933, Schaus Papers, Box 10, RU 7100, SIA.

125. Michael A. Salmon, *The Aurelian Legacy: British Butterflies and Their Collectors* (Berkeley: University of California Press, 2000), 219.

126. Riley to Jordan, 1946, KJC #5, NHM.

127. Jordan to Riley, Nov. 6, 1946, and Jan. 25, 1947, KJC #5, NHM.

128. Jordan to Frau Horn, Oct. 20, 1946, KJC #5, NHM.

129. Zimmerman to Gates Clarke, Nov. 19, 1955, USNM Division of Insects, Correspondence, 1909–1963, Box 17, RU 140, SIA.

130. Jordan to J. B. Corporaal, Sept. 25, 1946, KJC #5, NHM.

131. Jordan to Hans Eggers, Nov. 22, 1946, KJC #5, NHM.

132. *Society for Freedom in Science*, Bulletin no. 15 (Dec. 1955).

133. Riley to F. G. A. M. Smit, May 17, 1949, KJC #2, NHM.

134. Hoogstraal to Traub, May 1, 1964, Hoogstraal Papers, Box 61, RU 7454, SIA.

135. Zimmerman to J. F. Gates Clarke, Nov. 1, 1946, USNM Division of Insects Correspondence, Box 17, RU 140, SIA.

136. Zimmerman to Mayr, Feb. 6, 1955, Ernst Mayr Papers, General Correspondence, 1952–1987, HUGFP 74.7, Box 3, Folder 600, HUA.

137. Zimmerman to Mayr, Dec. 13, 1948, Ernst Mayr Papers, General Correspondence, 1931–1952, HUGFP 14.7, Box 6, Folder 281, HUA.

138. Zimmerman to Mayr, Nov. 11, 1953, Ernst Mayr Papers, General Correspondence, 1952–1987, HUGFP 74.7, Box 2, Folder 551, HUA.

139. Zimmerman to Mayr, Dec. 24, 1953, Ernst Mayr Papers, General Correspondence, 1952–1987, HUGFP 74.7 Box 2, Folder 551, HUA.

140. Zimmerman to Mayr, Sept. 22, 1957, Ernst Mayr Papers, General Correspondence, 1952–1987, HUGFP 74.7, Box 2, Folder 680, HUA.

141. Zimmerman to Gates-Clarke, Dec. 24, 1961, USNM Division of Insects Correspondence, Box 17, RU 140, SIA.

142. Zimmerman to Gates-Clarke, July 9, 1960, USNM Division of Insects Correspondence, Box 17, RU 140, SIA.

143. Zimmerman to Mayr, Feb. 21, 1961, Ernst Mayr Papers, General Correspondence, 1952–1987, Box 8, Folder 769, HUGFP 74.7, HUA.

144. Zimmerman to Gates-Clarke, July 9, 1960, USNM Division of Insects Correspondence, Box 17, RU 140, SIA.

145. Doubleday to Andrewes, May 29, 1938, Doubleday Papers, HLA.

146. E. O. Wilson, *The Naturalist* (Washington DC: Island Press, 1994), 218–219.

147. W. H. Bruford, *The German Tradition of Self-Cultivation: 'Bildung' from Humboldt to Thomas Mann* (Cambridge: Cambridge University Press, 1975), 222.

148. Thomas Mann, *The Magic Mountain* (New York: Vintage International edition, 1992; originally published 1924), 628.

149. Harry Clench to Mayr, Dec. 21, 1953, Ernst Mayr Papers, General Correspondence, 1952–1987, Box 2, Folder 531, HUGFP 74.7, HUA.

150. Mayr to David Lack, Nov. 24, 1954, Ernst Mayr Papers, General Correspondence, 1952–1987, Box 3, Folder 577, HUGFP 74.7, HUA.

151. Mayr to Miriam Rothschild, Mar. 21, 1955, Ernst Mayr Papers, General Correspondence, 1952–1987, Box 3, Folder 596, HUGFP 74.7, HUA.

152. Typed notes, Ernst Mayr Papers, Miscellaneous and Anonymous Correspondence, ca. 1920–1993, Box 1, HUGFP 74.10, HUA.

153. Ernst Mayr, "Where Are We?" *Cold Spring Harbor Symposia on Quantitative Biology* 24 (1959): 1–14, at 3.

154. Ernst Mayr, "Jordan's Contributions to Current Concepts in Systematics and Evolution," *TRESL* 107 (1955): 45–66.

155. Mayr to Meinertzhagen, Feb. 23, 1954, Ernst Mayr Papers, General Correspondence, 1952–1987, HUGFP 74.7, Box 2, Folder 540, HUA.

156. E. O. Wilson and W. L. Brown Jr., "The Subspecies Concept and Its Taxonomic Application," *Systematic Zoology* 2 (1953): 97–111, at 106 and 108.

157. Wilson and Brown, "The Subspecies Concept," 101.

158. William A. Gosline, "Further Thoughts on Subspecies and Trinomials," *Systematic Zoology* 3 (1954): 92–94.

159. Theodore H. Hubbell, "Entomology: The Naming of Geographically Variant Populations, or What Is All the Shooting About?" *Systematic Zoology* 3 (1954): 113–21, at 114.

160. Jordan, "The President's Address, 1929," 140.

161. For the use of *subspecies makers*, see W. L. McAtee, "Thoughts on Subspecies," *Scientific Monthly* 53 (1941): 368–371, at 368.

162. Robert Traub, "Jordan's Studies on Siphonaptera," *TRESL* 107 (1955): 33–42, at 33.

163. Malcolm Scoble, personal communication.

164. Ian J. Kitching and Jean-Marie Cadiou, *Hawkmoths of the World: An Annotated and Illustrated Revisionary Checklist* (*Lepidoptera: Sphingidae*) (Ithaca, NY: Cornell University Press, 2000), 14.

165. Ian Kitching, personal communication, July 2002.

166. Andrew Brower, personal communication.

167. Miriam Rothschild to Mayr, Mar. 2, 1955, Ernst Mayr Papers, General Correspondence, HUGFP 74.7, Box 3, Folder 596, HUA.

168. Francis Hemming, "Karl Jordan, 1860–1959," *Bulletin of Zoological Nomenclature* 17 (1960): 259–266 at 262.

169. Riley, "Heinrich Ernst Jordan," 262.

170. Francis Hemming, "Karl Jordan and Zoological Nomenclature," *TRESL* 107 (1955): 25–32, at 30–31.

171. August Grote, "Generic Names in the Noctuidae," *ERJV* 6 (1895): 77–81, at 81.

172. Zimmerman to Jordan, Mar. 13, 1955, KJC #2, NHM.

173. A. Ellen Prout to Jordan, Mar. 5, 1944, and Jordan to Prout, Feb. 29, 1944, KJC #4, NHM.

174. Jordan to Riley, Dec. 22, 1941, DF306/6, NHM.

175. Miriam Rothschild, personal communication, 2001.

176. Jordan, "The President's Address, 1929," 134.

177. Jordan to Riley, Dec. 15, 1951, DF306/9, NHM.

178. Jordan, "Presidential Address, 1953," 3.

179. From a letter dated Mar. 28, 1923, quoted in Garland E. Allen, *Thomas Hunt Morgan: The Man and His Science* (Princeton, NJ: Princeton University Press, 1978), 313.

Conclusion

1. See http://taxatoy.ubio.org/.

2. Jordan, "The Species Problem as Seen by a Systematist," *Proceedings of the Linnean Society of London* (1938): 241–247, at 241.

3. Niall Ferguson, *The House of Rothschild*, vol. 2: *The World's Banker, 1849–1999* (New York: Penguin, 2000), 456–459.

4. Ferguson, *The House of Rothschild*, 458.

5. Malcolm J. Scoble, "Unitary or Unified Taxonomy?" *Philosophical Transactions of the Royal Society B* 359 (2004): 699–710.

6. E. O. Wilson, "Systematics and the Future of Biology," *Proceedings of the National Academy of Sciences of the United States of America* 102 (2005): 6520–6521, at 6521.

7. "What Current Technologies Make This Project a More Realistic Proposition Than Before?" ALL Species Foundation website.

8. "Current Notes and Short Notices," *ERJV* (1915): 117.

9. E. O. Wilson, *Pheidole in the New World: A Dominant, Hyperdiverse Ant Genus* (Cambridge, MA: Harvard University Press, 2003), 13.

10. Transcript for "A Little Known Planet," National Public Radio, Apr. 30, 2004.

11. "ALL Species, the Sloan Foundation and the National Science Foundation Announce New $14 million Fund for Planetary Biodiversity Inventories," Oct. 24, 2002, ALL Species Foundation press release.

12. Circular in Mayr file of Alexander Wetmore Papers, Series 1, Box 39, Folders 4–5, RU 7006, SIA.

13. Edward H. Miller, "Biodiversity Research in Museums: A Return to Basics," in M. A. Fenger, E. H. Miller, J. A. Johnson, and E. J. R. Williams, eds., *Our Living Legacy: Proceedings of a Symposium on Biological Diversity* (Victoria, BC: Royal British Columbia Museum, 1993), 141–173.

14. Robert M. May, "Taxonomy as Destiny," *Nature* 347 (1990): 129–130, at 130.

15. Peter Davis, *Museums and the Natural Environment: The Role of Natural History Museums in Biological Conservation* (Leicester, UK: Leicester University Press, 1996), 151.

16. P. L. Forey, C. J. Humphries, and R. I. Vane-Wright, eds., *Systematics and Conservation Evaluation* (Oxford: Oxford University Press, 1994); Kevin J. Gaston, ed., *Biodiversity: A Biology of Numbers and Difference* (Oxford: Blackwell Science, 1996).

17. Quentin Wheeler, "Taxonomic Triage and the Poverty of Phylogeny," *Philosophical Transactions of the Royal Society of London* 359 (2004): 571–583, at 581. Also see Wheeler, ed., *The New Taxonomy*, Systematics Association special volume, series 76 (New York: CRC Press, 2008).

18. Quentin D. Wheeler, Peter H. Raven, and Edward O. Wilson, "Taxonomy: Impediment or Expedient?" *Science* 303 (2004): 285, and letters in response, idem 305 (2004): 1104–1107.

19. Constance Holden, "Entomologists Wane as Insects Wax," *Science* 246 (1989): 754–756.

20. Ghillean T. Prance, "Systematics: Relevance to the Twenty-First Century," *Encyclopedia of Life Sciences* (Chichester, UK: Wiley, 2001).

21. Jody Hey, Alan R. Templeton, Loren H. Rieseberg, Richard G. Harrison, Peter R. Grant, Roger K. Butlin, John A. Endler, Michael L. Arnold, Robert K. Wayne, John Avise, Douglas J. Futuyma, Donald M. Waller, James Mallett, and Steve Palumbi to Ryan Phelan and Members of the Science Advisory Board, All Species Foundation. Apr. 17, 2002. http://lifesci.rutgers.edu/~heylab/misc_stuff/Letter_to_All_Species_Foundation.pdf.

22. Russell Husted, "But Maybe Counting Is the Easiest Part," *Science* 294 (2001): 1834.

23. Lawler, "Up for the Count," *Science* 294 (2001): 769–770, at 769.

24. Christián Samper, "Counting All Species," *Science* 294 (2001): 1833.

25. R. Geeta, Andre Levy, J. Matt Hoch, and Melissa Mark, "Taxonomists and the CBD," *Science* 305 (2004): 1105.

26. E. O. Wilson, "The Linnaean Enterprise: Past, Present, and Future," *Proceedings of the American Philosophical Society of London* 149 (2005): 344–348, at 348 ($3 billion), and Lawler, "Up for the Count" ($20 billion).

27. Leslie R. Landrum, "What Has Happened to Descriptive Systematics? What Would Make It Thrive?" *Systematic Botany* 26 (2002): 438–442.

28. Wilson, "The Linnaean Enterprise," 347.

29. Ernst Mayr, "Methods and Strategies in Taxonomic Research," *Systematic Zoology* 20 (1971): 426–433, at 426.

30. Anthony Gill, personal communication, 2005.

31. Marcelo R. de Carvalho, Flávio A. Bockmann, Dalton S. Amorim, Mário de Vivo, Mônica de Toledo-Piza, Naércio A. Menezes, José L. de Figueiredo, Ricardo M. C. Castro, Anthony C. Gill, John D. McEachran, Leonard J. V. Compagno, Robert C. Schelly, Ralf Britz, John G. Lundberg, Richard P. Vari, and Gareth Nelson, "Revisiting the Taxonomic Impediment," *Science* 307 (2005): 353.

32. Hey et al. to Ryan Phelan and Members of the Science Advisory Board, All Species Foundation. Apr. 17, 2002.

33. Frank Newport, "On Darwin's Birthday, Only 4 in 10 Believe in Evolution," Feb. 11, 2009, GALLUP. http://www.gallup.com/poll/114544/darwin-birthday-believe-evolution.aspx.

34. Phil Ackery, Kim Goodger, and David Lees, "The Bürgermeister's Butterfly," *Journal of the History of Collections* 14 (2002): 225–230, at 225.

35. "A Leap for All Life: World's Leading Scientists Announce Creation of 'Encyclopedia of Life,'" May 2007, MacArthur Foundation, eNewsletters.

36. Jordan to Preston Clark, Sept. 19, 1932, TM2/69, NHM.

37. Holden, "Entomologists Wane as Insects Wax," 755.

Essay on Sources

Archives

The most important archival collection for the Tring naturalists is the Tring Museum Correspondence held at the Natural History Museum in London. Even in this wonderful collection, however, the traces of their life and work are limited in certain ways. Charles Rothschild commanded that his own correspondence be burned.* And the more than eighty thousand letters in the surviving collection, on which much of this study is based, is limited to incoming correspondence between 1892 and 1908, with the exception of four carefully bound letter books of outgoing correspondence for the first few years of the museum (after the 1908 scandal, all outgoing correspondence was carefully preserved as well). Legend has it that the rest of the letter books were burned in an effort to tidy up the museum after Jordan's death.†

I have only scratched the surface of this archive, and fascinating tales await future researchers, especially regarding Rothschild's collectors and the network of natural history agents. In particular, the archive is a tremendous resource for anyone interested in the collectors in the field, Ernst Hartert and Walter Rothschild's life and work, the hundreds of naturalists with whom the "Tring Triumvirate" exchanged information and specimens, and of course different perspectives on Jordan himself.

The Naturalist Tradition since Darwin

The history of the naturalist tradition is eloquently summarized in Paul Lawrence Farber's *Finding Order in Nature: The Naturalist Tradition from Linnaeus to E. O. Wilson* (John Hopkins University Press, 2000). Farber's various papers on the topic, includ-

---

* Miriam Rothschild to Harry Hoogstraal. October 27, 1964. Harry Hoogstraal Papers, Box 51, RU 7454, Smithsonian Institution Archives.

† Miriam Rothschild, *Dear Lord Rothschild: Birds, Butterflies, and History* (London: Hutchinson, 1983), 299–301.

ing "The Transformation of Natural History in the Nineteenth Century," in the *Journal of the History of Biology* 15 (1982):145–152, strongly influenced this book. In particular, Farber's insistence that "overly stressing the impact of general theories on taxonomy" ignores the "cumulative aspects of the history of systematics, and glosses over other changes that have occurred in natural history," has guided my choices in deciding what to focus on in Jordan's story; see his "Theories for the Birds: An Inquiry into the Significance of the Theory of Evolution for the History of Systematics," in *Religion, Science, and Worldview: Essays in Honor of Richard S. Westfall*, edited by Margaret Osler and Paul Farber (Cambridge University Press, 1985).

The work of Garland Allen, particularly his classic *Life Science in the Twentieth Century* (Cambridge University Press, 1978), is an important starting point for studying the history of the naturalist tradition in the twentieth century. Allen's work inspired a wealth of further research by scholars intent on testing his thesis that natural history declined amid the rise of experimental sciences and that the fields are opposed. See the contributions to *The American Development of Biology* (University of Pennsylvania Press, 1988) and *The Expansion of American Biology* (Rutgers University Press, 1991), both books edited by Ronald Rainger, Keith Benson, and Jane Maienschein. The work has continued in, among others, Bruno Strasser's "Collecting and Experimenting: the Moral Economies of Biological Research, 1960s–1980s," in *Preprints of the Max-Planck Institute for the History of Science*, 310 (2006): 105–123, and Lynn K. Nyhart's "Natural History and the 'New' Biology," in *Cultures of Natural History*, ed. N. Jardine, J. A. Secord, and E. C. Spary, 426–443 (Cambridge University Press, 1996).

A pioneer of relating natural history to its broader social context is David Elliston Allen. See his *The Naturalist in Britain: A Social History* (Princeton University Press, 1994, orig. published 1976) and "On Parallel Lines: Natural History and Biology from the Late Victorian Period," *Archives of Natural History* 25 (1998): 361–371. For further background on the history of natural history, see the papers in N. Jardine, J. A. Secord, and E. C. Spary, *Cultures of Natural History* (Cambridge University Press, 1996).

The histories of biology composed by the evolutionary biologist Ernst Mayr provide a fascinating perspective of one who lived and experienced much of this history. Mayr wrote his histories with the explicit aim of defending the role of naturalists in the history of biology. His *The Growth of Biological Thought: Diversity, Evolution, and Inheritance* (Harvard University Press, 1985) was simply the most ambitious of a wealth of writings on the history of evolution biology and systematics. Indeed, Mayr's first excursion into historical research seems to have been his tribute to Karl Jordan in 1955, where he stumbled across a very useful hero for his campaign. I have analyzed Mayr's use of Jordan in his histories of systematics in "Ernst Mayr, Karl Jordan, and the History of Systematics," *History of Science* 48 (2005): 1–35. Historians have since qualified

Mayr's narratives in various ways: see Mary P. Winsor, "The Creation of the Essentialism Story: An Exercise in Metahistory," *History and Philosophy of the Life Sciences* 28 (2006): 149–174.

I drew on Frederick Burkhardt, "England and Scotland: the Learned Societies," in T. F. Glick, ed. *The Comparative Reception of Darwinism* (University of Texas Press, 1974), 32–74, for information on the methodological ethos of British natural history societies. I also relied extensively on Lynn Nyhart's *Biology Takes Form* for the context of Jordan's education as a zoologist. Critical recent works on the history of the naturalist tradition in Germany and the United States include Nyhart's *Modern Nature: The Rise of the Biological Perspective in Germany* (University of Chicago Press, 2009) and Robert Kohler's *All Creatures: Naturalists, Collectors, and Biodiversity, 1850–1950* (Princeton University Press, 2006) and *Landscapes and Labscapes: Exploring the Lab-Field Border in Biology* (University of Chicago Press, 2002). Other important studies include work on specific disciplines since the mid-nineteenth century. See Mark Barrow, *A Passion for Birds: American Ornithology after Audubon* (Princeton University Press, 1998) and Erwin Stresemann's classic *Ornithology: From Aristotle to the Present* (Harvard University, 1975). A few studies have been done on entrepreneurial natural history in the nineteenth century. See Barrow, "The Specimen Dealer: Entrepreneurial Natural History in America's Gilded Age," in the *Journal for the History of Biology* 33 (2000): 493–534 and Kohlstedt, "Henry A. Ward: The Merchant Naturalist and American Museum Development," in the *Journal of the Society for the Bibliography of Natural History* 9 (1980): 647–661. Their focus is upon dealers in the United States. The Tring Correspondence provides an excellent source for adding a European and worldwide dimension to the story.

Heinrich Ernst Karl Jordan

Memorials to scientists must often be read with a critical eye, focusing as they do on the exciting points where a scientist "got it right," rather than missteps or administrative drudgery. But Jordan's memorialists composed a number of useful, entertaining, and highly informative tributes to their friend and colleague. Even allowing for the hyperbole of the genre, the admiration and respect his friends felt for Jordan is striking. The memorials easily bear rereading even half a century later, to capture both some of the personality his colleagues found so endearing and what they found so valuable in his life and work. See, for example, Francis Hemming, "Karl Jordan 1861–1959," in *Bulletin of Zoological Nomenclature* 17 (1960): 259–266. N. D. Riley's obituary in the *Memorials of the Fellows of the Royal Society* 6 (1960): 107–133, contains an almost complete list of Jordan's hundreds of publications.

A Festschrift for Jordan organized by the Entomological Society of London in 1955 is also an invaluable source. Published as volume 107 of the *Transactions of the Entomological Society of London*, the essays illustrate the range of Jordan's work and admirers. Most recently, biologist James Mallet has analyzed Jordan's work on the biological species concept in concert with the contributions of Alfred Russel Wallace and E. B. Poulton. See his "Poulton, Wallace, and Jordan: How discoveries in *Papilio* Butterflies Led to a New Species Concept 100 Years Ago," in *Systematics and Biodiversity* 1 (2004): 441–452.

This book is by no means a comprehensive biography. Even apart from the epistemological challenges that face the modern biographer, such a personal portrait would be difficult for a man who culled anything personal from his museum correspondence and whose habit of working "behind the blinds" was often noted. Fortunately we have Miriam Rothschild's "Biographical Sketch" in the 1955 *Festschrift* and her fascinating and often quite personal biography of her uncle, *Dear Lord Rothschild: Birds, Butterflies, and History* (Balaban, 1983), which has since been republished as *Walter Rothschild: The Man, the Museum, and the Menagerie* (Natural History Museum, 2008). Her writing confers more "life" to the men and women of Tring than the rather business-centered museum correspondence upon which this study is largely based.

Naturalists and Empire

For the ties between natural history and Empire, see Lucile Brockway, *Science and Colonial Expansion: The Rule of the British Royal Botanic Gardens* (Academic Press, 1979); Janet Browne, "Biogeography and Empire," in *Cultures of Natural History*, ed. N. J. Jardine, J. A. Secord, and E. C. Spary (Cambridge University Press, 1996), 305–321; Richard Drayton, *Nature's Government: Science, Imperial Britain, and the 'Improvement' of the World* (Yale University Press, 2000); Jim Endersby, *Imperial Nature: Joseph Hooker and the Practices of Victorian Science* (University of Chicago Press, 2008); Fa-ti Fan, *British Naturalists in Qing China: Science, Empire, and Cultural Encounter* (Harvard University Press, 2004); John M. MacKenzie, *The Empire of Nature: Hunting, Conservation, and British Imperialism* (Manchester University Press, 1988) and *Museums and Empire: Natural History, Human Cultures, and Colonial Identities* (Manchester University Press, 2009; Lynn K. Nyhart, "Essay Review: Biology and Imperialism," *Journal of the History of Biology* 28 (1995): 533–543; Michael A. Osborne, *Nature, the Exotic and the Science of French Colonialism* (Indiana University Press, 1994), and R. A. Stafford, *Scientist of Empire: Sir Roderick Murchison, Scientific Exploration, and Victorian Imperialism* (Cambridge University Press, 1989). For an introduction to the extensive amount of literature on science and empire more generally, see John M. MacKenzie, ed. *Imperialism and the Natural World* (Manchester University Press, 1990).

ESSAY ON SOURCES   361

Natural History Museums and Collections

On the importance of giving historical attention to museums as sites of science, see Sally Gregory Kohlstedt, "Museums: Revisiting Sites in the History of the Natural Sciences," *Journal of the History of Biology* 28 (1995): 151–166. For histories of particular museums, see W. T. Stern, *The Natural History Museum at South Kensington: A History of the British Museum (Natural History), 1753–1980* (Heinemann, 1981); Camille Limoges, "The Development of the Museum d'Histoire Naturelle of Paris, c. 1800–1914," in Robert Fox and George Weisz, eds., 211–240, *The Organization of Science and Technology in France, 1808–1914* (Cambridge University Press, 1980); R. T. Gunther, *A Century of Zoology: Through the Lives of Two Keepers* (Wm. Dawson, 1975); Mary P. Winsor, *Reading the Shape of Nature: Comparative Zoology at the Agassiz Museum* (University of Chicago Press, 1991); Ronald Rainger, *An Agenda for Antiquity: Henry Fairfield Osborn and Vertebrate Paleontology at the AMNH, 1890–1935* (University of Alabama Press, 1991), and James R. Griesemer and Elihu M. Gerson, "Collaboration in the Museum of Vertebrate Zoology," *Journal of the History of Biology* 26 (1993): 185–204.

For natural history museums that existed far away from the centralized institutions in Europe, see Susan Sheets-Pyenson, *Cathedrals of Science: The Development of Colonial Natural History Museums during the Late Nineteenth Century* (McGill-Queen's University Press, 1988); Maria Margaret Lopes and Irina Podgorny, "The Shaping of Latin American Museums of Natural History, 1850–1890," *Osiris* 15 (2000): 108–134; Maria Margaret Lopes, "Brazilian Museums of Natural History and International Exchanges in the Transition to the Twentieth Century," in Patrick Petitjean, Catherine Jami, and Annie Marie Moulin, eds. *Science and Empires: Historical Studies about Scientific Development and European Expansion* (Kluwer, 1992), 193–200.

Taxonomy, Systematics, and Biogeography

For studies on the history of systematics, see the above works on empire, museums, and natural history, as well as Mary P. Winsor, *Starfish, Jellyfish, and the Order of Life* (Yale University Press, 1976). David Hull's *Science as Process: An Evolutionary Account of the Social and Conceptual Development of Science* (University of Chicago Press, 1988) and Keith Vernon's "The Founding of Numerical Taxonomy," *British Journal for the History of Science* 21 (1988): 143–149 examine the history of systematics in the second half of the twentieth century. For another entomologist's life in taxonomy, see Pamela M. Henson, "Evolution and Taxonomy: John Henry Comstock's Research School in Evolutionary Entomology at Cornell University, 1874–1930," (PhD dissertation, University of Maryland, 1990). For useful studies of the history of taxonomy with a view to the present, see the work of Peter Stevens, including "Why Do We Name Organisms? Some Reminders from the Past," *Taxon* 51 (2002): 11–26; "Biological Systematics 1950–2000: Change, Progress, or Both?" *Taxon* 49 (2000): 635–659; and "J. D. Hooker, George Bentham, Asa Gray, and Ferdinand Mueller on Species Limits in Theory and Practice: A Mid-Nineteenth-Century Debate and Its Repercussions," in *Historical Records of Australian Science* 11 (1997): 345–370.

For studies of other taxonomists who navigated their way through changes in science in the twentieth century, see Joel Hagen, "Experimentalists and Naturalists in Twentieth Century Botany: Experimental Taxonomy, 1920–1950," *Journal of the History of Biology* 17 (1984): 249–270, and Keith Vernon, "Desperately Seeking Status: Evolutionary Systematics and the Taxonomists' Search for Respectability, 1940–1960," *British Journal for the History of Science* 26 (1993): 207–227. Taxonomy, systematics, and biogeography often appear in literature on the history of the evolution synthesis (listed below). Recently, Jim Endersby, in *Imperial Nature* (cited above), and Kurt Johnson and Steve Coates, in *Nabokov's Blues: The Scientific Odyssey of a Literary Genius* (McGraw-Hill, 2001) have written fascinating accounts of the lives of other naturalists.

Works on biogeography include Janet Brown, *The Secular Ark: Studies in the History of Biogeography* (Yale University Press, 1983); Jane R. Camerini, "Evolution, Biogeography, and Maps: An Early History of Wallace's Line," *Isis* 84 (1993): 700–727; and Robert Kohler's works cited above. William Kimler's work on E. B. Poulton in turn connects both the taxonomic and biogeographic traditions to evolutionary ecology; see "Advantage, Adaptiveness, and Evolutionary Ecology," *Journal of the History of Biology* 19 (1986): 215–233.

The specific issue of trinomial nomenclature and the study of geographical variation is examined by Stresemann (1975), Kohler (2006), and Barrow (1998). On the ornithologists within the trinomialist circle (and an ornithological counterpart to Jordan's tale), see both Stresemann (1975) and various papers by J. H. Haffer, including "The History of Species Concepts and Species Limits in Ornithology," *Bulletin of the British Ornithologists' Club Centenary Supplement* 112 (1992): 107–158; "Die Seebohm-Hartert-'Schule' der europäischen Ornithologie," *Journal für Ornithologie* 135 (1994): 37–54; "Ornithological Research Traditions in Central Europe during the 19th and 20th Centuries," *Journal für Ornithologie* 142 (2001): 27–93; and Haffer, Erich Rutschke, and Klaus Wunderlich, "Erwin Stresemann (1889–1972): Leben und Werk eines Pioniers der wissenschaftlichen Ornithologie," in *Acta Historica Leopoldina* 34 (2000): 399–427. Ernst Mayr's work connects these historical studies to modern biology.

Entomology, "Applied and Pure"

For historical studies under the broad rubric of entomology in the service of agricultural, medical, or other concerns, see John F. M. Clark, "Bugs in the System: Insects, Agricultural Science, and Professional Aspirations in Britain, 1890–1920," *Agricultural History* 75 (2001): 83–114, and Paolo Palladino, *Entomology, Ecology and Agriculture: The Making of Scientific Careers in North America, 1885–1985* (Harwood Academic, 1996). For studies on entomology in Britain and the United States in the nineteenth century, see John F. M. Clark, *Bugs and the Victorians* (Yale University Press, 2009) and W. Conner Sorensen, *Brethren of the Net: American Entomology, 1840–1880* (University of Alabama Press, 1995). For economic biology more generally, see Alison Kraft "Pragmatism, Patronage, and Politics in English Biology: The Rise and Fall of Economic Biology, 1904–1920," *Journal for the History of Biology* 37 (2004): 213–258. On France, see Stéphane Castonguay, "The Transformation of Agricultural Research in France: The Introduction of the American System," *Minerva* 43 (2005): 265–287; and on Germany, see Sarah Jansen, "Chemical-warfare Techniques for Insect Control: Insect 'Pests' in Germany before and after World War I," *Endeavour* 24 (2000): 28–33.

For the rise of state patronage of science in Britain more generally, see Peter Alter's *The Reluctant Patron: Science and the State in Britain, 1850–1920* (Berg, 1987), Robert Olby, "Social Imperialism and State Support for Agricultural Research in Edwardian Britain," *Annals of Science* 48 (1991): 509–526, and Frank Turner, "Public Science in Britain, 1880–1919," *Isis* 71 (1980): 589–608.

Evolution in the Twentieth Century

Much of the scholarship on the history of evolution theory in the twentieth century has focused on the "modern synthesis" of the 1930s and 1940s. For the original narrative and early critics, see the papers in Ernst Mayr and William B. Provine, *The Evolutionary Synthesis: Perpsectives on the Unification of Biology* (Harvard University Press, 1980). For examinations of the evolution synthesis as an institution and discipline-building enterprise, see Vassiliki Betty Smocovitis, "Organizing Evolution: Founding the Society for the Study of Evolution, 1939–1947," *Journal of the History of Biology* 27 (1994): 241–309, and Joseph Allen Cain, "Common Problems and Cooperative Solutions: Organizational Activity in Evolutionary Studies, 1936–1947," *Isis* 84 (1993): 1–25; Cain, "Ernst Mayr as Community Architect: Launching the Society for the Study of Evolution and the Journal Evolution," *Biology and Philosophy* 9 (1994): 387–427; Cain, "Towards a 'Greater Degree of Integration': The Society for the Study of Speciation, 1939–1941," *British Journal for the History of Science* 33 (2000): 85–108, and Cain, "For the 'Promotion' and 'Integration' of Various Fields: First Years of Evolution, 1947–1949," *Archives of Natural History* 27 (2000): 231–259. For other aspects of the synthesis, see Smocovitis, *Unifying Biology: the Evolutionary Synthesis and Evolutionary Biology* (Princeton University Press, 1996), Smocovitis, "Keeping up with Dobzhansky: G. Ledyard Stebbins, Plant Evolution, and the Evolutionary Synthesis," *History and Philosophy of the Life Sciences* 28 (2006): 11–50, and Jonathan Harwood, "Metaphysical Foundations of the Evolutionary Synthesis: A Historiographical Note," *Journal of the History of Biology* 27 (1994): 21–59. For the argument that the concept of an "evolutionary synthesis" be abandoned altogether, see Cain's recent "Rethinking the Synthesis Period in Evolutionary Studies," *Journal of the History of Biology* 42 (2009): 621–648.

For the background to why proponents thought an evolution synthesis was needed in the first place (and for developments in biology after Darwin published *On the Origin of Species*), see Peter J. Bowler, *The Eclipse of Darwinism: Anti-Darwinian Evolution Theories in the Decades around 1900* (Johns Hopkins University Press, 1983); and for a discussion of Jordan's work in this context, see John E. Lesch, "The Role of Isolation in Evolution: George J. Romanes and John T. Gulick," *Isis* 66 (1975): 483–503. For recent efforts to expand the study of the history of twentieth-century evolutionary biology beyond the synthesis, see Joe Cain and Michael Ruse, eds., *Descended from Darwin: Insights into the History of Evolutionary Studies, 1900–1970* (Diane, 2009).

Professionalization

There is an extensive literature available on the definitions, relationships, and dynamics of the elusive categories *amateur* and *professional* in natural history and the sciences more generally. For discussions on the subject in the naturalist tradition in both Britain and the United States, see works, cited above, by Clark (2009), Allen (1976 and 1998), Barrow (1998), and Sorensen (1995). The following studies are also useful: Elizabeth Keeney, *The Botanizers: Amateur Scientists in Nineteenth-Century America* (University of North Carolina Press, 1992); Richard Bellon, "Joseph Hooker's Ideals for a Professional Man of Science," *Journal of the History of Biology* 34 (2001): 51–82; Adrian Desmond, "Redefining the X Axis: 'Professionals,' 'Amateurs,' and the Making of Mid-Victorian Biology: A Progress Report," *Journal of the History of Biology* 34 (2001): 3–50; Peter C. Kjærgaard, "Competing Allies: Professionalisation and the Hierarchy of Science in Victorian Britain," *Centaurus* 44 (2002): 248–288; and Susan Leigh Star and James R. Griesemer, "Institutional Ecology, 'Translations,' and Boundary Objects: Amateurs and Professionals in Berkeley's Museum of Vertebrate Zoology, 1907–1939," *Social Studies of Science* 19 (1989): 387–420.

Although they were not aristocrats, the Rothschilds' wealth certainly placed them in the category of the extremely privileged. Historians have examined other cases where wealth and/or nobility increasingly placed one on the margins of the scientific establishment. In particular, see Mary Jo Nye, "Aristocratic Culture and the Pursuit of Science: The De Broglies in Modern France," *Isis* 88 (1997): 397–421. On the rise of professional society in general, see Harold Perkin, *The Rise of Professional Society: England since 1880* (Routledge, 1989).

Nationalism and Internationalism in Science

On nationalism and internationalism in science, see Paul Forman, "Scientific Internationalism and the Weimar Physicists: The Ideology and Its Manipulation in Germany after World War I," *Isis* 64 (1973): 151–180; Daniel J. Kevles, "'Into Hostile Political Camps': The Reorganization of International Science in World War I," *Isis* 62 (1970): 47–60; Brigitte Schroeder-Gudehus, "Nationalism and Internationalism," in G. N. Cantor et al., eds. *Companion to the History of Modern Science* (Routledge, 1989), 909–919; "Challenge to Transnational Loyalties: International Scientific Organizations after the First World War," *Science Studies* 3 (1973): 93–118; A. J. Cock, "Chauvinism and Internationalism in Science: The International Research Council, 1919–1926," *Notes and Records of the Royal Society of London* 37 (1983): 249–288; Elisabeth T. Crawford, "The Universe of International Science, 1880–1939," in Tore Frängsmyr, ed., *Solomon's House Revisited: The Organization and Institutionalization of Science*, Proceedings of Nobel Symposium 75 (Canton, MA: 1990), 251–269; Crawford, *Nationalism and Internationalism in Science, 1880–1939: Four Studies of the Nobel Population* (Cambridge University Press, 1992); and Elisabeth T. Crawford, Terry Shinn, and Sverker Sorlin, *Denationalizing Science: The Contexts of International Scientific Practice* (Kluwer Academic, 1993). On specific international congresses and organizations in the life sciences, see Helen M. Rozwadowski, *The Sea Knows No Boundaries: A Century of Marine Science under ICES* (University of Washington Press, 2003) and Debra Adrienne Everett-Lane, "International Scientific Congresses, 1878–1913: Community and Conflict in the Pursuit of Knowledge" (PhD diss., Columbia University, 2004).

Index

Addison, Christopher, 219
Africa: applied entomology and, 200; Jordan's expedition to, 234–35; Rothschild's collectors in, 232–33; scramble for, 22–23
Agassiz, Louis, 64
Aldrich, John Merton, 183, 215, 234
All Species Foundation (ASF), 1, 8, 300, 305, 307
Allen, Joseph Asaph, 63, 115
amateurs: as problematic, 47; tensions with professionals, 42, 143, 163; use by Tring naturalists, 107, 226
American Museum of Natural History, 221–22, 261, 291
Ansorge, W. J., 34
Anthribidae, 31, 54, 80, 255, 272–74, 276, 283–84, 288, 290, 297
anti-Semitism, 42
applied entomology: ambivalence toward, 157–60, 195; in Britain, 195–97; dependence on taxonomy, 215; forest entomology, 20, 194; in Germany, 20, 194; influence on taxonomy, 7, 198, 201–4, 209, 219, 227–28, 263, 288, 302; international congresses and, 126, 200, 252, 285; public's view of, 8; rise after World War I, 191–99, 207–12, 231–32; tension with taxonomists, 201–2, 204–5, 209, 214, 285, 290; in various countries, 195; wartime, 170, 174–75, 256
Arndt, Walther, 267
Association of Economic Biologists, 204
Avinoff, Andrey, 178, 243, 252

Baker, C. S., 174
Ball, Elmer Darwin, 201
Barclay-Smith, Phyllis, 225
Bates, Henry Walter, 52, 63, 66, 76, 238
Bateson, William, 100, 141, 238; campaign for the study of variation, 57–58; criticisms of taxonomists, 94, 116–17, 161; internationalism, 182; Jordan's criticism of, 99–100; support for discontinuous variation, 93–94, 115–16; support for experimentalism, 138–40, 186
Baylis, H. A., 204, 227
Beck, Rollo, 34
Berlepsch, Count Hans Hermann Carl Ludwig von, 21
Berlin Agreement, 217–18
Bethune-Baker, G. T., 168, 170
biodiversity crisis, 8, 306–7
biogeography: Jordan's early studies, 17–18; methodological framework for, 48–49; status outside academia, 19, 21; ties to nation-building, 11–12
biology: versus descriptive natural history, 5, 49–51, 75–77, 86, 101, 140, 143, 225, 230; as life history, 16, 51, 98
Birch, Fred, 134
Blair, K. G., 168
Blanchard, Raphael, 136, 137
Bloomfield, W., 89
Bonhote, John James Lewis, 36
Bonner, John Tyler, 248
Boulenger, G. A., 43
Bouvier, Eugène, 267
Braun, Rudolf Hans, 252
Brehm, Christian Ludwig, 58
Brierley, William B., 205, 228
British Museum (Natural History): and empire, 23; goals of, 108; growth of, 27; Jordan's relations with, 45–46, 225, 257, 269; resources of, 156–57, 198, 291; Walter Rothschild's relations with, 223–24; status of, 244, 259; wartime, 170, 173–74; workloads at, 286–87, 289

British Ornithologists' Union, 223, 225
British Science Guild, 136
Brooke, Charles Johnson, 132–33
Brown, William L., 293
Buchanan, Angus, 232–33, 235
Buckell, Edward, 135
Buckland, James, 131
Burr, Malcolm, 160–61, 184, 195
Busck, August, 173, 197
Butterfield, J., 112
Buxton, D. A. J., 168

Cadiou, Jean-Marie, 296
Calman, W. T., 198
capitalism: critiques of, 114–15, 118; dependence of collections upon, 114, 118, 122
Carnegie Museum of Natural History, 109, 121, 178, 206, 214, 243, 292
Carpenter, G. H., 51–52, 100, 161
Castle, W. E., 260
catalog of species: criticisms of, 87, 100, 108, 236, 262, 308–9; as goal of taxonomy, 161, 205, 309; Jordan on, 5, 20, 85, 103, 105, 108, 205; justifications of, 8, 10–12, 23–24, 50–52, 75, 170, 200; methodological framework for, 48–50, 75; progress of, 10, 22, 50, 283, 305; rate calculations, 230–31, 235, 309; recent efforts, 1, 8, 300, 305, 307
Chamberlain, Joseph, 25
Champion, G. C., 100
Chapman, Frank M., 224
Chapman, Royal, 208
Chapman, T. A., 112, 171
Cherrie, George K., 35
Cherrie, Stella, 35
Clark, Austin, 193–94, 226
class: and natural history, 113–15, 118, 132–33, 178, 192–94, 226, 240–41, 243–45, 292; of naturalists, 34, 37, 47, 52; rise of middle, 13–14
classification: difficulties of, 4–5, 16; evolution and, 4, 15, 49, 56, 75–77, 139, 144; justification of, 15, 20, 103; Linnaean system of, 10; naming and, 6, 59, 75, 79, 85, 94; as natural, 15–17, 87, 101–2, 105–6, 112
Claus, Carl, 55–56
Clench, Harry, 292
Cockerell, T. D. A., 139, 189, 190, 192
collections: aims of, 108; bequests of, 46, 223; destroyed during World War II, 267; and empire, 7, 21, 173, 188, 196, 247, 278, 302, 306; growth of, 27; Jordan's defense of, 106, 110, 143, 188; Jordan's early experience with, 17–18; methodological framework for, 48–50, 52; private, 7, 21, 34, 120, 126, 210, 226, 243; reform of, 46, 48, 57, 61, 63, 67–73, 124–26, 144–46, 189–90, 210–15; status of, 15–16, 49–52, 56–57, 77, 94, 116–17, 131–33, 138–40, 157, 189, 280–81. *See also* museums
collectors: dangers faced by, 34–35; decline after World War I, 231, 237; dependence on curators, 46; Depression and, 215; empire and, 25, 173, 247, 252–53; instructions to, 66, 68–73, 131, 146; need for more research on, 357; opposition to, 132–33; reform needed, 125–26, 144–46, 163; Rothschild's, 34, 68, 128; status of, 47, 52; World War II and, 254–55
Collin, J. E., 214–15
Comstock, John Henry, 100
conservation: as justification of systematics, 306–7; specimen collecting and, 52, 65, 131–34
Cope, Edward Drinker, 258
Corporaal, J. B., 211, 256, 288
cosmopolitanism, 24, 35; critiques of, 135–36; science and, 136. *See also* internationalism
Coues, Elliott, 58, 59
creation, theory of special, 3, 56–57, 63–64, 75
Crew, F. A., 187
Cunningham, J. T., 95
curators: administrative duties of, 27, 32–33, 35–36, 302; power of, 46, 70; private collectors and, 43; self-perception of, 38; status of, 47, 51

Dadd, E. M., 169
Dampf, Alfons, 201, 234
Darwin, Charles, 3–6, 13, 48, 51, 53, 56, 76, 91–92
dealers of natural history: in Britain, 44; in Germany, 21; as professionals, 47; Tring Archives and, 357
Depression: taxonomy and, 7, 215; Tring and, 221, 224–25
Deutsches Entomologisches Museum, 180, 210, 212, 254
Distant, W. L., 97, 112
Dixey, F. A., 49, 135, 159, 176
Dobzhansky, Theodosius, 187–88, 237–41, 260
Doherty, William, 34

Drosophila, 186–87, 229, 238–41, 274
Durrant, John Hartley, 44, 46, 48, 69, 81, 123, 173

ecology: applied entomology and, 232, 285; at congresses, 208, 285; German schoolteachers and, 19; taxonomy and, 13, 17, 20, 58, 85, 92
Edwards, W. H., 90
Eggers, Hans, 267
Ehlers, Ernst, 15–17, 58, 80, 88
Eichhorn, Albert, 34, 35
Eichler, Witold, 267
Eimer, Theodor, 95, 101–2, 107
Eltringham, Harry, 112
Elwes, H. J., 49–50, 61, 71–72, 89, 280
Empire: applied entomology and, 219, 231–32; British, 26; critics of, 115, 233–34, 278, 306, 308; Japan and natural history, 268; natural history and, 21–25, 32, 37, 115, 133, 173, 196, 231–35, 302
Engelhardt, G. P., 251
Enock, F., 195
Entomological Society of London: applied entomology and, 194, 197, 232, 256; evolution and, 162; genetics at, 187; scientific method and, 49–50, 100, 280–81; status of entomology and, 156–57, 214; World War I and, 168, 170
entomologists: British, 40–55; diversity of, 6, 119, 124–26, 144–46, 162–64, 294–95; nationalism among, 135–36, 181–82, 184; specialization among, 119, 252; status of, 74, 76–77, 93, 100, 141–46, 157, 170, 197, 285; World War I and, 168–75, 179–81; zoologists and, 138, 141, 145–46, 185, 209
Entomologist's Record and Journal of Variation, 49, 57, 168, 175, 186
entomology: justifications of, 8, 20, 159–60, 169–70, 191, 194, 196–97, 200, 208, 256, 285; need for revisions and monographs in, 78; reform of, 48–55, 57, 119–20, 124–26, 144–46, 201–15; shifts in journals of, 279–81; status of, 51, 125, 156–57, 203–6, 244. *See also* applied entomology
Escherich, Karl, 202
Esper, Eugenius Johann Christoph, 58
Essig, E. O., 239
Everett, Alfred, 34
evolution: biogeography and, 22; caution toward, 16, 48–50, 90–91; classification and, 16–17, 49, 101–2, 144; collections and, 63, 66; dependence on taxonomy, 5–6, 15, 50, 82, 94, 103, 106, 110, 126, 143, 160–61, 188; entomology and, 49–51, 76, 280; influence on taxonomy, 3–4, 6, 40, 49, 56–62, 75–77, 87, 96, 140, 144, 210; Jordan's writings on, 3, 91–100, 188–89, 236–41, 286, 295; mechanism of, 91–100, 162, 187–88
experimental method: congresses and, 185–86, 188, 208; as criterion of science, 5, 7, 94, 238–41
experimentalists: as critics of taxonomy, 16, 94, 115, 125, 138–40; Jordan on, 106
extinction, 131–32, 226

Farrell, Brian, 306
Faust, J., 36
Ferguson, Niall, 173, 175, 304–5
Ferris, Gordon Floyd, 210, 290
fleas, Jordan's work on, 2–3, 31–32, 158–59, 200, 211, 236–41, 256, 272, 274–75, 282, 289, 296
Flegel, Eduard Robert, 23
Flower, Sir William, 86, 298
Foetterle, F. G., 45, 109
Forbes, William, 220
Frohawk, Frederick W., 32
Frost, F., 119
Fruhstorfer, 34
Fryer, J. C. R., 250, 252

Gahan, Charles Joseph, 45, 170
Gardiner, John Stanley, 244, 269
Geddes, John M., 246
Genetical Society, 187
genetics, rise of, 6–7, 115–17, 138–40, 185–86, 238–41, 259, 262, 296
genital armature, 89–90, 99, 106–7, 202
geographical isolation, 91–92, 99–100
geographical variation: early studies of, 63; Jordan's writings on, 81–82, 90–100, 105–6, 110, 236–41, 248, 259; long series and, 66–67, 189, 239, 294–95; role in the modern synthesis, 262. *See also* subspecies; trinomials; variation
geography: natural history and, 12; specimen labels, 17–18, 69, 90, 108–9
Germany: empire of, 22–23, 179, 253; reputation of zoologists from, 41–42; and World War I, 179–80, 210, 212; zoological education in, 15–16
Gestro, Raphael, 125, 171
Gill, Anthony, 310
Girtanner, Cristoph, 97
Godman, Francis DuCane, 50, 64

Goldschmidt, Richard, 139, 186, 188, 230, 242, 245, 261
Goodson, Arthur, 32, 222
Goodson, Fred, 176
Gosline, William, 294
Graff, Ludwig von, 22
Great Depression. *See* Depression
Green, E. E., 144
Grote, Augustus Radcliffe, 90, 136, 297
Gulick, John T., 57–58, 63–64, 99
Günther, Albert, 24, 29, 36, 38, 43, 47, 60, 76

Haeckel, Ernst, 16
Hampson, George Francis, 108, 123, 171, 181
Handlirsch, Anton, 167, 170, 184
Harmer, Sidney, 298
Hartert, Claudia, 31, 179, 220
Hartert, Ernst: bird collection and, 220, 223, 226; collectors and, 47, 66, 68, 70–71, 177; on conservation, 131–32, 134; defense of Walter Rothschild, 43–44; on evolution, 124; as field-naturalist, 23, 37; international congresses of ornithology and, 123–24, 131; on nomenclature, 6; relations with Jordan, 21, 31; relations with Walter Rothschild, 29–30, 43, 129, 220; on trinomials, 60, 62, 124
Harwood, W. H., 135
Hellmayr, Carl Edward, 36
Hemming, Francis, 165, 172, 182, 297
Hering, Erich Martin, 250–51, 254, 266, 268, 271, 277–78, 283
Herrera, Alfonso L., 45, 77, 234, 278
Herrick, G. W., 206
Hescheler, Karl, 185, 200
Hoffman, Adolf, 253
Holland, W. J.: applied entomology and, 206–7, 254; at congresses, 145, 211, 213–14; opinion of Jordan, 278; and Tring, 109, 121
Hoogstraal, Harry, 279
Hopkins, G. H. E., 275, 279, 282
Horn, Walther: death of, 253–54; on Hitler's rise, 249, 251; internationalism of, 171, 270; nomenclature debates, 217–18; postwar Germany and, 179–80; taxonomic reform, 202, 210, 212–13, 230–31
Horn-Escherich-Nuttall Resolution, 202–5
Hose, Charles, 34
Houten, M. S. van, 137

Howard, Leslie Ossian: applied entomology and, 195–96, 200; at Ithaca congress, 206–7, 209
Hubbell, Theodore H., 294–95
Hubrecht, A. A. W., 137
Huxley, Julian, 260, 262–64
Huxley, Thomas Henry, 47, 52, 103, 193

Ibis, 43, 60, 225
Imperial Bureau of Entomology, 196
International Biological Program, 306
International Code of Zoological Nomenclature, 137, 165
International Commission of Zoological Nomenclature: controversies at, 216–18; entomologists and, 164; foundation of, 123; Jordan's role in, 164–66; trinomials and, 171; World War I and, 172, 181–83
International Commission on Entomological Nomenclature, 166
international congresses of entomology: aim of, 6, 118, 124–26, 138–40, 198–99, 216; applied entomology at, 158–60, 252, 285; cold war and, 271–72; genetics at, 241; Jordan's role in, 3, 118–27, 134–46, 182–84; world wars and, 167, 170–72, 270–71
international congresses of ornithology, 123, 131–32
international congresses of zoology: Berlin (1901), 122, 137, 139, 142, 165, 172; Boston (1907), 138–40, 195; Budapest (1927), 206, 217; Cambridge (1898), 72, 123, 137, 142; Graz (1910), 139; Leiden (1895), 121–23, 131, 137, 138; Lisbon (1935), 218; Monaco (1913), 139, 165; Moscow (1892), 122, 136, 138; motivations for, 122–23, 138; Padua (1930), 217; Paris (1889), 122; Paris (1932), 218, 230–31, 241
International Congress of Entomology: Berlin (1938), 251–52, 254; Brussels (1910), 140–46, 160–61, 191; Ithaca (1928), 206–13, 238; Madrid (1935), 237, 242; Montreal (1956), 299; Oxford (1912), 141, 161–62, 164, 166, 195, 284; Stockholm (1948), 251, 285; Zurich (1925), 182–84, 185–86, 199–205, 212
International Entomological Institute, 212–16, 218, 220, 230
internationalism: critics of, 272; practical motivations of, 137–38; prior to World War I, 121–23, 166; recovery after World War I, 181–84, 191, 199
International Research Council, 181–82

Jablonowski, Josef, 208
Jacoby, Martin, 36
Janse, A. J. T., 198, 243, 270
Japan, 268
Jeannel, René, 210, 266–67, 270
Jennings, Herbert Spencer, 187
Jordan, David Starr, 299
Jordan, Karl: amateurs, attitude toward, 163, 226; applied entomologists, interactions with, 126, 158–59, 203–6; applied entomology, attitude toward, 175, 194, 200, 204, 207, 211, 214, 219, 226, 268, 285, 287; applied entomology, early experience with, 20–21; British citizenship of, 168; British entomologists, relations with, 41–42, 44–45; British Museum, employee of, 269, 272, 286–87; childhood and early years, 10–14; class background, 12, 25, 31; classification methods of, 87, 99–102; as coleopterist, 3, 10, 19–20, 53, 79–80, 246, 273, 276; collectors and, 46–47, 68–69, 109; curatorial work of, 32, 53, 79–80, 96, 249; death of, 298; dependence on Walter Rothschild, 38–39, 46, 48, 79, 107, 301; *Don Quixote*, use of, 84; early work in museums, 17–18, 88; educational background, 14–18, 26; Entomological Society of London, member of, 45, 219, 246; evolution, interest in, 91, 94–111, 113, 143, 236–37, 258, 260, 281–84, 286; expedition to South-West Africa, 234–35; experimental method and, 16; genetics, attitude toward, 240, 259; geographical variation, work on, 90–100, 236–41; on the goal of taxonomy, 4–5, 15, 27, 53–55, 74, 79, 82, 103–4, 189, 236, 248; Hartert, relations with, 31, 177; international congresses of entomology and, 118–26, 134–46, 204–12, 270; international congresses of zoology, attendance at, 142; as internationalist, 136, 168, 182, 184, 199, 206, 216, 250, 256, 270; as lepidopterist, 3, 79–86, 130, 160–61, 245–46, 277; life in Tring and abroad during war, 167–68, 172, 176–78; Ernst Mayr, correspondence with, 261–64; *On Mechanical Selection*, 94–102, 105; methodological caution, 90–91, 94–95, 281–83; move to Britain, 26; naming practices of, 53–54, 81–86, 109; nationalism, attitude toward, 138, 171; network, importance of, 18, 33–35, 69, 106, 118–20, 130, 134, 140, 213, 276; new species described by, 2, 32, 45, 53–54; new systematics, hero of, 292–93, 297–99; nomenclature, work on, 163–66, 182–83, 216; Papilios, work on, 79–86, 101–2; personality of, 271, 297–98; plans for Tring after Walter Rothschild's death, 247–49, 255, 257–58, 259; pre-Tring employment, 19; priority rules, 82–84, 165–66; professional ethos of, 25, 28, 37; reform of entomology, campaign for, 103–11, 124–26, 189–90, 212–16, 218; relations with Charles Rothschild, 31–32, 176–77, 179, 183; relations with Walter Rothschild, 29–31, 79–80, 129, 222–23, 225, 245, 277; reputation of, 2–3, 100, 219, 277–78, 283, 296; salary of, 25; sale of Tring's birds and, 221–22; scandal of 1908 and, 128–31; species concept of, 96–97; on the importance of specimens, 54, 66, 80, 95, 97–98, 102–3, 106; attitude toward state patronage, 133, 258, 288–89; on status of entomology and taxonomy, 26, 35, 49, 53–54, 74, 79–80, 86–103, 110–11, 113, 124–26, 138–40, 142–43, 187, 189–90, 202, 245, 259; trinomials, support for, 58–62, 81–82, 87, 105; wartime breakdown in correspondence network of, 167, 169, 172–74, 188; World War II and, 254–55, 257–58
Jordan, Minna (Brünnig), 18, 31, 167, 183
Jourdain, F. C. R., 220
Junk, Wilhelm, 256

Kaye, W. J., 107, 112
Kirby, W. F., 36
Kitching, Ian, 296
Klages, S. M., 34
Kleinschmidt, Otto, 64
Kolbe, H. J., 143, 163
Kricheldorff, Adolph, 267
Kryger, J. P., 209
Kühn, Heinrich, 34, 35

Lack, David, 225, 292
Lamarckian evolution, 100
Lameere, Auguste, 140–44, 156, 159, 185, 203, 230, 238
Landrum, Les, 309
Lankester, C. H., 175
Lankester, E. Ray, 156–57, 192, 298
League of Nations, 250
Lecerf, Ferdinand, 267
Leigh, G. F., 131
Leiper, R. T., 201
Lepidopterists' Society, 286

Lesne, Pierre, 184, 267
Linnaeus, Carl, 10, 110, 310
Linnean Society of London, 134, 248, 260
Longstaff, George Blundell, 135
Lowe, Percy, 223
Lundblad, Olov, 270
Lydekker, R., 86
Lynes, Hubert, 36

Mach, Ernst, 91, 99
Major, C. I. Forsyth, 36
Manders, Neville, 169
Mann, Thomas, 292
Marchal, Paul, 241
Marshall, Guy K., 130, 196
Martini, Erich, 267
Mather, K., 261
Matteson, J. Harold, 218
Mayr, Ernst: campaign for evolution, 280; catalogs and, 309; Jordan as hero, 291–96, 306; memories of Tring, 30, 69; Walter Rothschild's death, 243; subspecies and trinomials, defense of, 293–94; on wartime correspondence with Jordan, 260–64
Mechanical Selection, On, 94–102, 105, 117
medical entomology, 158–59
Meek, A. T., 34, 35, 131
Meinertzhagen, Richard, 36, 244, 262, 278, 293
Meldola, Raphael, 50–52
Melou, G., 177
Mendelism, 115–18, 162, 185–87, 259
Merrifield, F., 159, 191
Metzger, A., 20
microscopes, 15–16, 106–7
Milne-Edwards, Alphonse, 138, 142
mimicry, 52, 143, 144, 160–61
Minall, Alfred, 32
Mitsukuri, Kakichi, 137
Möbius, Karl, 23, 58, 85
modern synthesis, 187–88, 230, 238, 259–64; Jordan as hero of, 292–93, 297–99
Moreau, R. E., 225, 280
Morgan, Thomas Hunt, 139–40, 186–87, 238, 260, 299
Moss, Miles: criticisms of collecting, 189; Jordan's advice to, 109, 163; during the wars, 173, 177, 257
Muir, Fred, 198, 202, 210–11, 215–16, 218
Murphy, Robert Cushman, 222–23, 227
Museum of Comparative Zoology, 291

museums: Jordan's early experience in, 17–18; relations with private collectors, 43, 46, 126. *See also* collections
mutationism, 92–94, 99–100, 110, 115, 129
Myhrvold, Nathan, 1

Nabokov, Vladimir, 178
National Socialism, 218, 250
nationalism: interwar period, 249–50; natural history and, 12, 24, 165–72, 216–18; opposition to collecting and, 233; prior to World War I, 135–37, 142, 165–72; World War I and, 177, 181–82
natural history dealers, 12, 70, 163
Natural History Museum (London). *See* British Museum (Natural History)
natural selection: eclipsed, 115–17, 139; as explaining naturalists' data, 56; Jordan's interest in, 91, 100, 143, 258, 260; studies by entomologists, 49; synthesis with genetics, 187, 240; Wallace as champion of, 63
natural theology, 10–11, 13, 257
Neave, S. A., 193, 198
networks: building collections through, 69, 106; cold war and, 271–72; importance to good taxonomy, 18, 33, 35, 45, 106, 130, 296, 301; international congresses and, 118–21, 134, 140, 213; Jordan and British, 41–42, 45; recovery after World War I, 179–84, 252–53; recovery after World War II, 266–72, 276; shifts in, 227, 231, 269, 290; wartime breakdown of, 167–74, 188, 242, 254–55, 264, 266
Neumann, Oscar, 36, 255
New Guinea, 35, 131
new species: Jordan's descriptions of, 2, 32, 45, 80, 235, 297; naming practices, 35, 45, 53–54, 82–83, 123, 144–46; priority rules and, 44, 165–66; process of describing, 55–56, 59, 82, 97–99; Rothschild Museum's descriptions of, 37, 53. *See also* catalog of species
new systematics, 262–63
Newton, Alfred, 23–24, 29, 43, 60, 65, 131
N. M. Rothschild & Sons: empire and, 25, 173; international network of, 120; during wartime, 168, 173; wealth of, 24–25
nomenclature: classification and, 6, 20; controversies over, 216–18; internationalism in, 136; Jordan's stance on, 81–86, 100; Linnaean system of, 10; nationalism in, 136–37, 165,

171; and postwar recovery, 182–83; reform of, 123, 144–46, 163–64, 181–83, 210–15, 230, 302; trinomials, 58
Novitates Zoologicae, 34, 53, 58, 62, 66, 162, 266

Oberthür, Charles, 64–65, 82, 107, 121, 169, 171, 173
Olivier, Ernest, 36
Olivier, Sir Sydney, 132
On Mechanical Selection, 94–102, 105, 117
Ormerod, Eleanor, 195
ornithologists, 59–60, 220–25
orthogenesis, 100, 101, 143
Osborn, Henry Fairfield, 221
Owen, Richard, 47
Oxford University Museum of Natural History, 49, 117, 130

Pagenstecher, A., 87
Papilionidae, 66, 79–86, 91, 110, 220
patronage: decline of, 241, 243–44; shifts in, 248–49, 301; state, 133, 156–61, 174–75, 191–93, 196–97, 287; upper class, 122, 192, 220, 226
Pendlebury, W., 168
phylogeny. *See* classification
physiological selection, 95
Pictet, Amé, 185–86, 199
Poche, Franz, 217
Poncelet, J. B., 249
Poulton, Edward Bagnall: applied entomology and, 160–61, 195; biology and, 50, 76–77, 100; Darwinism and, 49, 116–18, 139, 160–61; genetics and, 186–87, 259–60; knighthood, 298; Tring and, 130, 260
Preston-Clark, Benjamin, 162–63
Proceedings of the Zoological Society of London, 43, 48–49
professionals: applied entomologists, 21; ethos, 38, 52, 74; natural history dealers, 44; relations with non-professionals, 43, 46; rise of, 244–45; status of, 47, 133; zoologists, 19
Prout, Louis B., 97, 100, 119
Provine, William B., 238
Punnett, R. C., 162

Querci, Orazio, 231, 267

Radford, Charles, 259
Raven, Peter, 1

Rehn, James, 180
Reichenow, Anton, 36
Revision of the American Papilios, 86, 110, 113, 205
Revision of the Lepidopterous Family Sphingidae, 86, 99, 104–7, 112, 117
Revision of the Papilios of the Eastern Hemisphere, exclusive of Africa, 79–86
revisions, need for, 78, 80, 144, 295
Riggenbach, F. W., 36
Riley, N. D.: Berlin congress and, 252; Jordan and, 271, 284, 287–88; on Jordan's personality, 297; on the modern synthesis, 259; on museum administration, 287; Rothschild's bequest and, 246; state patronage and, 289; on Tring versus the British Museum, 108, 247; World War II and, 255; after World War II, 268–69, 271
Rockefeller Foundation, 219
Roepke, Walter, 267
Romanes, George, 64, 91, 95
Rothschild, Charles: anti-Semitism and, 42; conservation and, 131; father and, 30, 176; illness of, 176–77, 179, 183; Jordan and, 31–32, 176; organizational efforts, 119; scandal of 1908 and, 128–29; work on fleas, 158–59
Rothschild, Emma, 29–30, 114, 128, 129, 222, 243
Rothschild, Miriam: account of Walter Rothschild's scandals, 128–29; memories of Tring, 28, 30, 43, 79, 176, 183, 297–98; thoughts on Jordan, 225, 302; work on the bequest, 243, 246; work on fleas, 274–75, 282–83; World War II and, 253, 266
Rothschild, N. M., 25, 30, 37, 114, 176
Rothschild, Roszika, 176
Rothschild, Victor, 230, 248–49
Rothschild, Walter: bird collection, loss of, 220–27, 261; British naturalists, relations with, 43, 60, 223–24; collectors of, 34, 66; curators, dependence on, 38–39, 48; curators, relations with, 29–31, 79, 89, 220, 222–23, 225, 245; curators, selection of, 24; death of, 243–49; jealousy of, 43–44; Jordan's views of, 26, 30, 81; Miriam Rothschild's memories of, 28, 32; scandal of 1908, 127–31; service on ICZN, 165–66; status of, 47, 132; succession to title, 176; trinomials and, 58, 62, 162, 171; wealth of, 37, 44, 52, 85
Russian revolution, 178
Rutimeyer, Ernest, 277

Salvin, Osbert, 36
Samper, Christián, 308
Sanford, L. C., 221, 226
Schaus, William, 175, 181, 198, 252
Schlegel, Hermann, 58
Schulthess, Anton von, 184, 199
Schwartz, Martin, 254
scientific method: changing concepts of, 5, 49, 100–101, 132, 144, 247, 257, 280–82; at congresses, 208; entomologists and, 48–55, 99; experimentalism and, 94, 115–17, 138–40, 185–86, 238–41, 290; Germany versus Britain, 42; Jordan's views on the need for reform, 104–11; natural explanation in, 4, 13; physics, chemistry, and mathematics as criteria of, 26, 291; role of generalizations in, 27, 42, 49, 76, 90–91, 101
Sclater, Philip Lutley, 21, 60, 62, 137
Scoble, Malcolm, 296
Seebohm, Henry, 60
Seitz, Adalbert, 145, 174
Sennett, Noel, 168
Severin, Guillaume, 112, 130, 138, 184
Sharp, David, 45, 61, 93, 119, 126, 226
Sharp, William, 119, 195
Sharpe, Richard Bowdler, 59
Sherman, J. D., 255
Siphonaptera. *See* fleas
Sjöstedt, Yngve, 169, 170, 173, 254, 270, 284
Skinner, Henry, 184
Smart, John, 263–64, 281
Smit, F.G.A.M., 289
Smith, G. F. Herbert, 225
Smith, H. Grose, 36
Smithsonian Institution, 197, 215
socialism: Alfred Russel Wallace's, 114–15, 118, 133, 221, 247, 272; critiques of natural history, 132–33, 248–49; effect on science, 193–94, 196, 288; internationalism and, 121
Society for Freedom in Science, 288–89
Someren, V. G. L., 254
Spanish Civil War, 242
species: as arbitrary, 86–87, 204, 299; definitions of, 4, 56–57, 59, 95–97, 117, 262, 308; Jordan's definition of, 88–98, 236–37; methods of describing, 64, 77–78, 89, 98–99, 105–6
species-makers: criticisms of, 45, 51, 54, 57, 86, 162; defenses of, 5, 51–53, 93, 100; definition of, 3; Jordan's use of the phrase, 3–4, 55, 96, 98

species-mongers, 62, 76, 77–78, 80, 86, 124
specimens: dissection of, 88–90; duplicate, 63, 66, 70–71, 221, 247; exchange of, 33, 46, 106, 121, 130, 146, 276, 302; Jordan's early experience with, 17–18; justifications for collecting, 10–11, 23–24, 75, 95, 101, 196; labeling of, 17–18, 67–71; long series of, 55–57, 61–67, 84–86, 104–5, 110, 189, 239, 274; as old-fashioned, 5, 15, 51–52, 116, 225; opposition to collection of, 65, 131–34, 189, 233–34, 244; as reflecting living nature, 54, 67, 189; regulation of collecting of, 234; scientific natural history and, 45–46, 48–49, 95, 143; setting of, 135; as unifying, 34, 36, 44–46, 119, 120, 140; use in determining species, 97, 105–6; war and access to, 180–81, 231, 269
Speiser, Paul, 169, 178
Spengel, J. W., 102
Speyer, Adolf, 17
Speyer, August, 17
Sphingidae: Jordan's interests in, 246; Moss on need for life history, 189; Tring's revision of, 86, 104–6, 112, 296; variation within, 66
Standfuss, Otto, 129, 230
Staudinger, Otto, 58, 267
Stebbing, Thomas R. R., 77
Stejneger, Leonhard, 75, 165
Stewart, M. A., 275
Stiles, Charles, 138, 164–65, 217–18
Stresemann, Erwin, 226, 251, 291
Stromeyer collection, 17–18, 58, 67, 80, 126
Study, Eduard, 180
Sturtevant, A. H., 187, 230
subspecies: critics of, 45, 108, 171, 202, 216, 261, 293–94; descriptions of, 56; Jordan's stance on, 81–82, 87, 94–100, 105, 236–41, 248; nomenclature for, 58–62, 81–82; relation to biology, 67, 94–100, 262
Swynnerton, Charles, 232
synonymy: critiques of, 77, 123; debates over, 145, 164; as impetus to organizations, 125, 302; priority rules, 82–83, 165–66; type-specimens and, 270
systematics. *See* taxonomy
systematists. *See* taxonomists

taxonomists: debates between, 41–42, 50; diversity of, 8, 83–84, 119, 123–26, 144–46, 162–64, 294–95; goals of, 1, 4, 10, 12; influence of evolution on, 3–4, 6, 40, 56–62, 66, 87, 96, 124, 144, 210; positions for, 9, 18; status of, 3–7, 51–53, 56–57, 76, 116–18, 138–40, 143–44, 203–6, 258–59, 262; tension with applied workers, 201–2, 204–5, 209, 214, 285, 290

taxonomy: influence of rise of applied entomology on, 198, 201–4, 209, 219, 263, 288, 302; Jordan's justifications of, 74, 82, 94–100, 110, 143, 158, 160–61, 188, 197, 200, 205, 214, 248, 273, 295, 299; justifications of, 3, 5, 8, 10–11, 20, 23–24, 50, 51–52, 56, 59, 72, 75, 77, 159, 169–70, 196, 210; methodological framework for, 48–50, 52; need for revisions, 78; reform efforts, 202–3, 210–15, 230; subject to change in society, 7–8; universities and, 15, 18

Thomas, Oldfield, 36
Townsend, Charles, 233–34
trade, ties of natural history to, 34, 37, 172–73, 188, 192
Trägardh, Ivar, 270, 285
Transactions of the Entomological Society of London, 43, 48–55
Traub, Robert, 283, 289, 296
Trimen, Ronald, 130
Tring Museum. *See* Walter Rothschild Zoological Museum
Tring Park, 24–25
trinomials: criticisms of, 171, 202, 216, 293–94; Jordan's defense of, 87, 105, 110–11; support for, 58–62, 64, 66
Turner, H. J., 186, 251
Tutt, J. W.: biological entomology and, 49, 51–52, 54, 67, 119; criticism of British Museum, 157; description of British entomologists, 40, 41, 136; Jordan and, 100; priority rule and, 83; variation and, 57–58, 78

United Nations Convention on Biological Diversity, 308
United States: applied entomology in, 193–94, 197–98, 207–9; Department of Agriculture Bureau of Entomology, 198, 214; increase in influence in entomology, 175, 269
United States National Museum, 291

van Duzee, E. P., 202
variation: challenges posed by, 3, 55–56, 78; discontinuous versus continuous, 91–94, 110,

115–18, 139; explanations of, 4, 56–59; Jordan's conclusions regarding, 90–100; long series and, 66–67, 84, 107; need for more study of, 13–14, 57–58, 63–64, 72, 93, 144, 162; nonevolutionary views of, 64–65; Tring's study of, 82–86, 105, 117, 236–41, 262

Verity, Roger, 216, 267
Vinson, J., 272
Vries, Hugo de, 110, 129

Wagner, Moritz, 99, 267
Wallace, Alfred Russel: anti-imperialism, 142, 233; criticisms of Bateson, 93; defense of taxonomy, 5, 76; *The Geographical Distribution of Animals*, 22; Jordan reading, 91, 95; opposition to Bateson, 115–17; reform of collections needed, 63, 65; response to Tring's revisions, 113; as a socialist, 114–15, 118, 129, 133, 136, 194, 196, 221, 247, 257, 272; specimen labels, 68
Walsingham, Lord, 24, 44, 48, 61, 65, 81, 106, 156, 172, 173
Walter Rothschild Zoological Museum: aims of, 24, 35, 63, 65–67, 108, 247; applied entomology and, 159; bequest of, 223, 243, 246–47; bird collection, loss of, 220–27; collecting network of, 34, 130–31; criticisms of, 108; depression and, 221; end of the "hey-day," 127–31; growth of, 29–30, 37, 53; Jordan's plans for, 247–49, 255, 257–58; legacy of, 246–47; long series of specimens in, 62, 65–67, 71–72, 84–86, 92, 104–11; newspaper coverage of, 72; open-access of, 36; organization of, 36; resources of, 107, 109, 112, 175, 188, 219; status of, 53, 62, 69, 112, 117, 130, 225, 278; war and, 168; workers in, 32; work in Algeria, 129–30
war: influence on specimen exchange, 180–81, 269; influence on taxonomy, 7, 146, 172–75, 179, 181–83, 202; World War I, 166–79; World War II, 254–57, 261, 264, 266, 269
Warburg, J. C., 135
Warren, William, 36
Wasmann, Erich, 143
Watkins and Doncaster, 177
Watson, James D., 291
Weismann, August, 16, 49, 51, 91, 100, 103
Wells, H. G., 193
Wetmore, Alexander, 251
Wheeler, Quentin, 307
White, F. Buchanan, 89

Whitney, Gertrude Vanderbilt, 222
Whitney, Harry Payne, 222
Williams, C. B., 201
Wilson, E. B., 186
Wilson, Edward O., 1, 8, 262, 291, 293, 305–7, 311

Young, Fred, 32

Zeitschrift für wissenschaftliche Zoologie, 15
Zimmerman, Elwood, 264, 273, 276–77, 288, 290
Zoological Record, 181
Zopp, Johannes, 241